THE PRINCETON REVIEW

Cracking the SAT II: MATH Subject Tests

THE PRINCETON REVIEW

Cracking the SAT II: MATH Subject Tests

Jonathan Spaihts

1998–99 Edition

Random House, Inc.
New York 1998
www.randomhouse.com

Princeton Review Publishing, L.L.C.
2315 Broadway
New York, NY 10024
E-mail: info@review.com

Copyright © 1993, 1994, 1995, 1996, 1997, 1998 by Princeton Review Publishing, L.L.C.

All rights reserved under International and Pan-American Copyright Conventions. Published in the United States by Random House, Inc., New York, and simultaneously in Canada by Random House of Canada Limited, Toronto.

ISBN: 0-375-75100-9
ISSN: 1076-5387

SAT II is a registered trademark of the College Board.

Editor: Amy Bryant
Designer: Illeny Maaza
Illustrations by: The Production Department of The Princeton Review

Manufactured in the United States of America on partially recycled paper.

9 8 7 6 5 4 3 2 1

ACKNOWLEDGMENTS

The author would like to thank:

Bronwyn Collie, Chris Kensler, Margaret Balistreri, Amy Bryant, and Amy Zavatto for their managerial and editorial prowess; Jane Lacher and Jeannie Yoon for their sage counsel; Patricia Acero, John Bergdahl, Lee Elliott, Greta Englert, Mike Faivre, Effie Hadjiioannou, Scott Harris, Adam Hurwitz, Christine Lee, Robert McCormack, Chee Pae, John Pak, Carmine Raspaolo, Matthew Reilly, Chris Thomas, and Kirsten Ulve; for their acumen in the making of pages; Eric Payne for his advice and erudition; and Doug French for his gimlet eye.

CONTENTS

Acknowledgments   v

1   **Introduction**   1

2   **Strategy**   7

3   **Arithmetic**   15

4   **Algebra**   47

5   **Plane Geometry**   77

6   **Solid Geometry**   105

7   **Coordinate Geometry**   123

8   **Trigonometry**   145

9   **Functions**   171

10   **Statistics and Sets**   201

11   **Miscellaneous**   217

12   **Answers and Explanations**   235

13   **The Princeton Review Mathematics Subject Tests**   257

14   **Subject Test Answers and Explanations**   341

15   **Index**   389

About the Author   393

Introduction

WHAT ARE THE MATH SUBJECT TESTS?

The Math Subject Tests are standardized tests in mathematics. Colleges use these tests to assist them in admissions decisions and to place incoming students in classes of the right level. The Subject Tests are written by ETS, a company in the business of writing tests like these. ETS stays afloat by charging students to take the SAT I and SAT II tests, and charging again to send the scores to colleges. You'll also run into ETS if you ever apply to law school, medical school, business school, or graduate school. ETS is the enemy.

Each Math Subject Test has fifty multiple-choice questions and is one hour long. The tests are scored from 200 to 800 points. Math IC and Math IIC test a range of mathematical topics, from basic algebra to trigonometry and statistics. There is a substantial overlap between the subjects of the two tests, but they are nevertheless very different.

The Math IC Subject Test is requested by many colleges as part of the student's application. The Math IIC test is required by a few schools that emphasize mathematics, such as MIT and Cal Tech. If these are schools you might be interested in applying to, you'll want to consider taking the Math IIC Subject Test. As a general rule, it's a good idea to figure out as early as possible which schools you might be interested in. You should call or write these schools and find out exactly what test scores and information they require as part of an application.

WHAT'S ON THESE TESTS?

Here are the topics that are tested on the Math Subject Tests:

Topics on the Math IC Test:	Topics on the Math IIC Test:
algebra	algebra
simple algebraic functions	algebraic functions & graphs
plane geometry	solid geometry
solid geometry	coordinate geometry
coordinate geometry	trigonometry
basic trigonometry	statistics
elementary statistics	basic number theory
some miscellaneous topics	even more miscellaneous topics

As you can see, the tests cover very similar ranges of topics. The Math IIC, however, goes into these topics in much more depth than the Math IC does. The tests also emphasize different topics. One question in five on the Math IC is about plane geometry, which is barely dealt with on the Math IIC. On the Math IIC, about one question in five is about trigonometry, and roughly one in four is about algebraic functions and their graphs; the Math IC spends much less time on both of these topics.

Don't worry if you don't recognize some of the topic headings. Students taking the Math Subject Tests are not expected to have spent time on every one of these topics in school. What's more, you can do quite well on these tests even if you haven't studied *everything* on them.

WHEN SHOULD I TAKE A MATH SUBJECT TEST?

You want to take a Math Subject Test after you've studied trigonometry or precalculus in school. For most students, that means that the best time to take a Math Subject Test is at the June administration at the end of their junior year, when all their junior-year math is still fresh in their minds.

Keep in mind, however, that the Score Choice option is available on all SAT II Subject Tests. This means that if you request the Score Choice option, you can see your scores *before* you decide whether colleges will see them. So, if you really want to do well on a Math Subject Test, you can take it more than once. It's a good idea to take a Math Subject Test at an earlier administration (such as the December or January administration) as a practice run. At best, you'll get a good Subject Test score that much earlier. At worst, you'll get valuable practice and see which of your skills could use some work. And nobody needs to see your practice score.

The degree of emphasis you place on your Math Subject Test should depend on how colleges will use your score. Some schools use the Math Subject Tests to help them make admissions decisions. Others simply use the scores to place incoming students in appropriate math classes, or to fulfill school math requirements. If the test is used for admissions, then it is more important to you, and you should take it at the times suggested above, to give yourself a safety margin in case you feel the need to take it again. If the test is only used for placement, then you can safely take the test in the middle of your senior year, after you've taken care of your SAT I and other tests. Call the schools you're interested in and ask them how they use Subject Test scores.

One last note about Score Choice: Once you release a Subject Test score, that score will be seen by every school you apply to, even if those schools don't require a Math Subject Test. Your SAT I and SAT II scores are sent to schools as a block; you can't pick and choose. Keep this in mind as you decide whether or not to release each Subject Test score.

Which Test Should I Take?

Taking the Math IC is a fine idea for most students applying to more selective schools. You should base that decision on the admission requirements of the schools you're interested in. The Math IIC, on the other hand, is not for just anyone—it's a much harder test. The great majority of students who take a Math Subject Test choose to take the Math IC.

Taking the Math IIC test is appropriate for high school students who have had a year of trigonometry or precalculus and have done well in the class. You should also be comfortable using a scientific or graphing calculator. If you hate math, do poorly on math tests, or have not yet studied trigonometry or precalculus, the Math IIC test is probably not for you. It's worth noting, however, that while the Math IIC test is difficult, the test is scored on a comparatively generous curve. If you can cope with a significant fraction of the math on the Math IIC test, you might find it surprisingly easy to get a respectable score.

Many students who take the Math IIC test also take the Math IC. This is an easy way to add a high score to the list of scores that colleges will receive. If you feel you're not really cut out for the Math IIC, but must take it because you're applying to a school that requires the test, you'll probably want to take the Math IC as well. This will reflect your abilities in a better light.

The Calculator

The Math IC and Math IIC Subject Tests are designed to be taken with the aid of a calculator (hence the "C"). Students taking either test should have a scientific calculator and know how to use it. A "scientific" calculator is one which has keys for the following functions:

- the value of π
- square roots
- raising things to an exponent
- sine, cosine, and tangent
- logarithms

Calculators without these functions will be much less useful. Graphing calculators and programmable calculators are allowed on both Math Subject Tests, though they are not necessary for either one.

On the Math IC test, a calculator is helpful on about one question in four; this means that students who are comfortable with their calculators may save a little time on these questions, but that the questions can be answered without a calculator. Only a couple of questions on any Math IC test will *require* a calculator. It's possible to score well on the Math IC test without using a calculator at all, though it's better to have one.

On the Math IIC test, calculators are much more helpful. On average, a calculator is helpful on slightly more than half of the questions on a Math IIC test. As many as ten questions may *require* a calculator to get an exact answer—and these may include easy questions. Because algebraic functions and their graphs are emphasized heavily, graphing calculators are particularly useful on the Math IIC test. Students with graphing calculators who know how to use them may be able to shortcut some questions on this test.

Last notes on calculators: The calculator you take to the test must not have an alphabetic keyboard like a typewriter; it must not require a wall outlet for power; and it must not make noise or produce paper printouts. There will be no replacements at the test center for malfunctioning or forgotten calculators, though you're welcome to bring a spare. Oh, and laptop computers don't count as calculators. Leave the Power Book at home.

How to Use This Book

It's best to work through the chapters of this book in sequence, since the later chapters build on the techniques introduced in earlier chapters. If you want an overall review of Subject Test math, just start at the beginning and cruise through to the end. This book will give you all of the techniques and knowledge you need to do well on any of the Math Subject Tests. If you feel a little shaky in certain areas of math and want to review specific topics, the chapter headings and subheadings will also allow you to zero in on your own problem topics.

As with any subjects, pay particular attention to the math topics you don't like—otherwise, those are the ones that will burn you on the real test. We've put flashcard pages at the end of each chapter to help you study. Get some index cards and make real flashcards containing the rules on these pages. Flashcards are the best way to memorize the annoying rules you need to know to do well on the Math Subject Tests. Don't hesitate to add to that stack of flashcards whenever you find something important that you have trouble remembering.

If you really want to get your money's worth out of this book, you'll follow this study plan:

- Read through a lesson carefully until you feel that you understand it.
- Try the practice questions at the end of that lesson.
- Check your answers, and review any questions you got wrong until you understand your mistakes.
- At the end of each chapter, make flashcards for the rules and techniques on the flashcard page. Use them.
- When you've finished the book to your satisfaction, try a sample test at the end of the book.
- Score your test, and review it to see where your strengths and weaknesses lie.

Many study books for the Math Subject Tests are much thicker than this one, and contain lots of unnecessary material. Instead of making you wade through hundreds of extra pages, we've stripped our book down to the bare necessities. Each section contains just a few practice questions that focus on the rules and techniques tested by ETS—nothing extra. If you make sure you understand all of the practice questions, you'll understand the questions on the real test.

Specializing for your test

As we've said before, the Math IC Test is very different from the Math IIC Test, although they test some of the same subjects. One of the ways in which they differ is in the *number* of questions on the tests that deal with various topics. This chart gives you an idea of roughly how many questions of each type the two tests will contain:

Topic	Approximate Number of Questions	
	Math IC	Math IIC
Algebra	14-16	8-10
Plane Geometry	8-12	0
Solid Geometry	2-4	3-5
Coordinate Geometry	5-7	5-7
Trigonometry	3-5	8-12
Functions	5-7	10-14
Statistics	2-4	2-4
Miscellaneous	2-4	5-7

As you can see, the Math IC Test focuses mainly on algebra and plane geometry. The Math IIC Test focuses mainly on trigonometry and functions. Keep that in mind as you read through the book. Pay special attention to the chapters covering material that's big on *your* test.

Math IIC-only material

Because the Math IIC Test contains harder material than the Math IC Test, you'll sometimes run into material in this book that will never show up on the Math IC—it's too complicated. Such material will be marked with this button:

If you're planning to take only the Math IC Test (and that's most of you), ignore all sections and questions marked with the IIC-only button, and don't worry about them.

If you're planning to take the Math IIC Test, this whole book is for you. Do everything.

If you're still not sure whether you should be taking the Math IIC test, use the Math IIC-only material as a qualifying quiz. If you get more than half of the Math IIC-only questions wrong, the Math IIC Test is probably not for you. At the very least, you'll want to take the Math IC Test as well.

Question numbers

As you cruise through this strangely stimulating math book, you'll run into practice questions that seem to be numbered incorrectly. That's because the numbers of practice questions tell you what position those questions would occupy on a 50-question Math IC Test. The question number gives you an idea of how difficult ETS considers a given question.

Curious about where a question would fall on the Math IIC Test? Simple. Just subtract 15 from the given question number. You may notice that questions numbered 1-15 then seem not to exist on the Math IIC Test. You're right. There are no questions that easy on the Math IIC Test. They're still useful practice for you, but keep in mind that the Math IIC Test starts out tricky and stays that way.

Strategy

Cracking the Math Subject Tests

It's easy to get the impression that the only way to do well on the Math Subject Tests is to become a master of a huge number of math topics and practice endlessly. It's true that you've got to know some math to do well, but there's a great deal you can do to improve your score without staring into math books until you go blind.

Several important strategies will help you increase your scoring power. There are a few characteristics of the Math Subject Tests that you can use to your advantage:

- The questions on a Math Subject Test are arranged in order of difficulty. You can think of a test as being divided roughly into thirds, containing easy, medium, and difficult questions, in that order.

- The Math Subject Tests are multiple-choice tests. That means that every time you look at a question on the test, the correct answer is on the paper right in front of you.

- ETS writes incorrect answers on the Math Subject Tests by studying errors commonly made by students. These are common errors that you can learn to recognize.

The next few pages will introduce you to test-taking techniques that use these features of the Subject Tests to your advantage to increase your score.

PACING

The first step to improving your performance on a Math Subject Test is *slowing down*. That's right: you'll score better if you do fewer questions. It may sound strange, but it works. That's because the test-taking habits you've developed in high school are poorly suited to a Math Subject Test. It's a different kind of test.

Think about the way you take a test in school. On a test in class, you try to do every question; you BS when you're unsure of the answer, and whatever happens, you make sure you finish the test. This way of thinking will get you into trouble on the Math Subject Tests. Taking a Subject Test this way means rushing through the easy questions and making careless mistakes; lowering your score by guessing on questions you don't understand; and spending far too much time at the difficult end of the test, where the questions are hard to get right no matter how much time you spend.

A hard question on the Math Subject Tests isn't worth more points than an easy question. It just takes longer to do, and it's harder to get right. It makes no sense to rush through a test if all that's waiting for you are tougher and tougher questions—especially if rushing worsens your performance on the easy questions.

Instead of plowing through a Math Subject Test at full speed, just follow these simple guidelines:

- Take your time. Make sure you get easy questions right.
- Never let yourself get stuck. If a question takes you more than a couple of minutes, bail out and move on. People have different strong suits. Chances are you'll like the next question better.
- Toward the end of the test, pick and choose among the harder questions. Don't tackle the time-consuming ones.
- Don't do more questions than you have to in order to hit your target score.

These guidelines are the basic rules of correct *pacing*. Good pacing will ensure that you have more time to do each question, get fewer questions wrong, and reach your target score more reliably.

The following charts show you roughly how many questions you'd need to do in order to get a given score (the scoring curve will vary slightly from test to test). The charts assume that you get a few wrong answers along the way:

Your target score	Questions that you need to do (These numbers allow you a few mistakes)	
	Level IC	Level IIC
400	1 – 12	1 – 6
500	1 – 26	1 – 14
600	1 – 36	1 – 26
700	1 – 46	1 – 38
800	1 – 50	1 – 48

If you've already taken one of these Subject Tests, then your initial pacing goal is easy to set; just set your sights about 50 points higher than you scored last time. If you've never taken a Math Subject Test, you can use your Math score on the SAT I to set a goal; once again, aim about 50 points higher than your previous math score.

Setting a pacing target is a good way to improve your score, but it's important not to let an ambitious goal hurt your score. You may want a 700 Subject Test score on your application, but if that target makes you rush and get easy questions wrong, then you're hurting yourself. Force yourself to be realistic, and approach high goals gradually. Above all, memorize the pacing guidelines given above, and stick to them whenever you take a test.

The Tao of Testing: Seeking the Simple Way

It's true that the math on the Math Subject Tests gets difficult. But what exactly does that mean? Well, it *doesn't* mean that you'll be doing twenty-step calculations, or huge, crazy exponential expansions that your calculator can't handle. Difficult questions on the Math Subject Tests require you to understand some slippery mathematical *concepts*, and sometimes to recognize familiar math rules in strange situations.

The way to handle difficult questions on these tests is to learn and memorize the formulas we show you, and pay attention to the problem types in which they occur. Always look for the tricky, shortcut solution—ETS questions very often have simple, elegant answers. So, if you ever find yourself working with ten-digit numbers, or plodding through the thirteenth step of a long calculation, stop yourself. You can't afford to take that much time, and you can be sure that there's a simpler way.

The Process of Elimination

So, why is it helpful that the Math Subject Tests are multiple choice? Well, as we've pointed out, it means that every time you look at a question, the correct answer is written down right in front of you. The only things standing between you and victory are four *wrong* answer choices. They're pretending to be the right answer, too, but there's something wrong with each of them. Only one of the five answer choices will increase your score.

To prevent testers from taking advantage of the 1-in-5 odds and guessing blindly, ETS exacts a "guessing penalty" for wrong answers. When you get a question right, your raw score increases by one point. When you get a question wrong, your raw score decreases by 1/4 of a point. These numbers mean that the right and wrong answers of a tester who guesses randomly will cancel each other out. Guess all the way through a Subject Test, and your raw score will come out close to zero. Blind guessing gets you nowhere.

Because of the guessing penalty, many high school teachers advise students never to guess on the Math Subject Tests. But wait: It's true that blind guessing isn't helpful—the odds are against you. But if you can change the odds, then guessing becomes a powerful tool. You change the odds by *eliminating* wrong answer choices. The minute you eliminate even one wrong answer, then guessing is more likely to help you than to hurt you. The more wrong answers you eliminate, the more the odds are in your favor.

- ◆ To increase your score on the Math Subject Tests, eliminate wrong answer choices whenever possible, and guess aggressively whenever you can eliminate anything.

There are two major elimination techniques you should rely on as you move through a Math Subject Test: Approximating answers and identifying common errors.

APPROXIMATION

From time to time, you'll come to a question that you simply don't know how to answer. This is when you want to be careful not to get stuck or spend too much time fruitlessly. Don't immediately skip such problems, though. Often, it's possible to knock out a couple of answer choices quickly and guess. One of the most effective ways to knock out answer choices is approximation.

- You can eliminate answer choices by approximation whenever you have a general idea of the correct answer. Answer choices that aren't even in the right ballpark can be crossed out.

Take a look at the following questions. In each question, at least one answer choice can be eliminated by approximation. See whether you can make eliminations yourself. For now, don't worry about how to do these questions—just concentrate on eliminations.

21. If $x^{\frac{3}{5}} = 1.84$, then $x^2 =$

    (A) −10.40
    (B) −3.74
    (C)   7.63
    (D) 10.40
    (E) 21.15

You may not have been sure how to work with that ugly fractional exponent. But if you realized that x^2 can't be negative, no matter what x is, then you could eliminate (A) and (B)—the negative answers.

Here's another one:

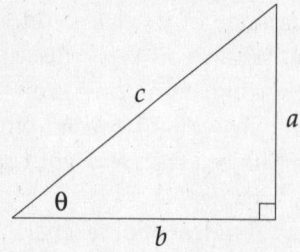

28. In Figure 1, if $c = 7$ and $\theta = 42°$, what is the value of a?

    (A) 0.3
    (B) 1.2
    (C) 4.7
    (D) 5.2
    (E) 6.0

Approximation using a diagram is a safe way to go whenever the diagram matches the description in the question. In this case, even if you weren't sure how to apply trigonometric functions to the question, you could still approximate based on the diagram provided. If c is 7, then a looks like, say, 5. That's not specific enough to let you decide between (C), (D), and (E), but you can eliminate (A) and (B). They're not even close to 5. At the very least, that gets you down to a 1-in-3 guess—much better odds.

One more:

37. The average (arithmetic mean) cost of Simon's math textbooks was $55.00, and the average cost of his history textbooks was $65.00. If Simon bought 3 math textbooks and 2 history textbooks, what was the average cost of the 5 textbooks?

   (A) $57.00
   (B) $59.00
   (C) $60.00
   (D) $63.50
   (E) $67.00

Here, once again, you might not be sure how to relate all those averages. However, you *could* realize that the average value of a group can't be bigger than the value of the biggest member of the group: so you could eliminate (E). You might also realize that, since there are more $55 books than $65 books, the average must be closer to $55.00 than to $65.00; so you could eliminate (C) and (D). That gets you down to only two answer choices, a 50-50 guess. Those are excellent odds.

These are all fairly basic questions. By the time you've finished this book, you won't need to rely on approximation to answer them. The technique of approximation will still work for you, however, whenever you're looking for an answer you can't figure out exactly.

JOE BLOGGS

As we mentioned before, ETS writes wrong answers by studying the common errors and careless mistakes made by thousands of students. They lay traps for the average student. We've nicknamed this average student Joe Bloggs. He's neither stupid nor brilliant; he does well on the easy questions, gets about half of the medium questions correct, and is suckered by the hard ones. ETS calls trap answers on difficult questions *distractors* (diabolical, isn't it?). We call them "Joe Bloggs answers."

This is a feature of the test that you can use to your advantage. Here's the idea: ETS stocks the difficult end of the test with questions that most students answered incorrectly in trial tests. So, the one thing you know about the hardest questions on a Math Subject Test is that ETS is planning on most students getting them wrong.

Anytime you find an answer choice immediately appealing on a hard question, stop and think again. If it looks that good to you, it probably looked good to many testers. That attractive answer choice is almost certainly a trap—it's a Joe Bloggs answer. The right answer won't be the one most people would pick. Remember:

♦ On hard questions, obvious answers are wrong. Eliminate them.

The following difficult questions have Joe Bloggs answers that you can eliminate. Again, don't worry about how to *do* the questions for now. Try to spot answers that are too appealing to be right—and eliminate them.

43. Ramona cycles from her house to school at 15 mph. Upon arriving, she realizes that it is Saturday and immediately cycles home at 25 mph. If the entire round-trip takes her 32 minutes, then what is her average speed, in miles per hour, for the entire round-trip?

    (A) 17.0
    (B) 18.75
    (C) 20.0
    (D) 21.25
    (E) 22.0

This is a tricky problem, and you may not be sure how to solve it. You can, however, see that there's a very tempting answer among the answer choices. If someone goes somewhere at 15 mph and returns at 25 mph, then it seems reasonable that the average speed for the trip should be 20 mph. On a #43, however, that's far too obvious to be right. You can eliminate (C). It's a Joe Bloggs answer.

Try one more:

49. If θ represents an angle such that $\sin^2 \theta = \tan \theta - \cos^2 \theta$, then $\sin \theta - \cos \theta =$

    (A) $-\sqrt{2}$
    (B) 0
    (C) 1
    (D) $2\sqrt{2}$
    (E) It cannot be determined from the information given.

On a question like this one, you might have no idea how to go about finding the answer. That "It cannot be determined" answer choice may look awfully tempting. You can be sure, however, that (E) will look tempting to *many* students. It's too tempting to be right on a question this hard. You can eliminate (E). It's a Joe Bloggs answer.

Keep Joe Bloggs elimination in mind whenever you're looking to eliminate and guess on hard questions.

FORMULAS

The techniques in this book will go a long way toward increasing your score, but there's a certain minimum amount of mathematical knowledge you'll need in order to do well on the Math Subject Tests. We've collected the most important rules and formulas into a list. As you move through the book, you'll find these lists at the end of each chapter.

The strategies in this chapter, and the techniques in the rest of this book, are powerful tools. They will make you a better test taker and improve your performance. Nevertheless, memorizing the formulas on our list is as important as learning techniques. Memorize those rules and formulas, and make sure you understand them.

Using That Calculator

Behold the First Rule of Intelligent Calculator Use:

> **Your calculator is only as smart as you are.**

It's worth remembering. Some test takers have a dangerous tendency to rely too much on their calculators. They try to use them on every question and start punching numbers in even before they've finished reading a question. That's a good way to make a question take twice as long as it has to.

The most important part of problem solving is done in your head. You need to read a question, decide which techniques will be helpful in answering it, and set the question up. Often, questions on the Math Subject Tests can be answered without much calculation—the setup itself often makes the answer clear. When a calculator *is* necessary, you'll know it. You'll need to produce an exact decimal value from a radical, a logarithm, a trig function, etc. Sometimes, using a calculator before you really need to will blind you to the shortcut solution to a problem.

When you do use your calculator, follow these simple procedures to avoid the most common calculator errors:

- Check your calculator's Operating Manual to make sure that you know how to use *all* of your calculator's scientific functions (such as the exponent and trigonometric functions).

- Clear the calculator at the beginning of each problem to make sure it's not still holding information from a previous calculation.

- Whenever possible, do long calculations one step at a time. It makes errors easier to catch.

- When you can, check your answers by approximating answers on paper or in your head. It'll help you catch careless errors, and it only takes a second.

Above all, remember that your brain is your main problem-solving tool. Your calculator is useful only when you've figured out exactly what you need to do to solve a problem.

Arithmetic

DEFINITIONS

To start with, there are a number of mathematical terms that will be thrown around freely on the test, and you'll want to recognize and understand them. Here are some of the most common terms:

Integers:	Positive and negative whole numbers, and zero. NOT fractions or decimals.
Prime Numbers:	Integers that are divisible only by themselves and 1. All prime numbers are positive; the smallest prime number is 2. Two is also the only even prime number.
Rational Numbers:	Integers, all fractions, and decimal numbers, positive and negative. Technically, any number that can be expressed as a fraction of two integers—which means everything except numbers containing radicals (like $\sqrt{2}$), π, or e.
Irrational Numbers:	All numbers with radicals that can't be simplified, such as $\sqrt{2}$ ($\sqrt{16}$ doesn't count because it can be simplified to 4). Also, all numbers containing π or e.

Real Numbers:	Any number on the number line. Everything except imaginary numbers (see below).
Imaginary Numbers:	The square roots of negative numbers, or any numbers containing i, which represents the square root of –1.
Consecutive Numbers:	The members of a set listed in order, without skipping any. Consecutive integers: 3, –2, –1, 0, 1, 2. Consecutive positive multiples of 3: 3, 6, 9, 12.
Positive Difference:	Just what it sounds like—the number you get by subtracting the smaller of two numbers from the bigger one. You can also think of it as the distance between two numbers on the number line.
Absolute Value:	The positive version of a number. You just strike the negative sign if there is one. You can also think of it as the distance on the number line between a number and zero.
Arithmetic Mean:	The average of a set of values. Also simply referred to as the "mean."
Median:	The middle value in a set when listed in order. In a set with an even number of members, the average of the *two* middle values.
Mode:	The value that occurs most often in a set. If no value appears more often than all the others in a set, then that set has no mode.

At the beginning of each chapter in this book, you may see additional definitions that pertain to the material in that chapter. Every time you see such definitions listed, be sure that you know them well.

FACTORIZATIONS

The "factors" of a number are all of the numbers by which it can be divided evenly. Some questions on the Math Subject Tests will specifically require you to identify the factors of a given number. You may find factorizations useful for solving other questions, even if they don't specifically talk about factorizations. There are two forms of factorization: plain old factorization, and prime factorization.

The factorization of a number is a complete list of its factors. The best way to compile a list of all of a number's factors is to write them in pairs, beginning with 1 and the number itself. Then count upward through the integers from 1, checking at each integer to see whether the number you're factoring is divisible by that integer. If it is, add that integer to the list of factors, and complete the pair.

Here is the factorization of 60:

1	60
2	30
3	10
4	15
5	12
6	10

Counting upward through the factors ensures that you won't miss any. You'll know your list is complete when the two columns of factors meet or pass each other. Here, the next integer after 6 that goes into 60 is 10, so you can be sure that the factorization is complete. This is the most efficient way to get a complete list of a number's factors.

The other kind of factorization is prime factorization. The prime factorization of a number is the unique group of prime numbers that can be multiplied together to produce that number. For example, the prime factorization of 8 is $2 \cdot 2 \cdot 2$. The prime factorization of 30 is $2 \cdot 3 \cdot 5$.

Prime factorizations are found by pulling the smallest possible prime number out of a number again and again until you can't anymore. The prime factorization of 75, for example, would be produced like this:

$$75 =$$
$$3 \cdot 25 =$$
$$3 \cdot 5 \cdot 5$$

When you've got nothing but prime numbers left, you're done. Here's the prime factorization of 78:

$$78 =$$
$$2 \cdot 39 =$$
$$2 \cdot 3 \cdot 13$$

Because they're often useful on the Math Subject Tests, you should be able to take prime factorizations quickly. Find the prime factorizations of the following numbers:

1. 64 = _____
2. 70 = _____
3. 18 = _____
4. 98 = _____
5. 68 = _____
6. 51 = _____

Prime factorizations are useful in many questions dealing with divisibility. For example:

> What is the smallest number divisible by both 14 and 12?

To find the *smallest* number that both numbers will go into, look at the prime factorizations of 12 and 14: $12 = 2 \cdot 2 \cdot 3$, and $14 = 2 \cdot 7$, so it's easy to build the factorization of the smallest multiple of both 12 and 14. It must contain, at least, two 2s, a 3, and a 7. That's $2 \cdot 2 \cdot 3 \cdot 7$, or 84. That's the smallest number you can divide evenly by 12 ($2 \cdot 2 \cdot 3$) and 14 ($2 \cdot 7$). Try this example:

> What is the largest factor of 180 that is NOT a multiple of 15?

Well, the prime factorization of 180 is $2 \cdot 2 \cdot 3 \cdot 3 \cdot 5$. All of the factors of 180 (except 1) will be made up of these prime factors. All you need to do is make the biggest number you can, using those prime factors, that doesn't have 15 (3 · 5) in it. The factor 2 · 2 · 5 may look tempting, but the largest number that fits the bill is 2 · 2 · 3 · 3, or 36.

Drill

Try the following practice questions. The answers to all drills are found in chapter 12.

3. What is the smallest integer divisible by both 21 and 18?

   (A) 42
   (B) 126
   (C) 189
   (D) 252
   (E) 378

7. If ¥$_x$ is defined as the largest prime factor of x, then for which of the following values of x would ¥$_x$ have the greatest value?

   (A) 170
   (B) 117
   (C) 88
   (D) 62
   (E) 53

9. If $x \, \Omega \, y$ is defined as the smallest integer of which both x and y are factors, then $10 \, \Omega \, 32$ is how much greater than $6 \, \Omega \, 20$?

   (A) 0
   (B) 70
   (C) 100
   (D) 160
   (E) 200

EVEN AND ODD, POSITIVE AND NEGATIVE

Some questions on the Math Subject Tests deal with the way numbers change when they're combined by addition and subtraction, or multiplication and division. The questions usually focus on changes in even and odd numbers, and positive and negative numbers.

Even and odd numbers

Even and odd numbers are governed by the following rules:

Addition and Subtraction
even + even = even even − even = even
odd + odd = even odd − odd = even
even + odd = odd even − odd = odd

Multiplication
even × even = even
even × odd = even
odd × odd = odd

Division of even and odd numbers follows the rules of multiplication, when division produces integers. Because many divisions don't produce whole numbers, division rules don't always apply. Only integers can be even or odd; fractions and decimals are neither even nor odd.

Positive and negative numbers

There are fewer firm rules for positive and negative numbers. Only the rules for multiplication and division are easily stated:

Multiplication and Division
positive × positive = positive   *positive ÷ positive = positive*
negative × negative = positive   *negative ÷ negative = positive*
positive × negative = negative   *positive ÷ negative = negative*

These rules are true for all numbers, because all real numbers except zero—including fractions, decimals, and even irrational numbers—are either positive or negative.

Addition and subtraction for positive and negative numbers are a little more complicated—it's best simply to use common sense. The one important rule to remember is that subtracting a negative is the same as adding a positive:

$$x - (-5) = x + 5$$
$$9 - (-6) = 9 + 6 = 15$$

If you remember this rule, adding and subtracting negative numbers should be simple.

Your understanding of these rules will be tested in questions that show you simple mathematical operations and ask you about the answers they'll produce.

Drill

Try the following practice questions. The answers to all drills are found in chapter 12.

15. If n and m are odd integers, then which of the following must also be an odd integer?

   I. mn
   II. $\dfrac{m}{n}$
   III. $(mn+1)^2$

   (A) I only
   (B) III only
   (C) I and II only
   (D) I and III only
   (E) I, II, and III

18. If c and d are integers and $cd < 0$, then which of the following statements must be true?

    (A) $\dfrac{cd}{d} > 0$

    (B) $c + d = 0$

    (C) $c^2 d > 0$

    (D) $3cd^2 \neq 0$

    (E) $cd(3+cd) < 0$

20. If x is an even positive integer and y is an odd negative integer, then which of the following must be an odd positive integer?

    (A) $x^3 y^2$

    (B) $(xy+2)^2$

    (C) $xy^2 - 1$

    (D) $x + y$

    (E) $\dfrac{x+y}{xy}$

Doing Arithmetic

This chapter deals with the basic manipulations of numbers: averages, word problems, exponents, and so on. Most of these operations can be greatly simplified by the use of a calculator, so you should practice them with your calculator in order to increase your speed and efficiency. Remember the points about calculator use from chapter 2, however: A badly used calculator will hurt you more than it helps you.

If you use your calculator incorrectly, you'll get questions wrong. If you use it on every question without thinking, it will slow you down. Keep your calculator near at hand, but *think* before you use it.

The Order of Operations

The order of operations is the sequence you've got to use to resolve complicated arithmetic expressions. It's symbolized by the acronym PEMDAS, which stands for Parentheses, Exponents, Multiplication and Division, Addition and Subtraction.

When using PEMDAS, it's important to remember that exponents and roots should be calculated at the same time, just as multiplication and division should be, followed by addition and subtraction. You can think of PEMDAS this way:

PEMDAS

Parentheses

Exponents and Roots

Multiplication and Division

Addition and Subtraction

Let's see how PEMDAS simplifies the following question. Try it yourself, and then compare your results to the explanation that follows:

5. What is the value of $2\sqrt{x^3 - 2}$ when $x = 3$?

First, plug that 3 into the expression:

$$2\sqrt{3^3 - 2}$$

Here, the radical includes several terms together: It's acting like a set of parentheses. You should resolve what's under the radical first. Start with the exponent:

$$2\sqrt{3^3 - 2} = 2\sqrt{27 - 2}$$

Then simplify:

$$2\sqrt{27 - 2} = 2\sqrt{25}$$

Once everything under the radical is done, you can take the square root:

$$2\sqrt{25} = 2 \times 5 = 10$$

Try one more. Simplify the expression yourself, and then compare your results to the explanation that follows:

3. Resolve $\dfrac{(3x^2 + 5)}{4}$ for $x = 5$.

First, plug the 5 into the expression:

$$\frac{[3(5)^2 + 5]}{4}$$

Now, you've got to simplify what's in the parentheses before you do the division.

Notice that the 5 is now in parentheses, separated from the 3; in the expression $3x^2$, the order of operations says that you must square x before you multiply by 3. Therefore, when you simplify what's in the parentheses, you do the exponent before the multiplication:

$$\frac{[3(5)^2 + 5]}{4} = \frac{[3(25) + 5]}{4} = \frac{(75 + 5)}{4} = \frac{80}{4}$$

When the parentheses are taken care of, *then* you can do the division:

$$\frac{80}{4} = 20$$

PEMDAS AND YOUR CALCULATOR

The safest way to do multistep problems like this on a calculator is one step at a time—just as they're done in the examples above.

On scientific calculators, it's possible to type complex expressions into your calculator all at once and let your calculator do the work of grinding out a number. But in order for your calculator to produce the right answer, the expression must be entered in exactly the right way—and that takes an understanding of the order of operations.

ARITHMETIC ◆ 21

For example, the expression $\dfrac{2\sqrt{3^3-2}}{5}$ would have to be typed into some calculators this way:

$$(2 \times \sqrt{(3\wedge 3 - 2)}) \div 5 =$$

On other calculators, it would have to look like this:

$$(2(3\wedge 3 - 2)\wedge(1/2)) \div 5 =$$

Any mistake in either pattern would produce an incorrect answer. On other calculators, the equation might have to be typed in still another way. If you intend to make your calculator do your work for you, check your calculator's operating manual and practice. And remember, the safest way to use your calculator is one step at a time.

Check your PEMDAS skills by working through the following complicated calculations with your calculator:

1. $0.2 \times \left[\dfrac{15^2 - 75}{6} \right] =$

2. $\dfrac{5\sqrt{6^3 - 20}}{2} =$

3. $\sqrt{\dfrac{(7^2 - 9)(.375 \times 16)^2}{10}} =$

4. $\sqrt{5\left[(13 \times 18) + \sqrt{121}\right]} =$

5. $\sqrt{\dfrac{2025^{0.5}}{0.2}} - \dfrac{5}{\frac{1}{3}} =$

FRACTIONS, DECIMALS, AND PERCENTAGES

On arithmetic questions, you will often be called upon to change fractions to decimal numbers, or decimal numbers to percentages, and so on. Be careful whenever you change the form of a number.

You turn fractions into decimals by doing the division represented by the fraction bar:

$$\dfrac{1}{8} = 1 \div 8 = .125$$

To turn a decimal number into a fraction, count the number of decimal places (digits to the right of the decimal point) in the number. Then place the number over a 1 with the same number of zeroes, get rid of the decimal point, and reduce:

$$.125 = \dfrac{125}{1000} = \dfrac{25}{200} = \dfrac{1}{8}$$

Decimal numbers and percentages are essentially the same. The difference is the percent sign (%), which means "÷ 100." To turn a decimal number into a percentage, just move the decimal point two places to the right, and add the percent sign:

$$.125 = 12.5\%$$

To turn percentages into decimal numbers, do the reverse; get rid of the percent sign and move the decimal point two places to the left.

$$0.3\% = 0.003$$

It's important to understand these conversions, and to be able to do them in your head as much as possible. Don't rely on the percent key on your calculator; it's far too easy to become confused and use it when converting in the wrong direction.

The chart below lists common values that you might be required to convert. Complete the chart for practice. Correct answers are at the end of chapter 12.

Fractions	Decimals	Percentages
		50%
	0.25	
		12.5%
	0.375	
$\frac{1}{3}$		
		$66\frac{2}{3}$ %
$\frac{1}{6}$		
$\frac{1}{10}$		
		1%
$\frac{3}{4}$		
	0.2	

Watch out for conversions between percentages and decimal numbers—especially ones involving percentages with decimal points already in them (like .15%). Converting these numbers is simple, but this step is still the source of many careless errors.

WORD-PROBLEM TRANSLATION

Most of the common careless errors made in answering math questions are committed in the very first step: Reading the question. All your skill in arithmetic does you no good if you're not solving the right problem; and all the power of your calculator can't help you if you've entered the wrong equation. Reading errors are particularly common in word problems.

The safest way to extract equations from long-winded word problems is to translate, word for word, from English to math. All of the following words have direct math equivalents:

English	Math
what what fraction how many	x, y, etc. (a variable)
a, an	1 (one)
percent	$\div 100$
of	$\times$ (multiplied by)
is, are, was, were	=

Using this table as a guide, you can translate any English sentence in a word problem into an equation. For example:

3. If the bar of a barbell weighs 15 pounds, and the entire barbell weighs 75 pounds, then *the weight of the bar is what percent of the weight of the entire barbell*?

The question at the end of the problem can be translated into

$$15 = \frac{x}{100} \times 75.$$

Solve this equation, and the question is answered. You'll find that x is equal to 20—and 20% is the correct answer.

Drill

For each of the following exercises, translate the information in English into an arithmetic equation. The answers to this drill are found in chapter 12.

1. 6.5 is what percent of 260?

2. If there are 20 honor students at Pittman High and 180 students at the school in all, then *the number of honors students at Pittman High is what percentage of the total number of students*?

3. Thirty percent of forty percent of 25 marbles is how many marbles?

4. What is the square root of one-third of 48?

5. The square root of what positive number is equal to one-eighth of that number?

Percent Change

"Percent change" is a way of talking about increasing or decreasing a number. The percent change is just the amount of the increase or decrease, expressed as a percentage of the starting amount.

For example, if you took a $100.00 item and increased its price by $2.00, that would be a 2% change—because the amount of the increase, $2.00, is 2% of the original amount, $100.00. At the same time, if you increased the price of a $5.00 item by the same $2.00, that would be a 40% increase—because $2.00 is 40% of $5.00. If you ever lose track of your numbers when computing a percent change, just use this formula:

$$\frac{change}{original} = \frac{x}{100}$$

Whenever you work with percent change, be careful not to confuse the amount of the change with the total after the change. Just concern yourself with the original amount and the amount of the increase or decrease. The new total doesn't matter.

Drill

Test your understanding of percent change with the following practice questions. The answers to all drills are found in chapter 12.

2. A 25-gallon addition to a pond containing 150 gallons constitutes an increase of what percent?

   (A) 14.29%
   (B) 16.67%
   (C) 17.25%
   (D) 20.00%
   (E) 25.00%

5. The percent decrease from 5 to 4 is how much less than the percent increase from 4 to 5?

   (A) 0%
   (B) 5%
   (C) 15%
   (D) 20%
   (E) 25%

12. Nicoletta deposits $150.00 in her savings account. If this deposit represents a 12% increase in Nicoletta's savings, then how much does her savings account contain after the deposit?

    (A) $1,100.00
    (B) $1,250.00
    (C) $1,400.00
    (D) $1,680.00
    (E) $1,800.00

Percent change shows up in many different problem types on the Math Subject Tests—it can be brought into almost any kind of math question. Here's one of the most common math question types that deal with percent change.

The change-up, change-down

It's a classic trick question to ask what happens if you increase something by a percent and then decrease it by the same percent, like this:

9. The price of a bicycle originally sold for $250.00 is marked up 30%. If this new price is subsequently discounted by 30%, then the final price of the bicycle is

    (A) $200.50
    (B) $216.75
    (C) $227.50
    (D) $250.00
    (E) $265.30

The easy mistake on this problem type is to assume that the price, after increasing by 30% and then decreasing by 30%, has returned to $250.00, the original amount. Nope! It doesn't actually work out that way, as you'll see if you try it step by step. First, you increase the original price by 30%:

$$\$250.00 + (\frac{30}{100} \times \$250.00) =$$

$$\$250.00 + \$75.00 =$$

$$\$325.00$$

Then discount this price by 30%:

$$\$325.00 - (\frac{30}{100} \times \$325.00) =$$

$$\$325.00 - \$97.50 =$$

$$\$227.50$$

The answer is (C). As you can see, the final amount isn't equal to the starting amount. The reason for the difference is that you're increasing the price by 30% of the starting number, and then decreasing by 30% of a *different* number—the new, higher price. The changes will never be of the same *amount*—just the same percent. You end up with a number smaller than your starting number, because the decrease was bigger than the increase. In fact, if you'd done the decrease *first* and then the increase, you would still have gotten the same number: $227.50.

Remember this tip whenever you *increase* a quantity by a percent and then *decrease* by the same percent: Your final result will always be a bit smaller than your original amount. The same thing is true if you *decrease* a quantity by a percent and then *increase* by the same percent: You'll get a number a bit lower than your starting number.

AVERAGES

You will be required to work with simple averages in a variety of question types. As you know, the average of a group of values is the sum of all the values divided by the number of values you're adding up.

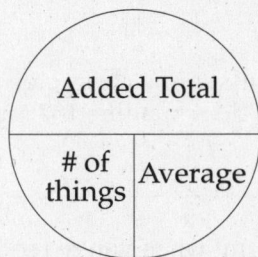

If you look again at the definition you just read, you'll see that every average involves three quantities: The added total, the number of things being added, and the average itself. The chart above is called an average wheel; it's The Princeton Review way of organizing the information in an average.

The added total, on top of the wheel, is the product of the two numbers on the bottom. Divide the top number by one of the bottom numbers, and you get the other bottom number.

When you run into an average in a Math Subject Test question, you'll be given two of the three numbers involved. Usually, solving the problem will depend on your supplying the missing number in the average.

Drill

Test your understanding of averages with the following questions. The answers to all drills are found in chapter 12.

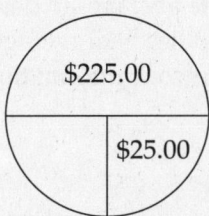

1. People at a dinner paid an average of $25.00 each. The total bill for dinner was $225.00.

What else do you know? _____

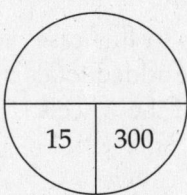

2. The average fruit picker on Wilbury Ranch picked 300 apples on Tuesday. There are 15 fruit pickers at Wilbury Ranch.

What else do you know? _____

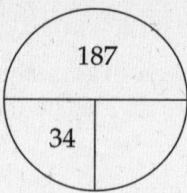

3. If the 34 students in the chess club lay down end to end, they would form a line 187 feet long.

What else do you know? _____

The average wheel becomes most useful when you're tackling a multiple-average question—one that requires you to manipulate several averages in order to find an answer. Here's an example:

32. Sydney's average score on the first 5 math tests of the year was 82. If she ended the year with a math test average of 88, and a total of 8 math tests were administered that year, what was her average on the last three math tests?

(A) 99.5
(B) 98.75
(C) 98.0
(D) 96.25
(E) 94.0

In this question, there are three separate averages to deal with: Sydney's average on the first five tests, her average on the last three tests, and her final average for all eight. In order to avoid confusion and careless errors, you've got to organize your information. Do this by drawing an average wheel for each average in the question:

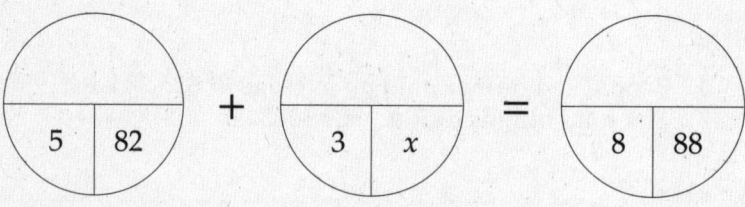

Then fill in the information you know. In this case, you know the average and the number of things added for each wheel; that leaves the added totals to be figured out. Fill in the added totals by multiplying the numbers on the bottoms of the wheels.

Then find the missing number by adding the totals together and solving:

$$410 + 3x = 704$$

$$3x = 294$$

$$x = 98$$

And as it turns out, Sydney averaged a 98 on her last three math tests; the answer is (C). Multiple-average questions are never terribly difficult. Just carefully count the averages in the question, and draw an average wheel for each one. Organization is everything on these questions; it's easy to make careless errors if you get your numbers scrambled.

33. At a charity fund-raiser, the average of the first 19 donations is $485.00. In order for the average of the first 20 donations to be $500.00, what must the amount of the twentieth donation be, in dollars?

   (A) 300
   (B) 515
   (C) 650
   (D) 785
   (E) 800

35. During the first 20 days of September, the Tribune received an average of 4 complaint letters per day. During the last 10 days of September, the Tribune received an average of 7 complaint letters per day. What was the Tribune's average number of complaint letters per day for the entire month of September?

   (A) 5.0
   (B) 5.33
   (C) 5.67
   (D) 6.0
   (E) 6.25

36. Over a year, Brendan sold an average of 12 umbrellas a day on rainy days, and an average of 3 umbrellas a day on clear days. If the weather was rainy one day in five, and this was not a leap year, what were Brendan's average daily umbrella sales for the year?

   (A) 4.8
   (B) 5.2
   (C) 6.75
   (D) 7.3
   (E) 9.0

EXPONENTS

An exponent is a simple way of expressing repeated multiplication. You can think of 5^3, for example, as three fives in a row, multiplied together. In this exponential expression, 5 is referred to as the "base," while 3 is the "exponent." Sometimes a third number is also present, called a "coefficient." In the expression $4b^2$, b is the base, 2 is the exponent, and 4 is the coefficient. Here, b is being squared, but the coefficient, 4, is not affected by the exponent.

There are a few important things to remember about the effects of exponents on various numbers:

- A positive number raised to any power remains positive. No exponent can make a positive number negative.
- A negative number raised to an *odd* power remains negative.
- A negative number raised to an *even* power becomes positive.

In other words, anything raised to an odd power keeps its sign. If a^3 is negative, then a is negative; if a^3 is positive, then a is positive. A term with an odd exponent has only one root (value that makes the equation true). For example, if $a^3 = -27$, there's only one value of a that makes it true: $a = -3$.

On the other hand, anything raised to an even power becomes positive, regardless of its original sign. This means that an equation with an even exponent has *two* roots. For example, if $b^2 = 25$, then b has two possible values: 5 and –5. The positive root is referred to as the "principal root," and questions on the Math Subject Test will often specify that they're looking for the positive or principal root of an exponential equation. Still, it's important to remember that two roots exist for any equation with an even exponent (the only exception is when $b^2 = 0$, in which case b can equal only 0, and b^2 has only one root).

One last thing to remember: since any real number becomes positive when raised to an even exponent, certain equations will have no real roots. For example, the equation $x^2 = -9$ has no real roots. There's no integer or fraction, positive or negative, that can be squared to produce a negative number. In this equation, x is said to be an *imaginary number*. The equation is considered to have no real solutions.

In the following exercises, find the roots of the exponential expression given. Specify whether each expression has one root, two roots, or no real roots.

1. $b^3 = 27$; $b =$
2. $x^2 = 121$; $x =$
3. $n^5 = 32$; $n =$
4. $c^2 = 10$; $c =$
5. $x^4 = 81$; $x =$
6. $x^3 = -8$; $x =$
7. $d^6 = 729$; $d =$
8. $n^0 = 1$; $n =$

For certain Math Subject Test questions, you'll need to do some algebraic calculations using exponents. To work with exponents in equations, you just need to remember a few basic rules:

Adding and subtracting when bases and exponents are the same

Terms with exponents can be added or subtracted only when they have the same base and exponent:

$$2a^3 + a^3 = 3a^3 \qquad\qquad 5x^2 - 4x^2 = x^2$$

If they don't have the same base *and* exponent, exponential terms can never be combined by addition or subtraction.

Multiplying exponents when bases are the same

Exponential terms can be multiplied when their bases are the same. Just leave the bases unchanged and add the exponents:

$$n^3 \times n^5 = n^8 \qquad\qquad 3 \times 3^4 = 3^5$$

Coefficients, if they are present, are multiplied normally:

$$2b \times 3b^5 = 6b^6 \qquad\qquad \frac{1}{2}c^3 \times 6c^5 = 3c^8$$

Dividing exponents when bases are the same

Exponential terms can also be divided when their bases are the same. Once again, the bases remain the same, and the exponents are subtracted:

$$x^8 \div x^6 = x^2 \qquad\qquad 7^5 \div 7 = 7^4$$

Coefficients, if they are present, are divided normally:

$$6b^5 \div 3b = 2b^4 \qquad\qquad 5a^8 \div 3a^2 = \frac{5}{3}a^6$$

Multiplying and dividing exponents when *exponents* are the same

There's one special case in which you can multiply and divide terms with different bases: when the *exponents* are the same. In this case you can multiply and divide under the exponents. Then the bases change and the exponents remain the same, for multiplication:

$$3^3 \times 5^3 = 15^3 \qquad\qquad x^8 \times y^8 = (xy)^8$$

And for division:

$$33^2 \div 3^2 = 11^2 \qquad\qquad x^{20} \div y^{20} = \left(\frac{x}{y}\right)^{20}$$

If exponential terms have different bases *and* different exponents, then there's no way to combine them by adding, subtracting, dividing, or multiplying.

One last rule:

Raising powers to powers

When an exponential term is raised to another power, the exponents are multiplied:

$$(x^2)^8 = x^{16} \qquad\qquad (7^5)^4 = 7^{20}$$

If there is a coefficient included in the term, then the coefficient is also raised to that power:

$$(3c^4)^3 = 27c^{12} \qquad\qquad (5g^3)^2 = 25g^6$$

Using these rules, you should be able to manipulate exponents wherever you find them.

Roots

Roots are exponents in reverse. For example: four times four is sixteen. That means that $4^2 = 16$. It also means that $\sqrt{16} = 4$. Square roots, or roots of the second power, are by far the most common roots on the Math Subject Tests. The square root of a number is simply whatever you would square to get that number.

You may also encounter other roots: cube roots, fourth roots, fifth roots, and so on. Each of these roots is represented by a radical with a number attached, like so: $\sqrt[3]{x}$, which means the cube root of x. Roots of higher degrees work just as square roots do. The expression $\sqrt[4]{81}$, for example, equals 3—the number that you'd raise to the 4th power to get 81. Similarly, $\sqrt[5]{32}$ is the number that, raised to the 5th power, equals 32—in this case, 2.

An important point: As you'll remember from the Exponents section, there are *two* numbers that equal 16 when raised to the 4th power—2 and –2. By definition, a radical refers only to the *principal* root of an expression. When there is only one root, that's the principal root. Where there's a positive root *and* a negative root, the positive root is considered the principal root and is the only root symbolized by the radical. Therefore $\sqrt[4]{16}$ means only 2, and not –2.

When the number under a radical has a factor whose root is an integer, then the radical can be *simplified*. This means that the root can be pulled out. For example, $\sqrt{48}$ is equal to $\sqrt{16 \times 3}$. Because 16 is a perfect square, its root can be pulled out, leaving the 3 inside: $4\sqrt{3}$. That's the simplified version of $\sqrt{48}$.

Working with Roots

The rules for manipulating roots when they appear in equations are the same as the rules for manipulating exponents. Roots can be combined by addition and subtraction only when they are roots of the same order *and* roots of the same number:

$$3\sqrt{5} - \sqrt{5} = 2\sqrt{5} \qquad\qquad 3\sqrt[3]{x} + 2\sqrt[3]{x} = 5\sqrt[3]{x}$$

Roots can be multiplied and divided freely as long as all the roots are of the same order—all square roots, or all cube roots, and so on. The answer must also be kept under the radical:

$$\sqrt{a} \times \sqrt{b} = \sqrt{ab} \qquad\qquad \sqrt[3]{24} \div \sqrt[3]{3} = \sqrt[3]{8} = 2$$

$$\sqrt{18} \times \sqrt{2} = \sqrt{36} = 6 \qquad\qquad \sqrt[4]{5} \div \sqrt[4]{2} = \sqrt[4]{\frac{5}{2}}$$

Be sure to memorize these rules before working with roots.

Special Exponents

There are some exponents on the Math Subject Tests that you've got to treat a little differently. Here are some unusual exponents you should be familiar with:

Zero

Any number raised to the power of zero is equal to 1, no matter what you start with. It's a pretty simple rule.

$$5^0 = 1 \qquad\qquad x^0 = 1$$

One

Any number raised to the first power is itself—it doesn't change. In fact, ordinary numbers, written without exponents, *are* numbers to the first power. You can think of then as having invisible exponents of 1. That's useful when using the basic exponent rules you've just reviewed. It means that $(x^4 \div x)$ can be written as $(x^4 \div x^1)$, which can prevent confusion when you're subtracting exponents.

$$x = x^1 \qquad\qquad 4^1 = 4$$

Fractional exponents

A fractional exponent is a way of raising a number to a power and taking a root of the number at the same time. The number on top is the normal exponent. The number on the bottom is the root—you can think of it as being in the "root cellar."

So, in order to raise a number to the $\frac{2}{3}$ power, you would square the number and then take the cube root of your result. You could also take the cube root first and then square the result—it doesn't matter which one you do first, as long as you realize that 2 is the exponent and 3 is the order of the root.

Remember that an exponent of 1 means the number itself, so that $x^{\frac{1}{2}}$ is equal to $\sqrt{x}$, the square root of x to the first power.

$$27^{\frac{1}{3}} = \sqrt[3]{27} = 3 \qquad\qquad b^{\frac{5}{2}} = \sqrt{b^5}$$

$$8^{\frac{2}{3}} = \sqrt[3]{8^2} = \sqrt[3]{64} = 4 \qquad\qquad x^{\frac{4}{3}} = \sqrt[3]{x^4}$$

Negative exponents

A negative exponent is treated exactly like a positive exponent, with one extra step: after you apply the exponent, you flip the number over—that is, you turn the number into its reciprocal.

$$a^{-4} = \frac{1}{a^4} \qquad\qquad 3^{-2} = \frac{1}{3^2} = \frac{1}{9}$$

$$x^{-1} = \frac{1}{x} \qquad\qquad \left(\frac{2}{3}\right)^{-1} = \frac{3}{2}$$

The negative sign works the same way on fractional exponents. First you apply the exponent as you would if it were positive, and then flip it over.

$$x^{-\frac{1}{2}} = \frac{1}{\sqrt{x}} \qquad\qquad a^{-\frac{3}{2}} = \frac{1}{\sqrt{a^3}}$$

Drill

In the following exercises, expand the exponential expressions. Where the bases are numbers, find the numerical values of the expressions. The answers to all drills are found in chapter 12.

1. $4^{\frac{3}{2}} =$

    (A) 2.52
    (B) 3.64
    (C) 8.00
    (D) 16.00
    (E) 18.67

2. $x^{-\frac{3}{4}} =$

    (A) $-\sqrt[5]{x} \cdot x^4$

    (B) $-\dfrac{x^3}{x^4}$

    (C) $\dfrac{x^4}{x^3}$

    (D) $\dfrac{1}{\sqrt[4]{x^3}}$

    (E) $-\sqrt[4]{x^3}$

3. $\left(\dfrac{2}{3}\right)^{-2} =$

    (A) 2.25
    (B) 1.67
    (C) 0.44
    (D) −1.50
    (E) −0.44

4. $\left(\dfrac{1}{a}\right)^{-\frac{1}{3}} =$

    (A) $-\dfrac{1}{\sqrt[3]{a}}$

    (B) $\sqrt[-3]{a}$

    (C) $\dfrac{1}{a^3}$

    (D) $-a^3$

    (E) $\sqrt[3]{a}$

5. $5^{\frac{2}{3}} =$

(A) 2.92
(B) 5.00
(C) 6.25
(D) 8.67
(E) 11.18

6. $\left(-\frac{5}{6}\right)^0 =$

(A) −1.2
(B) −0.8
(C) 0.0
(D) 1.0
(E) 1.2

Repeated Percent Change

On one common question type you'll have to work with percent change and exponents together. Occasionally, you'll be required to increase or decrease something by a percent *again and again*. Such questions often deal with growing populations, or bank accounts collecting interest. Here's an example:

40. Ruby had $1,250.00 in a bank account at the end of 1990. If Ruby deposits no further money in the account, and the money in the account earns 5% interest every year, then to the nearest dollar, how much money will be in the account at the end of 2000?

(A) $1,632.00
(B) $1,786.00
(C) $1,875.00
(D) $2,025.00
(E) $2,036.00

The easy mistake here is to find 5% of the original amount, which in this case would be $62.50. Add $62.50 for each of the ten years from 1990 to 2000 and you've got an increase of $625.00, right? Wrong. That would give you a final total of $1,875.00, but that's not the right answer. Here's the problem: The interest for the first year *is* $62.50, which is 5% of $1,250. But that means that now there's $1,312.50 in the bank account, so the interest for the second year will be something different. As you can see, this could get messy.

Here's the easy way. The first year's interest can be computed like any ordinary percent change, by adding the percent change to the original amount:

$$\$1{,}250.00 + \left(\frac{5}{100} \times \$1{,}250.00\right) = \text{total after one year}$$

But there's another way to write that. Just factor out the $1,250.00.

$$\$1,250.00 \times (1 + \frac{5}{100}) = \text{total after one year}$$

$$\$1,250.00 \times (1.05) = \text{total after one year}$$

We can get the total after one year by converting the percentage change to a decimal number, adding 1, and multiplying the original amount by this number. To get the total after two years, we just multiply by that number again:

$$\$1,250.00 \times (1.05) \times (1.05) = \text{total after two years}$$

And so on. So, to figure out how much money Ruby will have after ten years, all you have to do is multiply her original deposit by 1.05, ten times. That means multiplying Ruby's original deposit by 1.05 to the tenth power:

$$\$1,250.00 \times (1.05)^{10} = \text{total after ten years}$$

$$\$1,250.00 \times 1.629 = \text{total after ten years}$$

$$\$2,036.25 = \text{total after ten years}$$

So, to the nearest dollar, Ruby will have $2,036.00 after ten years. The answer is (E). There's a simple formula you can use to solve repeated percent-increase problems:

> Final amount = Original $\times (1 + \text{rate})^{\text{number of changes}}$

And the formula for repeated percent-*decrease* problems is almost identical. The only difference is that you'll be subtracting the percentage change from 1 rather than adding it:

> Final amount = Original $\times (1 - \text{rate})^{\text{number of changes}}$

Just remember that you've got to convert the rate of change (like an interest rate) from a percentage to a decimal number.

Here's another one. Try it yourself, and then check the explanation below.

43. The weight of a bar of hand soap decreases by 2.5% each time it is used. If the bar weighs 100 grams when it is new, what is its weight in grams after 20 uses?

   (A) 50.00
   (B) 52.52
   (C) 57.43
   (D) 60.27
   (E) 77.85

You've got all of your starting numbers. The original amount is 100 grams; the rate of change is 2.5%, or 0.025 (remember to subtract it, because it's a decrease). And you'll be going through 20 decreases, so the exponent will be 20. This is how you'd plug these numbers into the formula:

Final amount $= 100 \times (1 - .025)^{20}$

$= 100 \times (.975)^{20}$

$= 100 \times .60269$

Final amount $= 60.27$

The answer is (D). This is an excellent example of a question type that is difficult if you've never seen it before, and easy if you're prepared for it. Memorize the repeated percent-change formula and practice using it.

Drill

Try the following practice questions. The answers to all drills are found in chapter 12.

35. At a certain bank, savings accounts earn 5% interest per year. If a savings account is opened with a $1,000.00 deposit and no further deposits are made, how much money will the account contain after 12 years?

    (A) $1,333.33
    (B) $1,166.67
    (C) $1,600.00
    (D) $1,795.86
    (E) $12,600.00

40. In 1900, the population of Malthusia was 120,000. Since then, the population has increased by exactly 8% per year. If population growth continues at this rate, what will the population be in the year 2000?

    (A) 216,000
    (B) 2,599,070
    (C) 1,080,000
    (D) 5.4×10^7
    (E) 2.6×10^8

43. In 1995, Ebenezer Bosticle created a salt sculpture that weighed 2,000 pounds. If this sculpture loses 4% of its mass each year to rain erosion, what is the last year in which the statue will weigh more than 1,000 kilograms?

    (A) 2008
    (B) 2009
    (C) 2011
    (D) 2012
    (E) 2013

Scientific Notation

Scientific notation is a way of expressing numbers too large to be written out easily. A number is put into scientific notation by dividing by 10 until there is only one digit to the left of the decimal point, and then multiplying by a power of 10 equal to the number of tens that were divided out. The exponent on the ten will be equal to the number of places through which the decimal point moved.

For example, to write the number 1,204 in scientific notation, you would move the decimal point three places to the left and multiply by ten to the third power: 1.204×10^3. Five million (5,000,000) written in scientific notation would look like this: 5.0×10^6.

It's important to understand numbers written in scientific notation, because that's what your calculator will give you if the result of a calculation is too big for the calculator's display. Most calculators display numbers in scientific notation like this: 2.4657358 E7. That's equivalent to 2.4657358×10^7, which equals 24,657,358. Even though some calculations will produce numbers too big to be displayed in normal form, ETS may still expect you to work with them.

Drill

Try the following practice questions. The answers to all drills are found in chapter 12.

13. The numerical value of 3^{70} has how many digits?

    (A) 21
    (B) 28
    (C) 33
    (D) 34
    (E) 35

14. The number 6^n increases in length by how many digits when n goes from 44 to 55?

    (A) 12
    (B) 11
    (C) 10
    (D) 9
    (E) 8

19. Which of the following is the positive difference between 10^{18} and 9^{19}?

    (A) 5.3×10^{16}
    (B) 1.0×10^{17}
    (C) 3.5×10^{17}
    (D) 1.1×10^{18}
    (E) 7.3×10^{18}

Logarithms

Exponents can also be written in the form of logarithms. A logarithm is an exponent expressed in terms of the change it makes in a number. For example, $\log_2 8$ represents the exponent that turns 2 into 8. In this case, the "base" of the logarithm is 2. It's easy to make a logarithmic expression look like a normal exponential expression. Here we can say $\log_2 8 = x$, where x is the unknown exponent that turns 2 into 8. Then we can rewrite the equation this way: $2^x = 8$. Notice that in this equation, 2 is the base of the exponent, just as it was the base of the logarithm. Logarithms can be rearranged into exponential form using this definition:

> **Definition of a Logarithm**
>
> $\log_b n = x \iff b^x = n$

A logarithm that has no written base is assumed to be a base-10 logarithm. Base-10 logarithms are called "common logarithms," and are so frequently used that the base is often left off. Therefore, the expression "log 1,000" means $\log_{10} 1{,}000$. Most calculations involving logarithms are done in base 10 logs. When you punch a number into your calculator and hit the "log" button, the number it gives you is the exponent that would turn 10 into the number you entered.

Drill

Test your understanding of the definition of a logarithm with the following exercises. The answers to all drills are found in chapter 12.

1. $\log_2 32 =$ _____
2. $\log_3 x = 4: x =$ _____
3. $\log 1000 =$ _____
4. $\log_b 64 = 3: b =$ _____
5. $x^{\log_x y} =$ _____
6. $\log_7 1 =$ _____
7. $\log_2 x = \log_4$ _____
8. $\log_x x^{12} =$ _____
9. $\log 37 =$ _____
10. $\log 5 =$ _____

For the Math IC Test, that's about all you need to know about logarithms. As long as you can convert them from logarithmic form to exponential form, you should be able to handle any logarithm question you run into. For the IIC Test, however, you will need to work with logarithms in more complicated ways.

Logarithmic Rules

There are three properties of logarithms that are often useful on the IIC Test. These properties are very similar to the rules for working with exponents—which isn't surprising, because logarithms and exponents are the same thing. The first two properties deal with the logarithms of products and quotients:

> **The Product Rule**
>
> $\log_b (xy) = \log_b x + \log_b y$
>
> **The Quotient Rule**
>
> $\log_b \left(\dfrac{x}{y}\right) = \log_b x - \log_b y$

These rules are just another way of saying that when you multiply terms, you add exponents, and when you divide terms, you subtract exponents. Be sure to remember when you use them that the logarithms in these cases all have the same base.

ARITHMETIC ◆ 39

The third property of logarithms deals with the logarithms of terms raised to powers:

The Power Rule
$$\log_b (x^r) = r \log_b x$$

This means that whenever you take the logarithm of a term with an exponent, you can pull the exponent out and make it a coefficient:

$$\log (7^2) = 2 \log 7 = 2 (0.8451) = 1.6902$$

$$\log_3 (x^5) = 5 \log_3 x$$

These logarithm rules are often used in reverse, to simplify a string of logarithms into a single logarithm. Just as the sum and quotient rules can be used to expand a single logarithm into several logarithms, the same rules can be used to consolidate several logarithms that are being added or subtracted into a single logarithm. In the same way, the power rule can be used backwards to pull a coefficient into a logarithm, as an exponent. Take a look at how these rules can be used to simplify a string of logarithms with the same base:

$$\log 8 + 2 \log 5 - \log 2 =$$

$$\log 8 + \log 5^2 - \log 2 = \text{(Power Rule)}$$

$$\log 8 + \log 25 - \log 2 =$$

$$\log (8 \times 25) - \log 2 = \text{(Product Rule)}$$

$$\log 200 - \log 2 =$$

$$\log \left(\frac{200}{2}\right) = \text{(Quotient Rule)}$$

$$\log 100 = 2$$

Drill

In the following exercises, use the Product, Quotient, and Power rules of logarithms to simplify each logarithmic expression into a single logarithm with a coefficient of 1. The answers to all drills are found in chapter 12.

1. $\log 5 + 2 \log 6 - \log 9 =$
2. $2 \log_5 12 - \log_5 8 - 2 \log_5 3 =$
3. $4 \log 6 - 4 \log 2 - 3 \log 3 =$
4. $\log_4 320 - \log_4 20 =$
5. $2 \log 5 + \log 3 =$

LOGARITHMS IN EXPONENTIAL EQUATIONS

Logarithms can be used to solve many equations that would be very difficult or even impossible to solve any other way. The trick to using logarithms in solving equations is to convert all of the exponential expressions in the equation to base-10 logarithms, or *common logarithms*. Common logarithms are the numbers programmed into your calculator's logarithm function. Once you express exponential equations in term of common logarithms, you can run the equation through your calculator and get real numbers.

When using logarithms to solve equations, be sure to remember the meaning of the different numbers in a logarithm. Logarithms can be converted into exponential form following this guideline:

> **Definition of a Logarithm**
> $\log_b n = x \Leftrightarrow b^x = n$

Let's take a look at the kinds of tough exponential equations that can be solved using logarithms:

39. If $5^x = 2^{700}$, then what is the value of x?

This deceptively simple equation is practically impossible to solve using conventional algebra. Two to the 700th power is mind-bogglingly huge; there's no way to calculate that number. There's also no way to get x out of that awkward exponent position. This is where logarithms come in. Take the logarithm of each side of the equation:

$$\log 5^x = \log 2^{700}$$

Now use the Power Rule of logarithms to pull the exponents out:

$$x \log 5 = 700 \log 2$$

Then isolate x:

$$x = 700 \times \frac{\log 2}{\log 5}$$

Now you can use your calculator to get decimal values for log 2 and log 5, and plug them into the equation:

$$x = 700 \times \frac{.3010}{.6990}$$

$$x = 700 \times .4307$$

$$x = 301.47$$

And *voilà*, a numerical value for x. This is the usual way in which logarithms will prove useful on the Math Subject Tests (especially the Math IIC). Solving tough exponent equations will usually involve 1) taking the common log of both sides of the equation, and 2) using the Power Rule to bring exponents down. The same method can be used to find the values of logarithms with bases other than 10, even though logarithms with other bases aren't programmed into your calculator. For example:

25. What is the value of x if $\log_3 32 = x$?

You can't do this one in your head. The logarithm is asking, "What exponent turns 3 into 32?" Obviously, it's not an integer. You know that the answer will be between 3 and 4, because $3^3 = 27$ and $3^4 = 81$. That might be enough information to eliminate an answer choice or two, but it probably won't be enough to pick one answer choice. Here's how to get an exact answer. First, rewrite the logarithm in exponential form, using the definition of a logarithm:

$$\log_3 32 = x$$

$$3^x = 32$$

Then, as usual, take the common logarithm of both sides:

$$\log 3^x = \log 32$$

And then use the Power Rule to pull the exponents out:

$$x \log 3 = \log 32$$

$$x = \frac{\log 32}{\log 3}$$

$$x = \frac{1.5051}{0.4771}$$

$$x = 3.1546$$

And there's the exact value of x.

Drill

In the following examples, use the techniques you've just seen to solve these exponential and logarithmic equations. The answers to all drills are found in chapter 12.

1. If $2^4 = 3^x$, then $x =$
2. $\log_5 18 =$
3. If $10^n = 137$, then $n =$
4. $\log_{12} 6 =$
5. If $4^x = 5$, then $4^{x+2} =$
6. $\log_2 50 =$
7. If $3^x = 7$, then $3^{x+1} =$
8. If $\log_3 12 = \log_4 x$, then $x =$

Natural Logarithms
IIC ONLY

On the Math IIC Test, you may run into a special kind of logarithm, called a natural logarithm. Natural logarithms are logs with a base of e, a constant that is approximately equal to 2.718.

The constant e is a little like π; it's a decimal number that goes on forever without repeating itself, and, like π, it's a basic feature of the universe. Just as π is the ratio of a circle's circumference to its diameter, no matter what, e is a basic feature of growth and decay in economics, physics, and even in biology.

The role of e in the mathematics of growth and decay is a little complicated. Don't worry about that, because you don't need to know very much about e for the Math IIC test. Just memorize a few rules and you're ready to go.

Natural logarithms are so useful in math and science that there's a special notation for expressing them. The expression $\ln x$ (which is read as, "ell-enn x") means the log of x to the base e, or $\log_e x$. That means that there are three different ways to express a natural logarithm:

Definition of a Natural Logarithm

$$\ln n = x \qquad \log_e n = x \qquad e^x = n$$

You can use the definition of the natural logarithm to solve equations that contain an e^x term. Since e equals 2.718281828..., there's no easy way to raise it to a specific power. By rearranging the equation into a natural logarithm in "ln x" form, you can make your calculator do the hard work for you. Here's a simple example:

19. If $e^x = 6$, then $x =$

    (A) 0.45
    (B) 0.56
    (C) 1.18
    (D) 1.79
    (E) 2.56

The equation in the question, $e^x = 6$, can be converted directly into a logarithmic equation using the definition of a logarithm. It would then be written as $\log_e 6 = x$, or $\ln 6 = x$. To find the value of x, just punch 6 into your calculator and hit the "ln x" key. You'll find that $x = 1.791759$. The correct answer is (D).

For the Math IIC Test, you may also have to know the shapes of the two basic graphs associated with natural logs. Here they are:

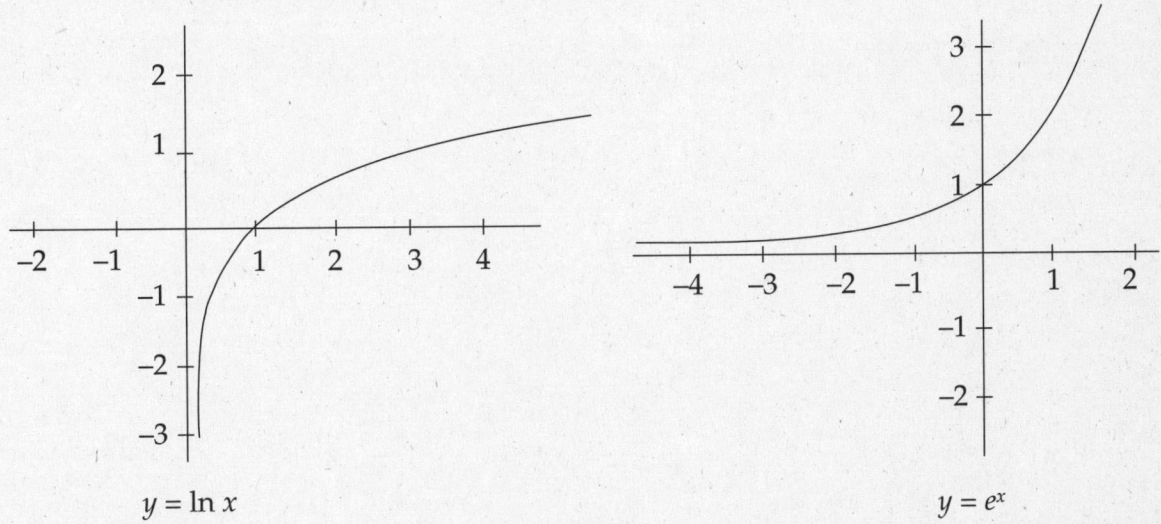

$y = \ln x \qquad\qquad\qquad y = e^x$

Finally, some questions may require you to estimate the value of e to answer a question. Just remember that $e \approx 2.718$. If you forget the value of e, you can always get your calculator to give it to you. Just punch in 1, and then hit the INVERSE key followed by the "ln x" key. The result will be e to the first power, which is just plain e.

Drill

18. If $e^z = 8$, then $z =$

    (A) 1.74
    (B) 2.08
    (C) 2.35
    (D) 2.94
    (E) 3.04

23. If Set $M = \{\pi, e, 3\}$, then which of the following shows the elements in set M in descending order?

    (A) $\{\pi, e, 3\}$
    (B) $\{e, 3, \pi\}$
    (C) $\{\pi, 3, e\}$
    (D) $\{3, \pi, e\}$
    (E) $\{3, e, \pi\}$

38. If $6e^{\frac{n}{3}} = 5$, then what is the value of n?

    (A) −0.55
    (B) −0.18
    (C) 0.26
    (D) 0.64
    (E) 1.19

FLASHCARDS

The rules and formulas listed below represent the most important points to study in this chapter. Memorize them by covering up the right column and testing yourself on the left column, or, better still, by making real flashcards as directed.

Front of Card	Back of Card
What does PEMDAS stand for?	• Parentheses • Exponents • Multiplication and Division • Addition and Subtraction
What is the formula for percent change?	$\dfrac{change}{original} = \dfrac{x}{100}$
The average of a set of values =	the sum of the values divided by the number of values.
What is the formula for a repeated percent increase?	Final = Original · (1 + rate)$^{\text{\# of changes}}$
What is the exponential equivalent of $\log_b n = x$?	$b^x = n$

ARITHMETIC ◆ 45

Front of Card   **Back of Card**

$\log_b(xy) =$ $\log_b x + \log_b y$

$\log_b \dfrac{x}{y} =$ $\log_b x - \log_b y$

$\log_b(x^r) =$ $r \log_b x$

Algebra

ALGEBRA ON THE SUBJECT TESTS

Algebra questions will make up about 30 percent of the questions on the Math IC Test, and about 20 percent of the questions on the Math IIC Test. Many of these questions are best answered by using the simple algebra rules outlined in this chapter; others can be shortcut by Princeton Review techniques, which you'll also find in the following pages.

Definitions

Here are some algebraic terms that will appear on the Math Subject Tests. Make sure you're familiar with them. If the meaning of any of these vocabulary words keeps slipping your mind, add those words to your flashcards.

Variable: An unknown quantity in an equation, represented by a letter.

Constant: An unchanging numerical quantity—in short, a number.

Term: An algebraic unit consisting of constants and variables multiplied together, such as $(5x)$ or $(9x^2)$.

Polynomial: An algebraic expression consisting of more than one term joined by addition or subtraction. For example: $x^3 - 3x^2 + 4x - 5$ is a polynomial with four terms.

Binomial: A polynomial with exactly two terms, like $(x - 5)$.

Quadratic: A quadratic expression is a polynomial with one variable whose largest exponent is a 2. For example: $x^2 - 5x + 6$. The equation $y = x^2 + 4$ is an example of a quadratic equation.

Root: A root of a polynomial is a value of the variable that makes the polynomial equal to zero. More generally, the roots of an equation are the values that make the equation true.

SETTING UP EQUATIONS

Many questions on the Math Subject Tests will require you to solve simple algebraic equations. Often these algebra questions are in the form of word problems. Setting up an equation from the information contained in a word problem is the first step to finding the solution—and the step at which many careless mistakes are made. The translation chart on page 24 is very useful for setting up equations from information given in English.

SOLVING EQUATIONS

An algebraic equation is an equation that contains at least one unknown—a variable. "Solving" for an unknown means figuring out what its value is. Generally, the way to solve for an unknown is to *isolate the variable*—that is, manipulate the equation until the unknown is alone on one side of the equal sign. Whatever's on the other side of the equal sign is the value of the unknown. Take a look at this example:

$$5(3x^3 - 16) - 22 = 18$$

In this equation, x is the unknown. To solve for x, you need to get x alone. You isolate x by undoing everything that's being done to x in the equation: If x is being squared, you need to take a square root; if x is being multiplied by 3, you need to divide by 3; if x is being decreased by 4, you need to add 4, and so on. The trick is doing these things in the right order. Basically, you should follow PEMDAS in reverse: Start by undoing addition and subtraction, then multiplication and division, then exponents, and, lastly, what's in parentheses.

The other thing to remember is that any time you do something to one side of an equation, you've got to do it to the other side also. Otherwise you'd be changing the equation, and you're trying to *rearrange* it, not change it. In this example, you'd start by undoing the subtraction:

$$5(3x^3 - 16) - 22 = 18$$
$$+ 22 + 22$$
$$5(3x^3 - 16) = 40$$

Then undo the multiplication by 5, saving what's in the parentheses for last:

$$
\begin{aligned}
5(3x^3 - 16) &= 40 \\
\div 5 \quad &\quad \div 5 \\
3x^3 - 16 &= 8
\end{aligned}
$$

Once you've gotten down to what's in the parentheses, follow PEMDAS in reverse again—first the subtraction, then the multiplication, and the exponent last:

$$
\begin{aligned}
3x^3 - 16 &= 8 \\
+ 16 \quad &\quad + 16 \\
3x^3 &= 24 \\
\div 3 \quad &\quad \div 3 \\
x^3 &= 8 \\
x &= 2
\end{aligned}
$$

At this point, you've solved the equation. You have found that the value of x must be 2. Another way of saying this is that 2 is the root of the equation $5(3x^3 - 16) - 22 = 18$. Equations containing exponents may have more than one root (see page 32 of the "Exponents" section of the last chapter).

Drill

Practice solving equations in the following examples. Remember that some equations may have more than one root. The answers to all drills are found in chapter 12.

1. If $\dfrac{(3x^2 - 7)}{17} = 4$, then $x =$

2. If $n^2 = 5n$, then $n =$

3. If $\dfrac{2a - 3}{3} = -\dfrac{1}{2}$, then $a =$

4. If $\dfrac{5s + 3}{3} = 21$, then $s =$

5. If $\dfrac{3(8x - 2) + 5}{5} = 4$, then $x =$

Factoring and Distributing

When manipulating algebraic equations, you'll need to use the tools of factoring and distributing. These are simply ways of rearranging equations to make them easier to work with.

Factoring

Factoring simply means finding some factor that is in every term of an expression and "pulling it out." By "pulling it out," we mean dividing each individual term by that factor, and then placing the whole expression in parentheses with that factor on the outside. Here's an example:

$$x^3 - 5x^2 + 6x = 0$$

On the left side of this equation, every term contains at least one x—that is, x is a factor of every term in the expression. That means you can factor out an x:

$$x^3 - 5x^2 + 6x = 0$$

$$x(x^2 - 5x + 6) = 0$$

The new expression has exactly the same value as the old one; it's just written differently, in a way that might make your calculations easier. Numbers as well as variables can be factored out, as seen in the example below:

$$17c - 51 = 0$$

On the left side of this equation, every term is a multiple of 17. Because 17 is a factor of each term, you can pull it out:

$$17c - 51 = 0$$
$$17(c - 3) = 0$$
$$c - 3 = 0$$
$$c = 3$$

As you can see, factoring can make equations easier to solve.

Distributing

Distributing is factoring in reverse. When an entire expression in parentheses is being multiplied by some factor, you can "distribute" the factor into each term, and get rid of the parentheses. For example:

$$3x(4 + 2x) = 6x^2 + 36$$

On the left side of this equation the parentheses make it difficult to combine terms and simplify the equation. You can get rid of the parentheses by distributing:

$$3x(4 + 2x) = 6x^2 + 36$$
$$12x + 6x^2 = 6x^2 + 36$$

And suddenly, the equation is much easier to solve:

$$12x^2 + 6x^2 = 6x^2 + 36$$

$$-6x^2 - 6x^2$$

$$12x = 36$$

$$x = 3$$

Drill

Practice a little factoring and distributing in the following examples, and keep an eye out for equations that could be simplified by this kind of rearrangement. The answers to all drills are found in chapter 12.

3. If $(11x)(50) + (50x)(29) = 4{,}000$, then $x =$

   (A) 2,000
   (B) 200
   (C) 20
   (D) 2
   (E) 0.2

17. $\dfrac{-3b(a+2)+6b}{-ab} =$

    (A) -3
    (B) -2
    (C) 0
    (D) 1
    (E) 3

36. $\dfrac{x^5 + x^4 + x^3 + x^2}{x^3 + x^2 + x + 1} =$

    (A) $4x^2$
    (B) x^2
    (C) $4x$
    (D) x
    (E) 4

Algebraic Functions

Algebra questions sometimes take the form of functions. A function is a set of algebraic instructions. On the Math IC Test, these functions are sometimes represented by symbols, like this:

13. If $\Diamond a \Diamond = a^2 - 5a + 4$, then $\Diamond 6 \Diamond =$

    (A) 6
    (B) 8
    (C) 10
    (D) 12
    (E) 14

Answer the question by putting the 6 into the definition of the function everywhere a is found:

$$(6)^2 - 5(6) + 4 =$$
$$36 - 30 + 4 =$$
$$10$$

The answer is (C). Don't be confused if a question requires you to plug something strange into a function. Just follow the instructions, and the answer will become clear:

17. If $y\S = y^2 - 6$, then which of the following equals $(y + 6)\S$?

    (A) y^2
    (B) $y^2 - 36$
    (C) $2y - 36$
    (D) $y^2 + 12y + 30$
    (E) $y^2 + 12y + 42$

To find the answer, just plug $(y + 6)$ into the definition of the function:

$$(y + 6)^2 - 6 =$$
$$(y^2 + 12y + 36) - 6 =$$
$$y^2 + 12y + 30$$

And (D) is the correct answer.

Drill

Practice your techniques on the following function questions. The answers to all drills are found in chapter 12.

27. If $[x] = -|x^3|$, then $[4] - [3] =$

    (A) -91
    (B) -37
    (C) -1
    (D) 37
    (E) 91

30. If $¥c$ is defined as $5(c - 2)^2$, then $¥5 + ¥6 =$

    (A) $¥7$
    (B) $¥8$
    (C) $¥9$
    (D) $¥10$
    (E) $¥11$

$$\S a = \begin{cases} a \text{ if } a \text{ is even} \\ -a \text{ if } a \text{ is odd} \end{cases}$$

34. $\S1 + \S2 + \S3 \ldots \S100 + \S101 =$

    (A) -151
    (B) -51
    (C) 0
    (D) 50
    (E) 51

PLUGGING IN

Plugging in is a technique for short-cutting algebra questions. It works on a certain class of algebra questions in which relationships are defined, but no real numbers are introduced. For example:

11. The use of a neighborhood car wash costs n dollars for a membership, and p cents for each wash. If a membership includes a bonus of 4 free washes, which of the following reflects the cost, in dollars, of getting a membership at the car wash and washing a car q times, if q is greater than 4?

    (A) $100n + pq - 4p$
    (B) $n + 100pq - 25p$
    (C) $n + pq - \dfrac{p}{25}$
    (D) $n + \dfrac{pq}{100} - \dfrac{p}{25}$
    (E) $n + \dfrac{p}{100} - \dfrac{q}{4}$

Here, there are no numbers assigned to most of the quantities in the question. All the question tells you is how those quantities are related. Assembling an algebraic equation to describe the relationship between these variables can be tricky—especially since, in this question, you also have to convert from dollars to cents in there somewhere.

The idea behind "plugging in" is that if these relationships are true, then it doesn't matter *what* numbers you put into the question; you'll always arrive at the same answer choice. So the easiest way to get through the question is to plug in easy numbers, follow them through the question, and see which answer choice they lead you to. Let's see how plugging in would work on this example:

Let's start with n, the membership fee: Plug in an easy number like 3, so that a membership costs $3.00.

Then, plug in a number for p, the charge per wash. Since this number is in cents, and we'll need to convert it to dollars in the answers, choose a number that can be converted easily to dollars, like 200. Let's make $p = 200$, so a wash costs $2.00.

Lastly, let's say that q, the number of washes, is 5. That's as easy as it gets—with 4 free washes, you're paying for only 1.

Then, just work out the answer to the question using *your numbers*. How much does it cost for a membership and 5 washes? Well, that's $3.00 for a membership, 4 washes free, and 1 wash for $2.00. The total is $5.00. That means that if you plug your numbers into the answer choices, the right answer should give you 5. We call that your target number—the number you go looking for in the answer choices.

When you plug $n = 3$, $p = 200$, and $q = 5$ into the answer choices, the only answer choice that gives you 5 is (D). That's the right answer, and you're done.

Take a look at one more:

13. The size of an art collection is tripled, and then 70% of the collection is sold. Acquisitions then increase the size of the collection by 10%. The size of the art collection is then what percent of its size before these three changes?

    (A) 240 percent
    (B) 210 percent
    (C) 111 percent
    (D) 99 percent
    (E) 21 percent

Here again, you're given relationships between quantities (before and after, in this case), but no actual numbers. That's a sign that you can plug in whatever numbers you like. The unknown quantity here is the size of the collection to start with. Since you're working with percentages, 100 is a good number to plug in—it'll make your math easy.

You start with a collection of 100 items. It's tripled, meaning it increases to 300. Then it's decreased by 70%. That's a decrease of 210, so the collection's size decreases to 90. Then, finally, it increases by 10%. That's an increase of 9, for a final collection size of 99. Since the collection began at 100, it's now at 99% of its original size. The answer is (D). It doesn't matter what number you choose for the original size of the collection—you'll always get the right answer. The trick to choosing numbers is picking ones that makes your math easy.

Occasionally, more than one answer choice will produce the correct answer. When that happens, eliminate the answer choices that didn't work out, and plug in a different set of numbers. The new numbers will produce a new target number. Use this new target number to eliminate the remaining incorrect answer choices. You will rarely have to plug in more than two sets of numbers.

You can plug in whenever:

- the answer choices contain variables, percentages, or ratios
- there are unknowns in the questions
- the question seems to call for an algebraic equation

When plugging in, keep a few simple rules in mind:

- Avoid plugging in 1 or 0, which often make more than one answer choice produce the same number. For the same reason, avoid plugging in numbers that appear in the answer choices—they're more likely to cause several answer choices to produce your target number.

- Plug in numbers that make your math easy: 2, 3, and 5 are good choices in ordinary algebra. Multiples of 100 are good in percentage questions, and multiples of 60 are good in questions dealing with seconds, minutes, and hours.

Plugging in offers several notable advantages. By plugging in numbers, you're checking your math as you do it; with algebra, it takes an extra step to check your work with numbers. You can also plug in even when you don't know how to set up an algebraic equation. When algebra confuses you, always see if plugging in is an option. Finally, plugging in offers fewer opportunities for careless errors than algebra does. Plugging in is safer.

Drill

Try solving the following practice questions by plugging in. Remember to check all your answer choices, and plug in a second set of numbers if more than one answer choice produces your target number. The answers to all drills are found in chapter 12.

15. The price of an item in a store is p dollars. If the tax on the item is $t\%$, what is the total cost in dollars of n such items, including tax?

    (A) npt
    (B) $npt+1$
    (C) $\dfrac{np(t+1)}{100}$
    (D) $100n(p+pt)$
    (E) $\dfrac{np(t+100)}{100}$

23. Vehicle A travels at x miles per hour for x hours. Vehicle B travels a miles per hour faster than Vehicle A, and travels b hours longer than Vehicle A. Vehicle B travels how much farther than Vehicle A, in miles?

    (A) $x^2 - ab$
    (B) $a^2 + b^2$
    (C) $ax + bx + ab$
    (D) $x^2 + abx + ab$
    (E) $2x^2 + (a+b)x + ab$

33. For any real number n, $|5-n| - |n-5| =$

    (A) -2
    (B) -1
    (C) 0
    (D) 1
    (E) 2

38. If Company A builds a skateboards per week, and Company B builds b skateboards per day, then in m weeks, Company A builds how many more skateboards than Company B?

    (A) $7bm$
    (B) $m(a-7b)$
    (C) $7(ma-mb)$
    (D) $7m(a-b)$
    (E) $\dfrac{m(a-b)}{7}$

BACKSOLVING

Backsolving is another approach to algebra questions. It uses numbers instead of algebra to find the answer. As you've just seen, "plugging in" is useful on questions whose answer choices contain variables, percentages, or ratios—not actual numbers. Backsolving, on the other hand, is useful on questions whose answer choices *do* contain actual numbers.

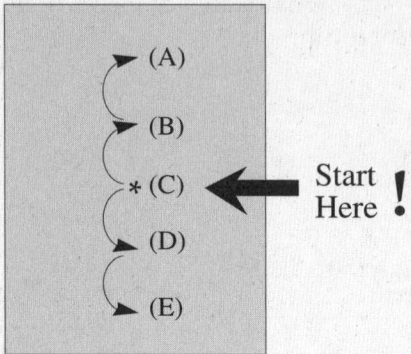

To backsolve an algebra question, take (C), the middle answer choice, and stick it back into the problem. If it makes all of the statements in the question true, then it's the right answer. If it doesn't, eliminate (C) and try another answer choice. Usually, you'll know from your experience with (C) whether you want to try a smaller or larger answer choice. If (C) is too small, you can eliminate answer choices (A) and (B) since the answer choices are arranged in numerical order. If (C) is too large, try again from choices (A) and (B).

Like plugging in, backsolving can open doors for you when you're unsure how to approach a question with algebra. Also like plugging in, it checks your answers as you pick them, eliminating careless errors. Particularly at the tough end of a Math Subject Test, where you're getting into hard material, plugging in and backsolving can enable you to solve problems that would otherwise stump you.

Let's take a look at a backsolving example:

10. A duck travels from point A to point B. If the duck flies $\frac{3}{4}$ of the way, walks $\frac{1}{9}$ of the way, and swims the remaining 10 kilometers of her trip, what is the total distance traveled by the duck?

    (A) 36
    (B) 45
    (C) 56
    (D) 72
    (E) 108

To backsolve this question, you'd start with answer choice (C). The answer choices represent the quantity asked for in the question—in this case, the total distance traveled by the duck. Answer choice (C), therefore, means that the duck traveled a total distance of 56 kilometers. Follow this information through the problem:

The duck flies $\frac{3}{4}$ of the way: $\frac{3}{4}$ of 56 is 42 kilometers.

The duck walks $\frac{1}{9}$ of the way: $\frac{1}{9}$ of 56 is 6.22 kilometers.

That makes 48.22 kilometers, which leaves 7.78 kilometers in the trip.

BUT the duck swims 10 kilometers!

That means that (C) isn't the right answer. It also tells you that you need a longer trip to fit in all that flying, walking, and swimming; move down to (D), the next largest answer, and try again.

The duck flies $\frac{3}{4}$ of the way: $\frac{3}{4}$ of 72 is 54 kilometers.

The duck walks $\frac{1}{9}$ of the way: $\frac{1}{9}$ of 72 is 8 kilometers.

That makes 62 kilometers, which leaves 10 kilometers in the trip.

And THAT'S exactly how far the duck swims, 10 kilometers.

Right answer. Generally, you'll never have to try more than three answer choices when backsolving—and often, the first or second answer you try will be correct. Keep your eyes open for backsolving opportunities on the Math Subject Tests, particularly when you run into an algebra question that you're not sure how to solve.

Drill

Solve the following questions by backsolving. Remember to start with (C), the middle answer choice. The answers to all drills are found in chapter 12.

11. Matt has 4 more hats than Aaron and half as many hats as Michael. If the three together have 24 hats, how many hats does Michael have?

    (A) 7
    (B) 9
    (C) 12
    (D) 14
    (E) 18

17. A shipment of 3,200 items is divided into 2 portions so that the difference between the portions is one half of their average. What is the ratio of the smaller to the larger portion?

    (A) 1 : 2
    (B) 1 : 3
    (C) 2 : 5
    (D) 3 : 5
    (E) 5 : 8

27. Three distinct positive integers have a sum of 15 and a product of 45. What is the largest of these integers?

   (A) 1
   (B) 3
   (C) 5
   (D) 9
   (E) 15

INEQUALITIES

Inequalities are equations that aren't *quite* equations. They're not statements of equality, but of inequality. There are four basic inequality signs, which should be read like this:

$a < b$	a is less than b.
$a > b$	a is greater than b.
$a \leq b$	a is less than or equal to b.
$a \geq b$	a is greater than or equal to b.

Apart from the different meanings of the inequality signs, inequalities can generally be treated just like equations: You can add, subtract, multiply, and divide on both sides of the inequality sign, and you still solve by isolating the variable.

There's just one important point to remember:

> Whenever you multiply or divide both sides of an inequality by a negative, the inequality sign switches directions.

Multiplying across an inequality by a negative flips the signs of all of the terms in the inequality. The inequality sign itself must also flip:

$$4n - 20 > -3n + 15 \qquad\qquad x \geq 5$$
$$-4n + 20 < 3n - 15 \qquad\qquad -x \leq -5$$

As long as you remember this rule, you can treat inequalities just like equations and use all of your usual algebra tools in solving them.

Drill

Practice solving inequalities in the following exercises. The answers to all drills are found in chapter 12.

1. If $\dfrac{6(5-n)}{4} \leq 3$, then _____

2. If $\dfrac{r+3}{2} < 5$, then _____

3. If $\dfrac{4(1-x)+9}{3} \leq 5$, then _____

4. If $8(3x + 1) + 4 < 15$, then _____

5. If $23 - 4t \geq 11$, then _____

6. If $4n - 25 \leq 19 - 7n$, then _____

7. If $-5(p+2) < 10p - 13$, then _____

8. If $\dfrac{23s+7}{10} \geq 2s + 1$, then _____

9. If $-3x - 16 \leq 2x + 19$, then _____

10. If $\dfrac{14s-11}{9} \geq s - 1$, then _____

Working with Ranges

A variable's value may sometimes be defined by a range of values, instead of a single numerical value. For example:

> At a certain amusement park, anyone under 12 years of age is not permitted to ride the Stupendous Hurlcoaster, because the person could easily lose his or her mind due to the ride's extreme funkiness. Anyone over 60 years of age is also prohibited from the ride, as the incredible velocity of the Hurlcoaster may cause spontaneous coronary explosion. If x is the age of a rider of the Stupendous Hurlcoaster, what is the range of possible values of x?

The end values of the range are obviously 12 and 60. But are 12 and 60 included in the range themselves, or not? If you read carefully, you'll see that only those *under* 12 or *over* 60 are barred from riding the Hurlcoaster. If you're 12 or 60, you're perfectly legal. The range of possible values of x is therefore given by $12 \leq x \leq 60$. Noticing the difference between "greater than" and "greater than or equal to" is crucial to many range questions.

You can manipulate ranges in a couple of ways. You can add and subtract ranges, as long as their inequality signs point the same way. You can also multiply or divide across a range to produce new information, as long as you obey that basic rule of inequalities: Flip the sign if you multiply or divide by a negative.

> If the range of possible values for x is given by $-5 < x < 8$, find the range of possible values for the range of possible values for each of the following:
>
> 1. $-x$: _____
>
> 2. $4x$: _____
>
> 3. $x + 6$: _____
>
> 4. $(2-x)$: _____
>
> 5. $\dfrac{x}{2}$: _____

Adding ranges

Occasionally, a question on the Math Subject Tests will require you to add or subtract ranges. Take a look at this example:

> If $3 < a < 10$ and $-6 < b < 3$, what is the range of possible values of $a + b$?

Here, the range of $(a + b)$ will be the sum of the range of a and the range of b. The lower bound of $(a + b)$ is the lower bound of a plus the lower bound of b. The upper bound of $(a + b)$ is the sum of the upper bounds of a and b:

$$\begin{array}{r} 3 < a < 10 \\ + \; -6 < b < 3 \\ \hline -3 < a + b < 13 \end{array}$$

You can add any two ranges, but you must be careful about which way the signs are pointing in order to be sure that you're adding the upper bounds together and the lower bounds together. For example, in the following question, the ranges of x and y are given in different forms:

> If $7 < x < 13$ and $4 > y > -1$, and $z = x + y$, then what is the range of possible values of z?

In this case, in order to add the ranges of x and y, you can still simply add the two lower bounds to get the lower bound of z, and add the two upper bounds to get the upper bound of z. Just be sure not to get the upper and lower bounds confused.

$$\begin{array}{r} 7 < x < 13 \\ + \; -1 < y < 4 \\ \hline 6 < z < 17 \end{array}$$

Notice that in order to add these ranges by stacking them, it was necessary to turn the range-definition of y around, so that the signs of both ranges pointed in the same direction.

Subtracting ranges

To subtract one range from another, take the range that is being subtracted and multiply it by -1: This will flip the signs of all the terms in the inequality, as well as flipping the inequality signs themselves. Then, add the ranges normally:

> If $-4 < a < 5$ and $2 < b < 12$, then what is the range of possible values of $a - b$?

$$\begin{array}{r} -4 < a < 5 \\ + \; -12 < -b < -2 \\ \hline -16 < a - b < 3 \end{array} \qquad \begin{array}{c} 2 < b < 12 \\ -2 > -b > -12 \end{array}$$

Never stack the ranges and *subtract* them. You'll get an incorrect answer. Ranges can only be *added* in this way. In order to subtract, you must always invert the term being subtracted (multiply by -1) and then *add*.

Multiplying ranges

Although it's uncommon, you may get a question requiring you to *multiply* ranges. If you're asked to find the *product* of two ranges, find all of the products it's possible to get by multiplying one of the bounds of one limit by a bound of the other. The largest of these numbers will be the upper limit of the product; the smallest will be the lower limit:

If $-3 < f < 4$ and $-7 < g < 2$, then what is the range of possible values of fg?

These are the four possible products of the bounds of f and g:

$(-3)(-7) = 21$ $(-3)(2) = -6$
$(4)(-7) = -28$ $(4)(2) = 8$

The greatest of these values is 21; the least is –28; and so the range of possible values of fg is:

$$-28 < fg < 21$$

And that's all there is to multiplying ranges.

Drill

Try the following range questions. The answers to all drills are found in chapter 12.

1. If $-2 \leq a \leq 7$ and $3 \leq b \leq 9$, then what is the range of possible values of $b - a$?
2. If $2 \leq x \leq 11$ and $6 \geq y \geq -4$, then what is the range of possible values of $x + y$?
3. If $-3 \leq n \leq 8$, then what is the range of possible values of n^2?
4. If $0 < x < 5$ and $-9 < y < -3$, then what is the range of possible values of $x - y$?
5. If $-3 \leq r \leq 10$ and $-10 \leq s \leq 3$, then what is the range of possible values of $r + s$?
6. If $-6 < c < 0$ and $13 < d < 21$, then what is the range of possible values of cd?

DIRECT AND INDIRECT VARIATION

Direct and indirect variation are specific relationships between quantities. Quantities that vary directly are said to be *in proportion* or *proportional*. Quantities that vary indirectly are said to be *inversely proportional*.

Direct variation

If x and y are in direct variation, that can be said in several ways: x and y are in proportion; x and y change proportionally; or x varies directly as y. All of these descriptions come down to the same thing: x and y increase and decrease together. Specifically, they mean that the quantity $\frac{x}{y}$ will always have the same numerical value. That's all there is to it. Take a look at a question based on this idea:

17. If n varies directly as m, and n is 3 when m is 24, then what is the value of n when m is 11?

    (A) 1.375
    (B) 1.775
    (C) 1.95
    (D) 2.0
    (E) 2.125

ALGEBRA ◆ 61

To solve the problem, use the definition of direct variation: $\frac{n}{m}$ must always have the same numerical value. Set up a proportion:

$$\frac{3}{24} = \frac{n}{11}$$

And solve by cross multiplying and isolating n:

$$24n = 33$$

$$n = 33 \div 24$$

$$n = 1.375$$

And that's all there is to it. The correct answer is (A). All direct variation questions can be answered this way.

Indirect variation

If x and y are in indirect variation, that can be said in several ways as well: x and y are in inverse proportion; x and y are inversely proportionate; or x varies indirectly as y. All of these descriptions come down to the same thing: x increases when y decreases, and decreases when y increases. Specifically, they mean that the quantity xy will always have the same numerical value.

Take a look at this question based on indirect variation:

15. If a varies indirectly as b and $a = 3$ when $b = 5$, then what is the value of a when $b = 7$?

    (A) 2.14
    (B) 2.76
    (C) 3.28
    (D) 4.2
    (E) 11.67

To answer the question, use the definition of indirect variation: The quantity ab must always have the same value. Therefore, you can set up this simple equation:

$$3 \times 5 = a \times 7$$
$$7a = 15$$
$$a = 15 \div 7$$
$$a = 2.142857$$

So the correct answer is (A). All indirect variation questions on the Math Subject Tests can be handled this way.

Drill

Try these practice exercises using the definitions of direct and indirect variation. The answers to all drills are found in chapter 12.

15. If a varies indirectly as b, and $a = 3$ when $b = 5$, then what is the value of a when $b = x$?

    (A) $\dfrac{3}{x}$
    (B) $\dfrac{5}{x}$
    (C) $\dfrac{15}{x}$
    (D) $3x$
    (E) $3x^2$

18. If n varies directly as m, and $n = 5$ when $m = 4$, then what is the value of n when $m = 5$?

    (A) 4.0
    (B) 4.75
    (C) 5.5
    (D) 6.25
    (E) 7.75

24. If p varies directly as q, and $p = 3$ when $q = 10$, then what is the value of p when $q = 1$?

    (A) 0.3
    (B) 0.33
    (C) 0.5
    (D) 3.3
    (E) 3.33

26. If r and s are inversely proportional, and $r = 7$ when $s = 3$, then what is r when $s = 4$?

    (A) 2.50
    (B) 3.75
    (C) 4.10
    (D) 5.25
    (E) 6.5

WORK AND TRAVEL QUESTIONS

Word problems dealing with work and travel tend to cause a lot of careless mistakes, because the relationships among distance, time, and speed—or among work-rate, work, and time—sometimes confuse test takers. When working with questions about travel, just remember this:

$$\text{distance} = \text{rate} \times \text{time}$$

When working with questions about work being done, remember this:

work done = rate of work × time

Drill

Answer the following practice questions using these formulas. The answers to all drills are found in chapter 12.

11. A factory contains a series of water tanks, all of the same size. If Pump 1 can fill 12 of these tanks in a 12-hour shift, and Pump 2 can fill 11 tanks in the same time, then how many tanks can the two pumps fill, working together, in 1 hour?

    (A) 0.13
    (B) 0.35
    (C) 1.92
    (D) 2.88
    (E) 3.33

12. A projectile travels 227 feet in one second. If there are 5,280 feet in 1 mile, then which of the following best approximates the projectile's speed in miles per hour?

    (A) 155
    (B) 170
    (C) 194
    (D) 252
    (E) 333

18. A train travels from Langston to Hughesville and back in 5.5 hours. If the two towns are 200 miles apart, what is the average speed of the train in miles per hour?

    (A) 36.36
    (B) 72.73
    (C) 109.09
    (D) 110.10
    (E) 120.21

25. Jules can make m muffins in s minutes. Alice can make n muffins in t minutes. Which of the following gives the number of muffins that Jules and Alice can make together in 30 minutes?

    (A) $\dfrac{m+n}{30st}$
    (B) $\dfrac{30(m+n)}{st}$
    (C) $30(mt+ns)$
    (D) $\dfrac{30(mt+ns)}{st}$
    (E) $\dfrac{mt+ns}{30st}$

AVERAGE SPEED

The "average speed" question is a specialized breed of travel question. Here's what a basic "average speed" question might look like:

15. Roberto travels from his home to the beach, driving at 30 mph. He returns along the same route at 50 mph. If the distance from Roberto's house to the beach is 10 miles, then what is Roberto's average speed for the round trip in miles per hour?

    (A) 32.5
    (B) 37.5
    (C) 40.0
    (D) 42.5
    (E) 45.0

The easy mistake to make on this question is to simply choose answer choice (C), the average of the two speeds. Average speed isn't found by averaging speeds, however. Instead, you have to use this formula:

$$\text{average speed} = \frac{\text{total distance}}{\text{total time}}$$

The total distance is easy to figure out: 10 miles each way is a total of 20 miles. Total time is a little trickier. For that, you have to use the "distance = rate × time" formula. Here, it's useful to rearrange this equation to read this way:

$$\text{time} = \frac{\text{distance}}{\text{rate}}$$

On the way to the beach, Roberto traveled 10 miles at 30 mph, which took 0.333 hours, according to the formula. On the way home, he traveled 10 miles at 50 mph, which took 0.2 hours. That makes for 20 miles in a total of .533 hours. Plug those numbers into the average-speed formula, and you get an average speed of 37.5 mph. The answer is (B).

Here's a general tip for "average speed" questions: On any round-trip in which the traveler moves at one speed heading out and another speed returning, the traveler's average speed will be a little lower than the average of the two speeds.

Drill

Try these "average speed" questions. The answers to all drills are found in chapter 12.

19. Alexandra jogs from her house to the lake at 12 mph, and jogs back by the same route at 9 mph. If the path from her house to the lake is 6 miles long, what is her average speed for the round trip?

    (A) 11.3
    (B) 11.0
    (C) 10.5
    (D) 10.3
    (E) 10.1

24. A truck travels 50 miles from town S to town T in 50 minutes, and then immediately drives 40 miles from town T to town U in 40 minutes. What is the truck's average speed in miles per hour, from town S to town U?

    (A) 1
    (B) 10
    (C) 45
    (D) 60
    (E) 90

33. Ben travels a certain distance at 25 mph and returns across the same distance at 50 mph. What is his average speed in miles per hour for the round-trip?

    (A) 37.5
    (B) 33.3
    (C) 32.0
    (D) 29.5
    (E) It cannot be determined from the information given.

SIMULTANEOUS EQUATIONS

It's possible to have a set of equations that can't be solved individually but can be solved in combination. A good example of such a set of equations would be:

$$x + 6 = y$$

$$y = 8$$

Neither of these equations by itself is enough to give you the value of x. But when you combine the equations, finding the answer is easy:

$$x + 6 = 8$$

$$x = 2$$

Solving systems of equations this way is called solving equations *simultaneously*. This term applies to any way of solving equations that works by combining multiple equations into one, but on the Math Subject Tests it has a more specific meaning. For the purposes of the Math Subject Tests, "simultaneous equations" are equations with very similar forms that can be solved only when they are added together or subtracted from one another. Here's an example of a system of simultaneous equations as they might appear on a Math Subject Test question:

7. If x and y are real numbers such that $3x + 4y = 10$ and $2x - 4y = 5$, then what is the value of x?

Each equation has two variables, and so cannot be solved alone. But if the two equations are added, the solution becomes easy:

$$\begin{aligned} 3x + 4y &= 10 \\ +\ 2x - 4y &= 5 \\ \hline 5x &= 15 \\ x &= 3 \end{aligned}$$

66 ♦ CRACKING THE SAT II: MATH SUBJECT TESTS

On the Math Subject Tests, solving simultaneous equations will generally involve adding or subtracting equations in order to cause one term to drop out (as the $4y$ did above), or rearranging the terms into the form demanded by the question:

8. If $12a - 3b = 131$ and $5a - 10b = 61$, then what is the value of $a + b$?

Here, the goal is not to isolate a variable, as we did last time, but to produce an equation that will show the value of $a + b$. In a sense, you're trying to isolate $a + b$. If you can produce an equation that has only a and b on one side, then the number on the other side will be the answer. Try to answer the question yourself by solving simultaneously, and then check the explanation below.

$$\begin{array}{r} 12a - 3b = 131 \\ -5a - 10b = 61 \\ \hline 7a + 7b = 70 \\ a + b = 10 \end{array}$$

So the value of $a + b$ is 10. In this case, it was necessary to subtract one equation from the other. A little practice will enable you to see quickly when adding is a good idea, or when subtracting is probably better. Sometimes it may be necessary to multiply one of the equations by a convenient factor to make terms that will cancel out properly. For example:

6. If $4n - 8m = 6$, and $-5n + 4m = 3$, then $n =$

$$4n - 8m = 6$$
$$-5n + 4m = 3$$

Here, it quickly becomes apparent that neither adding nor subtracting will combine these two equations very usefully. However, things look a little brighter when the second equation is multiplied by 2:

$$\begin{array}{r} 4n - 8m = 6 \\ -10n + 8m = 6 \\ \hline -6n = 12 \\ n = -2 \end{array} \qquad 2(-5n + 4m = 3)$$

Occasionally, a simultaneous equation can be solved only by *multiplying* all of the pieces together. This will generally be the case only when the equations themselves involve multiplication alone, and not the kind of addition and subtraction that the previous equations have contained. Take a look at this example:

$$ab = 3 \qquad bc = \frac{5}{9} \qquad ac = 15$$

34. If the above statements are true, what is one possible value of *abc*?

    (A) 5.0
    (B) 8.33
    (C) 9.28
    (D) 18.54
    (E) 25.0

This is a tough one. No single one of the three small equations can be solved by itself. In fact, no two of them together can be solved. It takes all three to solve the system, and here's how it's done:

$$ab \times bc \times ac = 3 \times \frac{5}{9} \times 15$$
$$aabbcc = 25$$
$$a^2b^2c^2 = 25$$

Once you've multiplied all three equations together, all you have to do is take the square roots of both sides, and you've got a value for *abc*:

$$a^2b^2c^2 = 25$$
$$abc = 5, -5$$

And so (A) is the correct answer.

Drill

Try answering the following practice questions by solving equations simultaneously. The answers to all drills are found in chapter 12.

26. If $a + 3b = 6$, and $4a - 3b = 14$, $a =$

    (A) –4
    (B) 2
    (C) 4
    (D) 10
    (E) 20

31. If $2x - 7y = 12$ and $-8x + 3y = 2$, which of the following is the value of $x - y$?

    (A) 12.0
    (B) 8.0
    (C) 5.5
    (D) 1.0
    (E) 0.8

$$ab = \frac{1}{8}, bc = 6, ac = 3$$

34. If all of the above statements are true, what is one possible value of *abc*?

    (A) 3.75
    (B) 2.25
    (C) 2.0
    (D) 1.5
    (E) 0.25

37. If $xyz = 4$ and $y^2z = 5$, what is the value of $\frac{x}{y}$?

    (A) 20.0
    (B) 10.0
    (C) 1.25
    (D) 1.0
    (E) 0.8

FOIL

A binomial is an algebraic expression that has two terms (pieces connected by addition or subtraction). FOIL is how to multiply two binomials together. The letters of FOIL stand for:

First
$$(x - 3)(x + 2) = x^2$$

Outside
$$(x - 3)(x + 2) = x^2 + 2x$$

Inside
$$(x - 3)(x + 2) = x^2 + 2x - 3x$$

Last
$$(x - 3)(x + 2) = x^2 - x - 6$$

Suppose you wanted to do the following multiplication:

$$(x + 5)(x - 2)$$

You would multiply the two *first* terms together: $x \times x = x^2$

And then the *outside* terms: $x \times -2 = -2x$

And then the *inside* terms: $5 \times x = 5x$

And finally the two *last* terms: $5 \times -2 = -10$

String the four products together and simplify them to produce an answer:

$$x^2 - 2x + 5x - 10$$
$$x^2 + 3x - 10$$

And that's the product of $(x + 5)$ and $(x - 2)$.

Drill

Practice using FOIL on the following binomial multiplications. The answers to all drills are found in chapter 12.

1. $(x - 2)(x + 11) =$
2. $(b + 5)(b + 7) =$
3. $(x - 3)(x - 9) =$
4. $(2x - 5)(x + 1) =$
5. $(n^2 + 5)(n - 3) =$
6. $(3a + 5)(2a - 7) =$
7. $(x - 3)(x - 6) =$
8. $(c - 2)(c + 9) =$
9. $(d + 5)(d - 1) =$

Factoring Quadratics

An expression like $x^2 + 3x + 10$ is a *quadratic* polynomial. A quadratic is an expression that fits into the general form $ax^2 + bx + c$, with a, b, and c as constants. An equation in general quadratic form looks like this:

General Form of a Quadratic Equation

$$ax^2 + bx + c = 0$$

Often, the best way to solve a quadratic equation is to factor it into two binomials—basically FOIL in reverse. Let's take a look at the quadratic we worked with in the previous section, and the binomials that are its factors.

$$x^2 + 3x - 10 = (x + 5)(x - 2)$$

Notice that the coefficient of the quadratic's middle term (3) is the sum of the constants in the binomials (5 and –2), and that the third term of the quadratic (–10) is the product of those constants. That relationship between a quadratic expression and its factors will always be true. To factor a quadratic, look for a pair of constants whose sum equals the coefficient of the middle term, and whose product equals the last term of the quadratic. Suppose you had to solve this equation:

$$x^2 - 6x + 8 = 0$$

Your first step would be to factor the quadratic polynomial. That means looking for a pair of numbers that add up to –6 and multiply to 8. Because their sum is negative but their product is positive, you know that the numbers are both negative. And as always, there's only one pair of numbers that fits the bill—in this case, –2 and –4:

$$x^2 - 6x + 8 = 0$$
$$(x - 2)(x - 4) = 0$$
$$x = \{2, 4\}$$

Two and four are the *roots* of the equation.

Once a quadratic is factored, it's easy to solve for x. The product of the binomials can be zero only if one of the binomials is equal to zero—and there are only two values of x that will make one of the binomials equal to zero: 2 and 4. The equation is solved.

Drill

Solve the following equations by factoring the quadratic polynomials. Write down all of the roots of each equation (values of x that make the equations true). The answers to all drills are found in chapter 12.

1. $a^2 - 3a + 2 = 0$
2. $d^2 + 8d + 7 = 0$
3. $x^2 + 4x - 21 = 0$
4. $3x^2 + 9x - 30 = 0$
5. $2x^2 + 40x + 198 = 0$
6. $p^2 + 10p = 39$
7. $c^2 + 9c + 20 = 0$
8. $s^2 + 4s - 12 = 0$
9. $x^2 - 3x - 4 = 0$
10. $n^4 - 3n^2 - 10 = 0$

Special Quadratic Identities

There are a few quadratic expressions that you should be able to factor at a glance. Because they are useful mathematically, and above all, because ETS likes to put them on the Math Subject Tests, you should memorize the following identities:

$$(x + y)^2 = x^2 + 2xy + y^2$$
$$(x - y)^2 = x^2 - 2xy + y^2$$
$$(x + y)(x - y) = x^2 - y^2$$

Here are some examples of these quadratic identities in action:

$$n^2 + 10n + 25 = (n + 5)^2 = (n + 5)(n + 5)$$
$$r^2 - 16 = (r + 4)(r - 4)$$
$$n^2 - 4n + 4 = (n - 2)^2 = (n - 2)(n - 2)$$

But knowing the quadratic identities will do more for you than just allow you to factor some expressions quickly. ETS writes questions based specifically on these identities. Such question are easy to solve if you remember these equations and use them, and quite tricky (or even impossible) if you don't. Here's an example:

36. If $a + b = 7$, and $a^2 + b^2 = 37$, then what is the value of ab?

    (A) 6
    (B) 12
    (C) 15
    (D) 22
    (E) 30

Algebraically, this is a tough problem to crack. You can't divide $a^2 + b^2$ by $a + b$ and get anything useful. In fact, most of the usual algebraic approaches to questions like these don't work here. Even plugging the answer choices back into the question (backsolving) isn't very helpful. What you *can* do is recognize that the question is giving you all of the pieces you need to build one of the quadratic identities: $(x + y)^2 = x^2 + 2xy + y^2$. To solve the problem, just rearrange the identity a little and plug in the values given by the question:

$$(a + b)^2 = a^2 + b^2 + 2ab$$

$$(7)^2 = 37 + 2ab$$

$$49 = 37 + 2ab$$

$$12 = 2ab$$

$$6 = ab$$

And presto, the answer appears. It's not easy to figure out what x or y is specifically—and you don't need to. Just find the value asked for in the question. If you remember the quadratic identities, solving the problem is easy.

Drill

Try solving the following questions using the quadratic identities, and take note of the clues that tell you when the identities will be useful. The answers to all drills are found in chapter 12.

17. If $n - m = -3$ and $n^2 - m^2 = 24$, then which of the following is the sum of n and m?

    (A) −8
    (B) −6
    (C) −4
    (D) 6
    (E) 8

19. If $x + y = 3$ and $x^2 + y^2 = 8$, then $xy =$

    (A) 0.25
    (B) 0.5
    (C) 1.5
    (D) 2.0
    (E) 2.25

24. If the sum of two nonzero integers is 9 and the sum of their squares is 36, then what is the product of the two integers?

    (A) 9.0
    (B) 13.5
    (C) 18.0
    (D) 22.5
    (E) 45.0

The Quadratic Formula

Unfortunately, not all quadratic equations can be factored by the reverse–FOIL method. The reverse–FOIL method is only practical when the roots of the equation are integers. Sometimes, however, the roots of a quadratic equation will be non–integer decimal numbers, and sometimes a quadratic equation will have no real roots at all. Consider this quadratic equation:

$$x^2 - 7x + 8 = 0$$

There *are* no integers that add up to –7 and multiply to 8. This quadratic cannot be factored in the usual way. To solve this equation, it's necessary to use the *quadratic formula*, a formula that produces the root or roots of any equation in the general quadratic form $ax^2 + bx + c = 0$.

The Quadratic Formula

$$x = \frac{-b \pm \sqrt{b^2 - 4ac}}{2a}$$

The a, b, and c in the formula refer to the coefficients of an expression in the form $ax^2 + bx + c$. For the equation $x^2 - 7x + 8 = 0$, $a = 1$, $b = -7$, and $c = 8$. Plug these values into the quadratic formula, and you get the roots of the equation:

$$x = \frac{-(-7) + \sqrt{(-7)^2 - 4(1)(8)}}{2(1)} \qquad x = \frac{-(-7) - \sqrt{(-7)^2 - 4(1)(8)}}{2(1)}$$

$$x = \frac{7 + \sqrt{49 - 32}}{2} \qquad x = \frac{7 - \sqrt{49 - 32}}{2}$$

$$x = \frac{7 + \sqrt{17}}{2} \qquad x = \frac{7 - \sqrt{17}}{2}$$

$$x = 5.56 \qquad x = 1.44$$

So the equation $x^2 - 7x + 8 = 0$ has two real roots, 5.56 and 1.44.

It's possible to tell quickly, without going all the way through the quadratic formula, how many roots an equation has. The part of the quadratic formula under the radical, $b^2 - 4ac$, is called the *discriminant*. The value of the discriminant gives you the following information about a quadratic equation:

- If $b^2 - 4ac > 0$, then the equation has two real roots.
- If $b^2 - 4ac = 0$, then the equation has one real root and is a perfect square.
- If $b^2 - 4ac < 0$, then the equation has no real roots. Both of its roots are imaginary.

Drill

In the following exercises, use the discriminant to find out how many roots each equation has, and whether the roots are real or imaginary. For equations with real roots, find the exact value of those roots using the quadratic formula. The answers to all drills are found in chapter 12.

1. $x^2 - 7x + 5 = 0$
2. $3a^2 - 3a + 7 = 0$
3. $s^2 - 6s + 4 = 0$
4. $x^2 - 2 = 0$
5. $n^2 + 5n + 6.25 = 0$

FLASHCARDS

The rules and formulas listed below represent the most important points to study in this chapter. Memorize them by covering up the right column and testing yourself on the left column, or, better still, by making real flashcards as directed.

Front of Card

distance =

Average speed =

What does "FOIL" stand for?

$(x + y)^2 =$

$(x - y)^2 =$

Back of Card

rate × time

$\dfrac{\text{total distance}}{\text{total time}}$

First, Outside, Inside, Last

$x^2 + 2xy + y^2$

$x^2 - 2xy + y^2$

Front of Card

$(x + y)(x - y) =$

What is the quadratic formula for solving equations in the form $y = ax^2 + bx + c$?

Back of Card

$x^2 - y^2$

$x = \dfrac{-b \pm \sqrt{b^2 - 4ac}}{2a}$

Plane Geometry

Plane Geometry on the Subject Tests

ETS uses the term "plane geometry" to refer to the kind of geometry that is commonly tested on the SAT I: Questions about lines and angles, triangles and other polygons, and circles. Questions testing plane geometry appear almost exclusively on the Math IC Test. About 20 percent of the questions on the Math IC Test concern plane geometry. None of the questions on the Math IIC test will focus on plane geometry, but the tools in this chapter will be needed to answer some Math IIC questions about coordinate geometry and solid geometry.

Definitions

Here are some geometry terms that appear on the Math Subject Tests. Make sure you're familiar with them. If the meaning of any of these vocabulary words keeps slipping your mind, add that word to your flashcards.

Line: A "line" in plane geometry is perfectly straight, and extends infinitely in both directions.

Line Segment: A line segment is a section of a line—still perfectly straight, but having limited length.

Bisector:	Any line that cuts a line segment, angle, or polygon exactly in half. It *bisects* another shape.
Midpoint:	The point that divides a line segment into two equal halves.
Plane:	A "plane" in plane geometry is a perfectly flat surface that extends infinitely in two dimensions.
Complementary Angles:	Angles whose measures add up to 90 degrees.
Supplementary Angles:	Angles whose measures add up to 180 degrees.
Parallel Lines:	Lines that run in exactly the same direction—they are separated by a constant distance, never growing closer together or farther apart. Parallel lines never intersect.
Polygon:	A flat shape formed by straight line segments, such as a rectangle or triangle.
Quadrilateral:	A four-sided polygon.
Altitude:	A vertical line drawn from the highest point of a polygon to the polygon's base.
Perimeter:	The sum of the lengths of a polygon's sides.
Radius:	A line segment extending from the center of a circle to a point on that circle.
Arc:	A portion of a circle's edge.
Chord:	A line segment connecting two distinct points on a circle.
Sector:	A portion of a circle's area between two radii, like a slice of pie.
"Inscribed":	A shape that is *inscribed* in another shape is placed inside that shape with the tightest possible fit. A circle inscribed in a square is the largest circle that can be placed inside that square.
"Circumscribed":	A *circumscribed* shape is drawn around another shape with the tightest fit possible. A circle circumscribed around a square is the smallest circle that can be drawn around that square.
"Perpendicular":	Perpendicular lines are at right angles to one another.
"Tangent":	Something that is tangent to a curve touches that curve at only one point without crossing it. A shape may be "internally" or "externally" tangent to a curve, meaning that it may touch the inside or outside of the curve.

Basic Rules of Lines and Angles

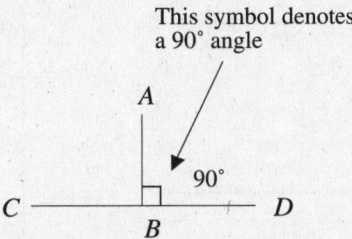

A right angle has a measure of 90°. The angles formed by perpendicular lines are right angles.

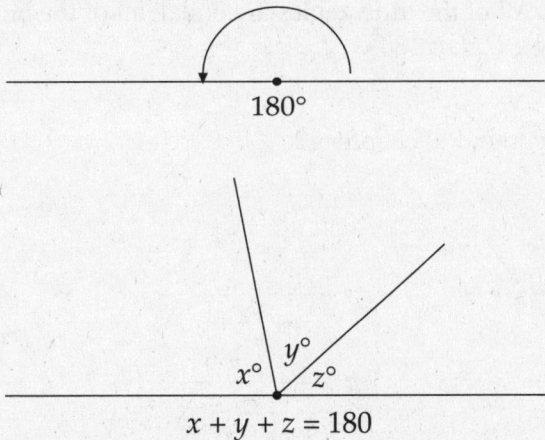

An angle opened up into a straight line (called a "straight angle") has a measure of 180°. If a number of angles make up a straight line, then the measures of those angles add up to 180°.

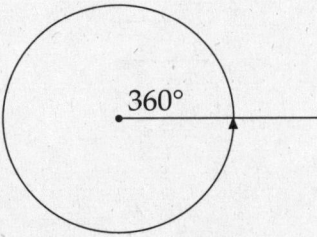

Any line rotated through a full circle moves through 360°. If a group of angles makes up a full circle, then the measures of those angles add up to 360°.

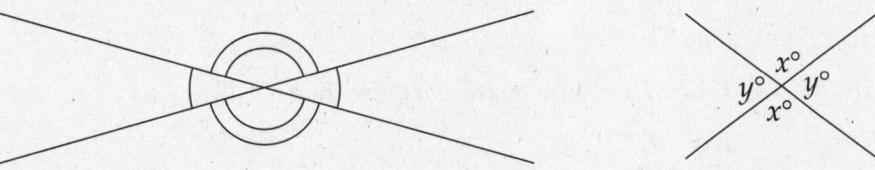

When two lines cross, opposite angles are equal (these are called "vertical angles"). Adjacent angles form straight lines and, therefore, add up to 180°.

PLANE GEOMETRY ◆ 79

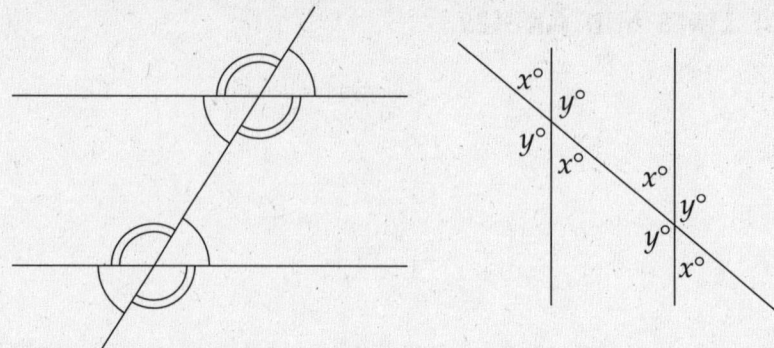

Fred's Theorem: When parallel lines are crossed by a third line, two kinds of angles are formed—little angles and big angles. All of the little angles are equal, all of the big angles are equal, and any little angle plus any big angle equals 180°.

Drill

The answers to all drills are found in chapter 12.

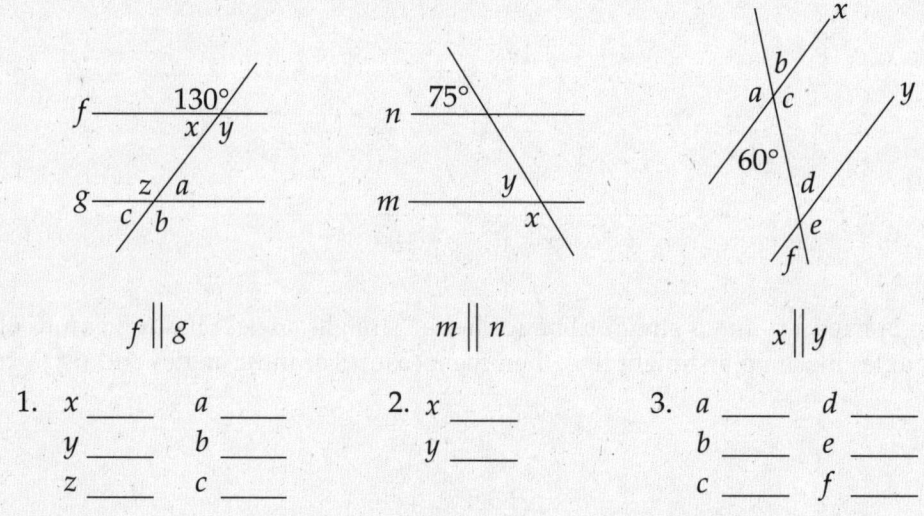

1. x _____ a _____
   y _____ b _____
   z _____ c _____

2. x _____
   y _____

3. a _____ d _____
   b _____ e _____
   c _____ f _____

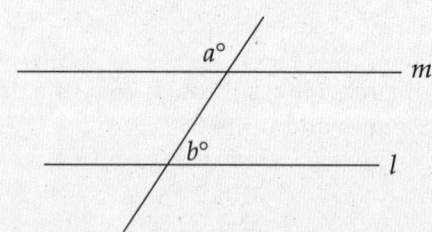

8. If line l and line m are parallel, then $a + b =$

    (A) 90°
    (B) 180°
    (C) 270°
    (D) 360°
    (E) It cannot be determined from the information given.

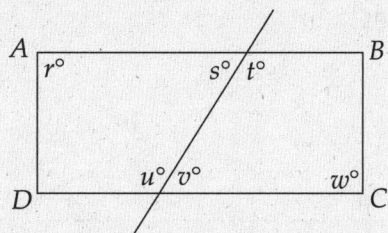

13. If rectangle ABCD is crossed by a line as shown, then which of the following is equal to t?

    (A) v
    (B) w
    (C) r + s
    (D) w − v
    (E) r + w − s

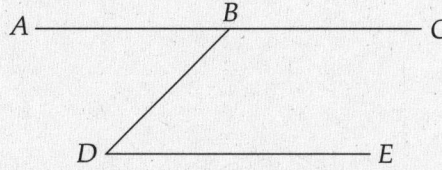

16. If AC ∥ DE, the which of the following is the difference between ∠DBC and ∠BDE ?

    (A)   0°
    (B) 45°
    (C) 90°
    (D) 180°
    (E) It cannot be determined from the information given.

TRIANGLES

Triangles appear in the majority of plane-geometry questions on the Math Subject Tests. What's more, triangle techniques are useful in solving questions that don't obviously relate to triangles, such as coordinate-geometry and solid-geometry questions. The following rules are some of the most important in plane geometry.

The Rule of 180°

For starters: The three angles of any triangle add up to 180°. This rule helps to solve a great many plane-geometry questions.

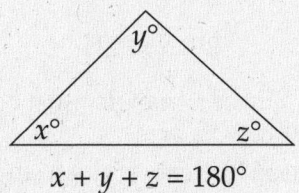

$x + y + z = 180°$

The proportionality of triangles

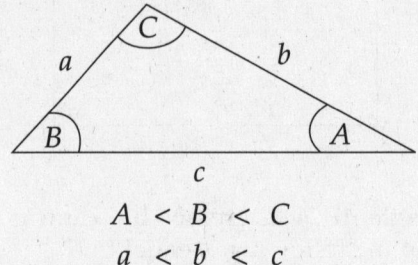

$A < B < C$
$a < b < c$

In a triangle, the smallest angle is always opposite the shortest side; the middle angle is opposite the middle side; and the largest angle is opposite the longest side. If a triangle has sides of equal length, then the opposite angles will have equal measures.

If you have trouble figuring out which side is opposite a certain angle in a triangle, remember this simple rule: the opposite side is the side that doesn't touch a given angle.

Isosceles triangles

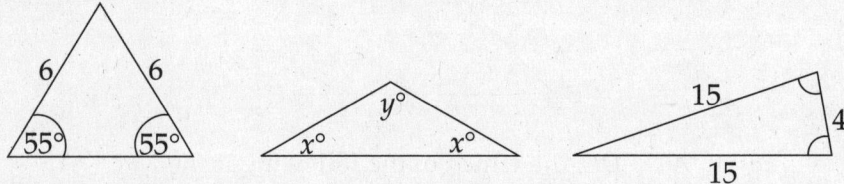

An isosceles triangle has two equal sides and two equal angles.

Equilateral triangles

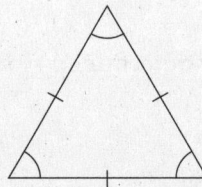

An equilateral triangle has three equal sides and three equal angles. Each angle has a measure of 60°.

The Third Side Rule

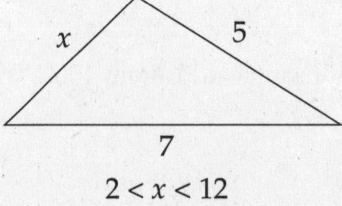

$2 < x < 12$

The Third Side Rule: The length of any side of a triangle must be between the sum and the difference of the lengths of the other two sides.

The Third Side Rule is commonly used to create tricky triangle questions. Watch out for questions that employ the Third Side Rule in isosceles triangles.

Drill
The answers to all drills are found in chapter 12.

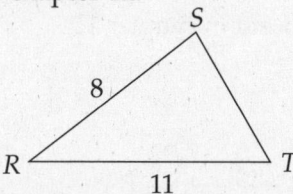

12. Which of the following expresses the possible values of p, if p is the perimeter of RST?

    (A) $3 < p < 19$
    (B) $3 < p < 22$
    (C) $19 < p < 22$
    (D) $19 < p < 38$
    (E) $22 < p < 38$

17. An isosceles triangle has sides of 5, 11, and x. How many possible values of x exist?

    (A) one
    (B) two
    (C) three
    (D) four
    (E) more than four

18. The distance from A to D is 6, and the distance from D to F is 4. Which of the following is NOT a possible value of the distance from F to A?

    (A) 2
    (B) 4
    (C) 7
    (D) 9
    (E) 11

RIGHT TRIANGLES

Right triangles are, not surprisingly, triangles that contain right angles. The sides of right triangles are referred to by special names: The sides that form the right angle are called the legs of the triangle; the longest side, opposite the right angle, is called the hypotenuse. There are many techniques and rules for right triangles that won't work on just any triangle. The most important of these rules is the relationship between angles described by the Pythagorean Theorem:

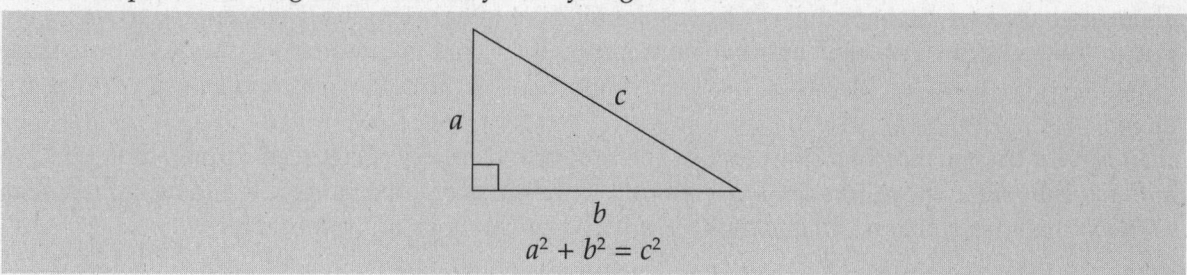

Keep in mind when you use the Pythagorean Theorem that the c always represents the hypotenuse.

Drill

In the following triangles, use the Pythagorean Theorem to fill in the missing sides of the triangles shown. The answers to all drills are found in chapter 12.

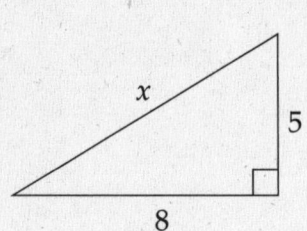 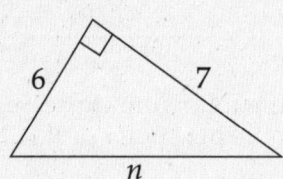 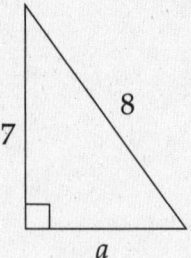

1. $x =$ _____ 2. $n =$ _____ 3. $a =$ _____

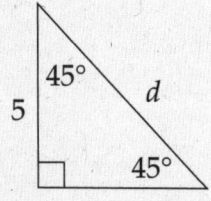 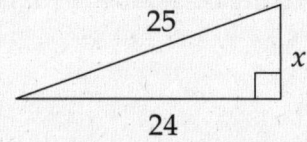 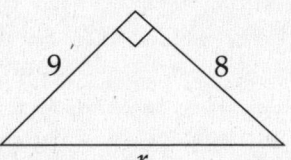

4. $d =$ _____ 5. $x =$ _____ 6. $r =$ _____

Pythagorean Triplets

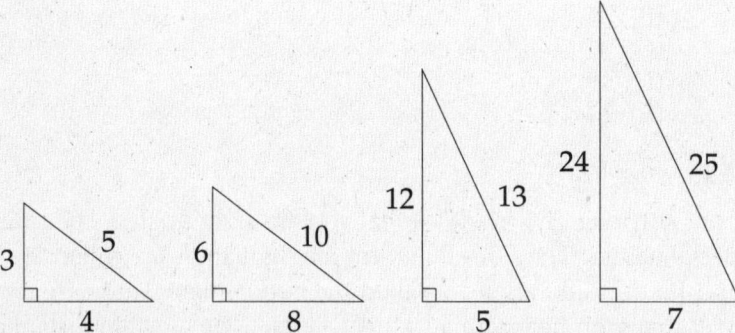

There are only a few right triangles whose sides all have integer lengths. These special triangles are called "Pythagorean Triplets," but that's not important. What is important is that ETS puts these triangles on the test a lot. Memorize the basic proportions, 3:4:5 and 5:12:13, and keep an eye out for them.

If a right triangle has two sides that fit the proportions of a Pythagorean Triplet, then you can automatically fill in the third side. The multiples of these basic proportions will also be Pythagorean Triplets; that means that 6:8:10 and 30:40:50 are also proportions of right triangles.

Drill

Use the proportions of the Pythagorean Triplets to complete the triangles below. The answers to all drills are found in chapter 12.

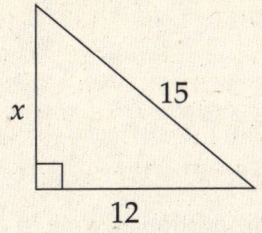 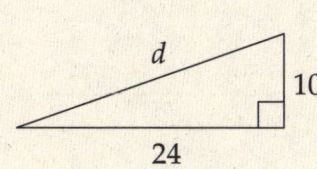 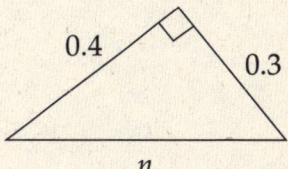

1. $x =$ _____ 2. $d =$ _____ 3. $n =$ _____

The 45-45-90 Triangle

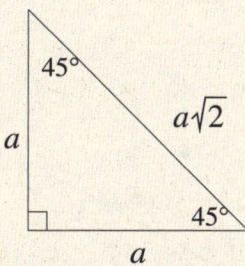

A right triangle with angles of 45°, 45°, and 90° has sides in definite proportions. The legs of the triangle are of equal length (the 45-45-90 is the only isosceles right triangle). The length of the hypotenuse is equal to the length of a leg multiplied by $\sqrt{2}$.

Drill

Use the proportions of the 45-45-90 triangle to complete the dimensions of the following triangles. The answers to all drills are found in chapter 12.

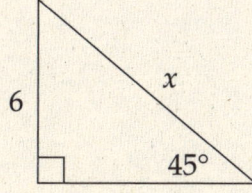 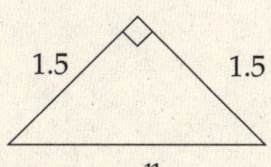 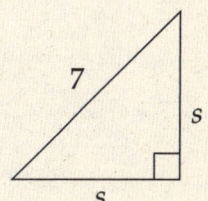

1. $x =$ _____ 2. $n =$ _____ 3. $s =$ _____

The 30-60-90 triangle

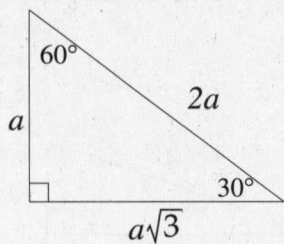

A right triangle with angles of 30°, 60°, and 90° has sides in a definite proportion as well. In a 30-60-90 triangle, the hypotenuse is twice as long as the shorter leg. The length of the longer leg is equal to the length of the shorter leg times $\sqrt{3}$.

Drill

Use the proportions of the 30-60-90 triangles to complete the dimensions of the following triangles. The answers to all drills are found in chapter 12.

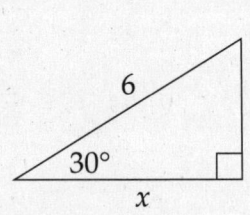   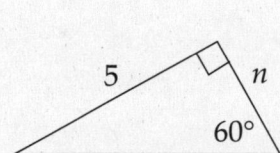   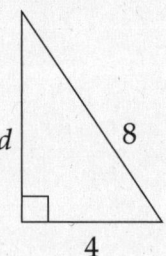

1. $x =$ _____   2. $n =$ _____   3. $d =$ _____

Drill

Use all of your right-triangle techniques to answer the following questions. The answers to all drills are found in chapter 12.

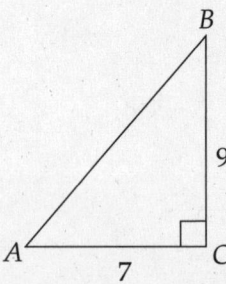

7. The perimeter of triangle ABC has how many possible values?

   (A) one
   (B) two
   (C) three
   (D) four
   (E) infinitely many

13. A right triangle with a side of length 6 and a side of length 8 also has a side of length x. What is x?

    (A) 7
    (B) 10
    (C) 12
    (D) 14
    (E) It cannot be determined from the information given.

16. A straight 32-foot ladder is leaned against a vertical wall so that it forms a 30° angle with the wall. To what height does the ladder reach?

    (A) 9.24
    (B) 16.00
    (C) 27.71
    (D) 43.71
    (E) 54.43

19. An isosceles right triangle has a perimeter of 23.9. What is the area of this triangle?

    (A) 16.9
    (B) 24.5
    (C) 25.0
    (D) 33.8
    (E) 49.0

Similar Triangles

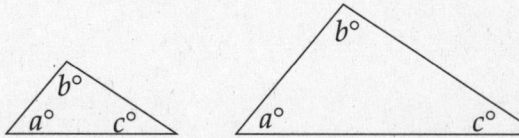

Triangles are said to be "similar" when they have the same angle measures. Basically, similar triangles have exactly the same shape, although they may be of different sizes. Their sides, therefore, are in the same proportion:

Corresponding sides of similar triangles are proportional.

For example: two 30-60-90 triangles of different sizes would be similar. If the short side of one triangle were twice as long as the short side of the other, then you could expect all of the larger triangle's dimensions to be twice the smaller triangle's dimensions. Similar triangles don't have to be right triangles, however. Triangles will be related proportionally whenever they have identical angle measures.

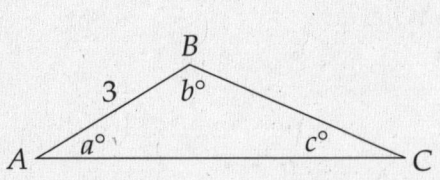 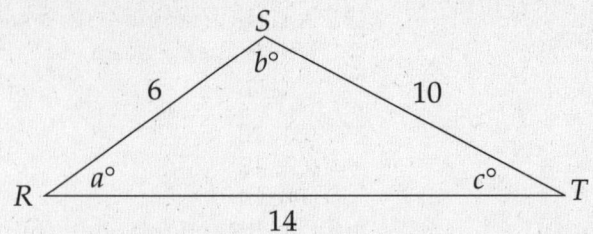

In the figure above, both triangles have angles measuring $a°$, $b°$, and $c°$. Because they have the same angles, you know they're similar triangles. Side AB of the large triangle and side RS of the smaller triangle are *corresponding* sides; each is the short side of its triangle. You can use the lengths of those two sides to figure out what the proportion between the triangles is. The length of AB is 3 and the length of RS is 6; in other words, RS is twice as long as AB. You can expect every side of RST to be twice as long as the corresponding side of ABC. That makes $BC = 5$ and $AC = 7$.

Drill
Use the proportionality of similar triangles to complete the dimensions of the following triangles:

1.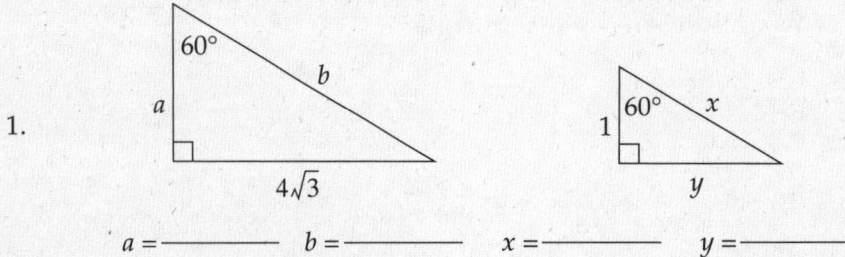

   $a = $ _____ $b = $ _____ $x = $ _____ $y = $ _____

2.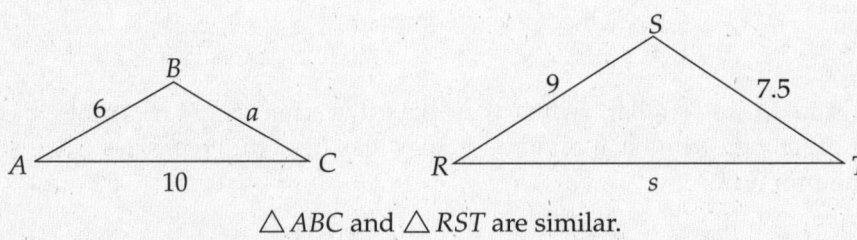

   $\triangle ABC$ and $\triangle RST$ are similar.

   $a = $ _____ $s = $ _____

3.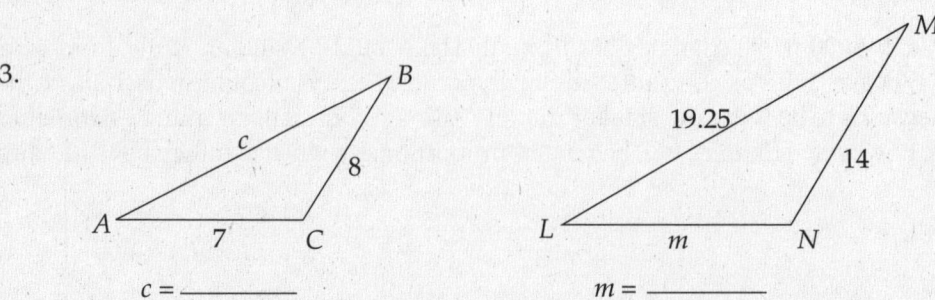

   $c = $ _____ $m = $ _____

88 ◆ CRACKING THE SAT II: MATH SUBJECT TESTS

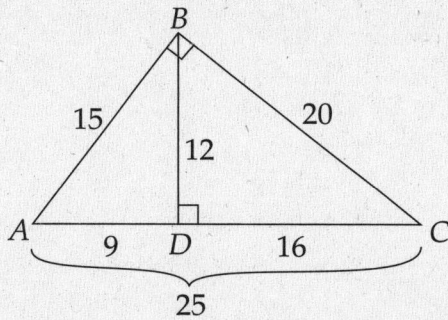

A useful note: Whenever a right triangle is divided in two by an altitude drawn to the hypotenuse, the result is three similar triangles of different sizes. The three triangles will be related proportionally.

Drill
The answers to all drills are found in chapter 12.

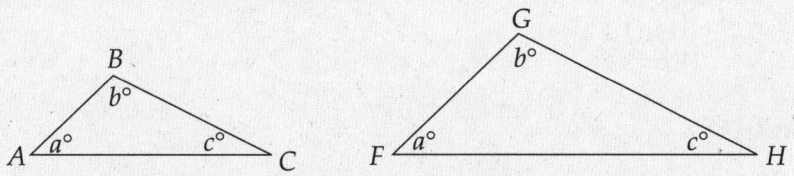

37. FG is twice as long as AB. If the area of triangle FGH is 0.5, what is the area of triangle ABC?

   (A) 0.13
   (B) 0.25
   (C) 0.50
   (D) 1.00
   (E) 2.00

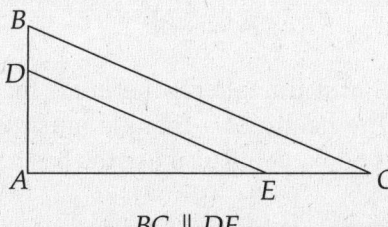

BC ∥ DE

40. If the length of DB is half of the length of AD, then the area of triangle ADE is what fraction of the area of triangle ABC?

   (A) $\frac{5}{9}$
   (B) $\frac{1}{2}$
   (C) $\frac{4}{9}$
   (D) $\frac{1}{4}$
   (E) $\frac{1}{9}$

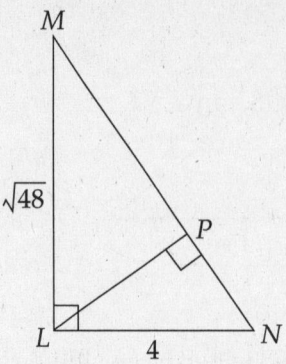

45. What is the area of $\triangle LPN$?

 (A) 3.46
 (B) 6.93
 (C) 8.00
 (D) 11.31
 (E) 13.86

Area of Triangles

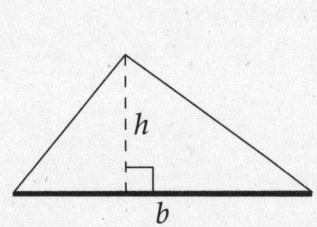 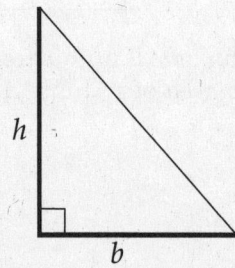 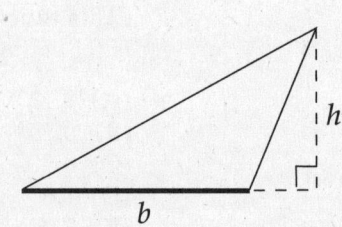

The Area of a Triangle:

$$A = \frac{1}{2}bh$$

Notice that the height, or altitude, of a triangle can be *inside* the triangle, *outside* the triangle, or formed by a *side* of the triangle. The height of a triangle must sometimes be computed with the Pythagorean Theorem. Later in this book, you'll also use the basic functions of trigonometry to find the height of a triangle.

The height of an equilateral triangle can be found by dividing it into two 30-60-90 triangles; but you can save yourself the time and trouble if you memorize the following formula:

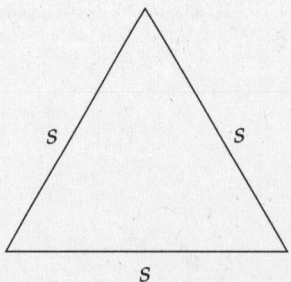

The Area of an Equilateral Triangle:

$$A = \frac{s^2\sqrt{3}}{4}$$

Drill

Try the following practice questions about the areas of triangles. The answers to all drills are found in chapter 12.

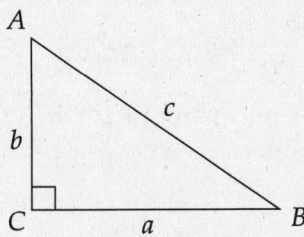

9. If the area of △ABC is equal to 3b, then a =

    (A) $\dfrac{3}{4}$

    (B) $\dfrac{3}{2}$

    (C) 3

    (D) 4

    (E) 6

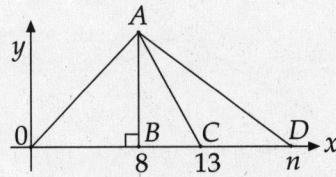

15. If △OAB and △ADC are of equal area, then n =

    (A) 8
    (B) 16
    (C) 18
    (D) 21
    (E) 24

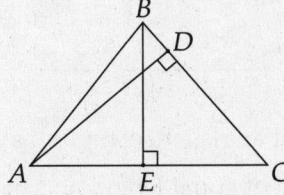

37. If AC = 12, BC = 10, and AD = 9, then BE =

    (A) 7.0
    (B) 7.5
    (C) 8.0
    (D) 8.5
    (E) 9.0

PLANE GEOMETRY ◆ 91

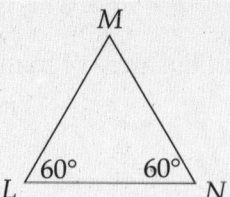

38. If △LMN has a perimeter of 24, then what is the area of △LMN?

(A) 13.86
(B) 20.78
(C) 27.71
(D) 36.95
(E) 41.57

44. An equilateral triangle with an area of 12 has what perimeter?

(A) 12.00
(B) 13.39
(C) 15.59
(D) 15.79
(E) 18.66

Quadrilaterals

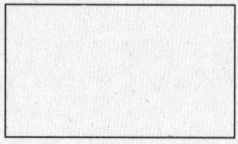
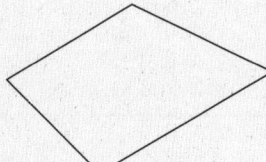
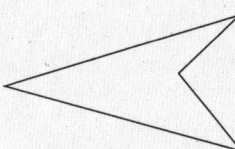

A quadrilateral is any shape formed by four straight lines in a plane. The internal angle measures of a quadrilateral always add up to 360°.

Rectangles

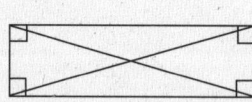
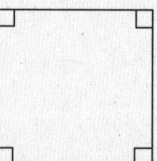

A rectangle is a quadrilateral with four right angles. Rectangles have the following properties:

- Opposite sides of a rectangle are of equal length and parallel to one another.
- The diagonals of a rectangle are of equal length and cut each other exactly in half.

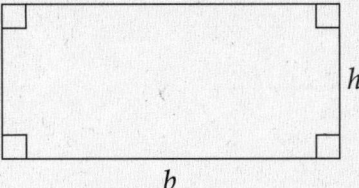

The area of a rectangle is given by this formula:

The Area of a Rectangle:
$A = bh$

Squares

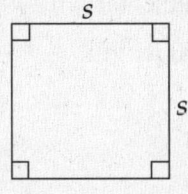 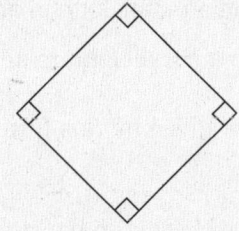

A square is a rectangle with four sides of equal length. (If you are asked to draw a rectangle, drawing a square is a legitimate response.)

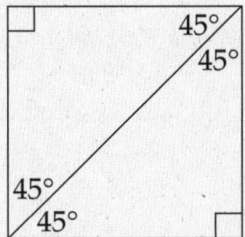

A diagonal in a square divides the square into two 45-45-90 triangles. The area of a square is given by either one of these formulas:

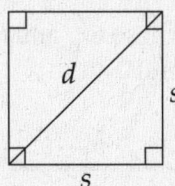

Area of a Square
$A = s^2$
or
$A = \dfrac{d^2}{2}$

The second, less known formula for the area of a square can be used to shortcut questions that would otherwise take many more steps, and require you to use a 45-45-90 triangle.

Parallelograms

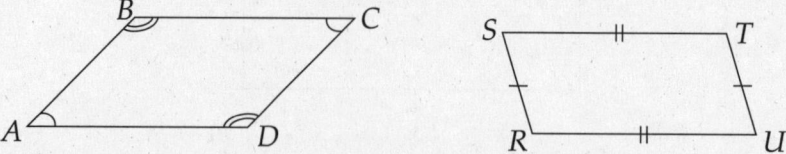

A parallelogram is a quadrilateral whose opposite sides are parallel. Rectangles are parallelograms, but a parallelogram does not need to have right angles. Note, however, that since the sides are parallel, "Fred's Theorem" applies to the angles of a parallelogram. Parallelograms have the following characteristics:

- Opposite angles in a parallelogram are equal.
- Adjacent angles in a parallelogram are supplementary; they add up to 180°.
- Opposite sides in a parallelogram are of equal length.

The area of a parallelogram is given by this formula:

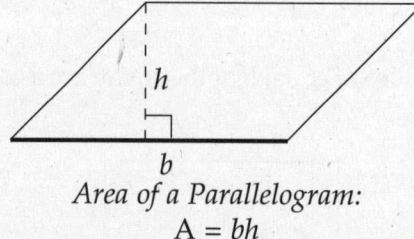

Area of a Parallelogram:
$A = bh$

You may need to use the Pythagorean Theorem to find the height of a parallelogram.

Trapezoids

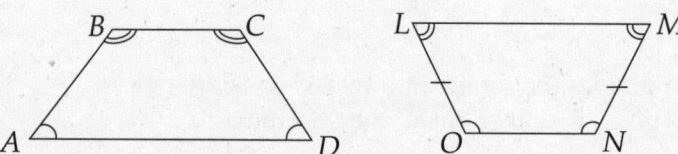

A trapezoid is a symmetrical quadrilateral whose top and bottom are parallel but different in length. The area of a trapezoid is given by this formula:

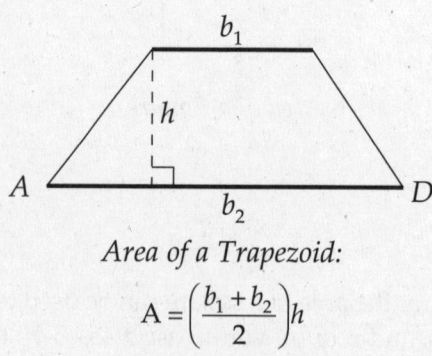

Area of a Trapezoid:
$$A = \left(\frac{b_1 + b_2}{2}\right)h$$

Drill

Try the following practice questions using quadrilateral formulas. The answers to all drills are found in chapter 12.

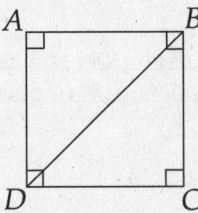

22. If $AB = BC$ and $DB = 5$, then the area of $ABCD =$

   (A) 12.50
   (B) 14.43
   (C) 17.68
   (D) 35.36
   (E) 43.30

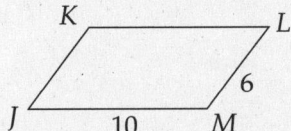

34. If $\angle KJM = 60°$, what is the area of parallelogram $JKLM$?

   (A) 18.34
   (B) 25.98
   (C) 34.64
   (D) 51.96
   (E) 103.92

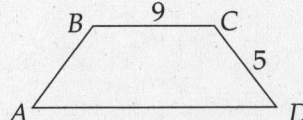

40. The bases of trapezoid $ABCD$ differ in length by 6, and the perimeter of the trapezoid is 34. What is the trapezoid's area?

   (A) 45.0
   (B) 48.0
   (C) 54.0
   (D) 60.0
   (E) 62.5

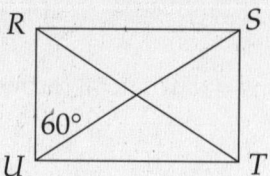

45. If the area of rectangle *RSTU* is 62.35, what is the sum of the rectangle's diagonals?

   (A) 18.8
   (B) 24.0
   (C) 32.0
   (D) 36.0
   (E) 40.8

OTHER POLYGONS

The Math Subject Tests may occasionally require you to deal with polygons other than triangles and quadrilaterals. Here are the names of the other polygons you're likely to see:

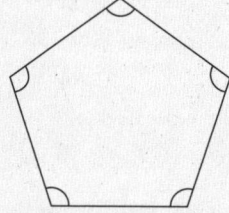

pentagon: a five-sided polygon

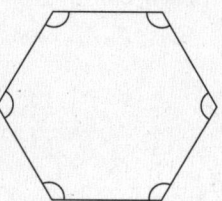

hexagon: a six-sided polygon

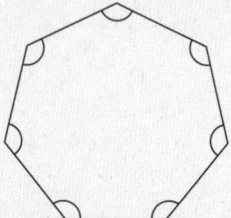

heptagon: a seven-sided figure

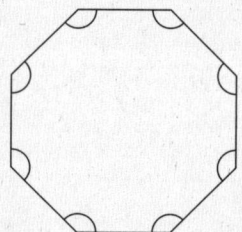

octagon: an eight-sided figure

All of the polygons pictured above are *regular* polygons. That means that their sides and angles are all of the same size.

You know that the internal angles of a triangle add up to 180° and that the internal angles of a quadrilateral add up to 360°. But what about the angles of a hexagon, or an octagon? You can compute the sum of a polygon's internal angles using this formula:

Sum of the Angles of an *n*-sided Polygon

Sum of Angles = $(n - 2) \times 180°$

Using this formula, we can figure out that the angles of a hexagon (a 6-sided figure) would have a sum of (4×180) degrees. That's 720°. If you know that the figure is a *regular* hexagon, then you can even figure out the measure of each angle: $720° \div 6 = 120°$.

Circles

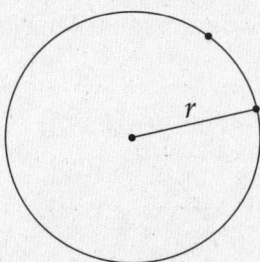

A circle is defined as the set of all the points located at a certain distance from a given center point. A point that is said to be *on* a circle is a point on the edge of the circle, not contained within the circle.

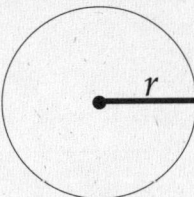

The *radius* is the distance from the center to the edge of the circle.

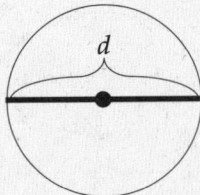

The *diameter* of a circle is the distance from edge to edge through the circle's center. The diameter is twice as long as the radius.

The *circumference* of a circle is the distance around the circle—essentially, the circle's perimeter. The circumference is given by this formula:

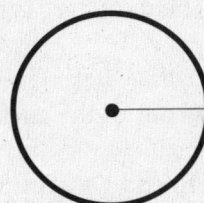

Circumference of a Circle:
$C = 2\pi r$

The *area* of a circle is given by this formula:

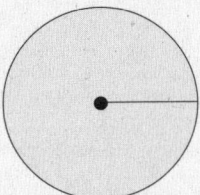

Area of a Circle:
$A = \pi r^2$

Drill
Use these circle formulas to complete the dimensions of the following circles. The answers to all drills are found in chapter 12.

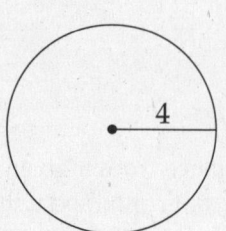

Area = 20

Circumference = 8

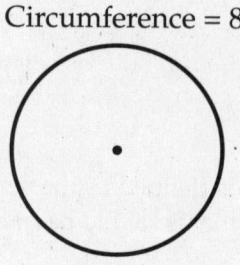

1. C = _____

   A = _____

2. C = _____

   r = _____

3. A = _____

   r = _____

A Slice of Pie

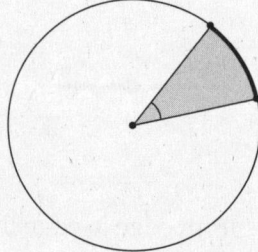

The portion of a circle's area between two radii is called a *sector*. The portion of the circle's circumference that falls between the radii is called an *arc*. Between any two points on a circle's edge, there are two arcs, a major arc and a minor arc. The minor arc is the shorter of the two, and it's usually the one ETS is concerned about.

The angle between two radii is called a *central angle*. The degree measure of a central angle is equal to the degree measure of the arc that it cuts out of the circle's circumference. The minor arc formed by a 40° central angle is said to be a 40° arc.

To put it simply, the piece of a circle defined by a central angle (like a slice of pie) takes the same fraction of everything. A 60° central angle, for instance, takes one-sixth of the circle's 360°; the arc that is formed will be one-sixth of the circumference; and the sector that is formed will be one-sixth of the circle's area.

Inscribed Angles

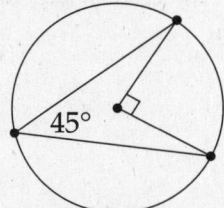 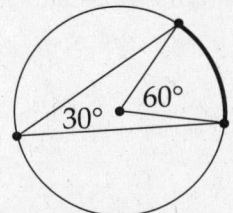 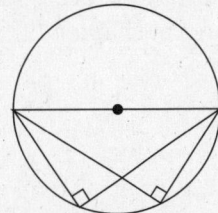

An angle formed by two chords is called an *inscribed angle*. While a central angle with a certain degree measure intercepts an arc of the same degree measure, an inscribed angle intercepts an arc with *twice* the degree measure of the angle. For example, a 30° central angle intercepts a 30° arc, while a 30° inscribed angle intercepts a 60° arc. In the same way, a right angle inscribed in a circle always intercepts a 180° arc, which means that it also marks the ends of a diameter.

All angles inscribed in the same segment of a circle (or identical circles) are equal.

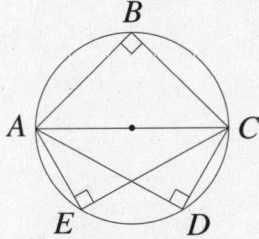

Any angle inscribed in a semicircle is a right angle.

Tangent Lines

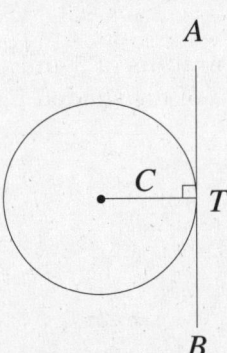

A *tangent line* to a circle is a line that touches the circle at only one point. A tangent line is always perpendicular to the radius touching the same point.

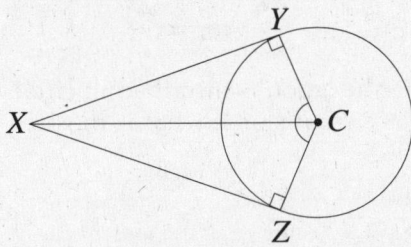

From any point outside a circle, there are two separate tangent lines to that circle. The distances to the two points of tangency are equal, and the radii to the points of tangency make equal angles with the line connecting the external point to the circle's center.

Drill

Try the following practice questions using the rules and techniques for circles. The answers to all drills are found in chapter 12.

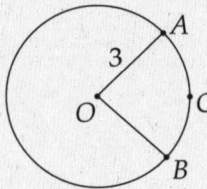

Note: Figure not drawn to scale.

12. If the length of arc ACB is 4.71, which of the following best approximates the measure of $\angle AOB$?

   (A) 60.0°
   (B) 72.0°
   (C) 86.4°
   (D) 90.0°
   (E) 98.6°

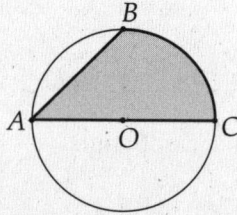

29. If circle O has a radius of 5 and $\angle BAC = 45°$, then what is the area of the shaded figure?

   (A) 32.13
   (B) 31.52
   (C) 26.70
   (D) 25.41
   (E) 24.26

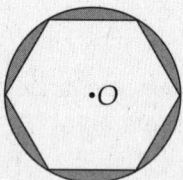

31. A regular hexagon is inscribed in circle O. If the circle has a radius of 4, what is the area of the shaded region?

   (A) 8.3
   (B) 8.7
   (C) 9.0
   (D) 9.4
   (E) 10.2

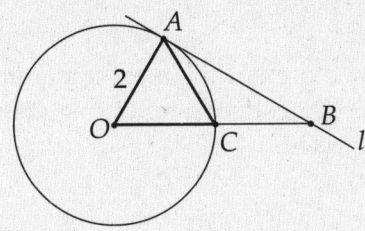

43. Line *l* is tangent to circle O at A, and OA = AC. What is the length of AB?

    (A) 1.73
    (B) 2.83
    (C) 3.46
    (D) 4.74
    (E) 5.20

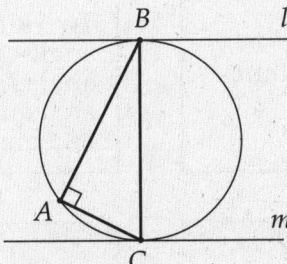

45. The right angle ∠CAB is inscribed in a circle. Lines *l* and *m* are tangent to the circle at B and C, respectively. Which of the following must be true?

    I. AB<BC
    II. ∠ACB = 60°
    III. *l* and *m* meet when extended to the right.

    (A) I only
    (B) III only
    (C) I and II only
    (D) II and III only
    (E) I, II, and III

FLASHCARDS

The rules and formulas listed below represent the most important points to study in this chapter. Memorize them by covering up the right column and testing yourself on the left column, or, better still, by making real flashcards as directed.

Front of Card	Back of Card
The internal angles of a triangle add up to...	180°
"Fred's Theorem" states that when parallel lines are crossed by a third line...	Two kinds of angles are formed, little angles and big angles. All of the little angles are equal, all of the big angles are equal, and any little angle plus any big angle equals 180°.
What is the Third Side Rule for triangles?	The length of any side of a triangle must be between the sum and difference of the other two sides.
What is the Pythagorean Theorem?	$a^2 + b^2 = c^2$
What are the proportions of the sides of a 30-60-90 triangle?	The lengths of the sides are in a ratio of $1 : \sqrt{3} : 2$.

Front of Card

What are the proportions of the sides of a 45-45-90 triangle?

What is the formula for the area of a triangle?

What is the area of an equilateral triangle?

What is the area of a rectangle?

What is the area of a square? (two ways)

What is the area of a parallelogram?

Back of Card

The lengths of the sides are in a ratio of $1:1:\sqrt{2}$.

$A = \frac{1}{2}bh$

$A = \frac{s^2\sqrt{3}}{4}$

$A = bh$

$A = s^2$ or $A = \frac{d^2}{2}$

$A = bh$
(The height is found as in a triangle)

Front of Card | **Back of Card**

| What is the area of a trapezoid? | $A = \left(\dfrac{b_1 + b_2}{2}\right)h$ |

| What is the sum of the internal angles of a polygon with n sides? | Sum of Angles $= (n-2) \cdot 180°$ |

| What is the circumference of a circle? | $C = 2\pi r$ or $C = \pi d$ |

| What is the area of a circle? | $A = \pi r^2$ |

Solid Geometry

SOLID GEOMETRY ON THE SUBJECT TESTS

Questions about solid geometry frequently test plane-geometry techniques. They're difficult mostly because the added third dimension makes them harder to visualize. You're likely to run into 3 or 4 solid-geometry questions on either one of the Math Subject Tests, however, so it's important to practice. If you're not the artistic type and have trouble drawing cubes, cylinders, and so on, it's worthwhile to practice sketching the shapes in the following pages. The ability to make your own drawing is often helpful.

RECTANGULAR SOLIDS

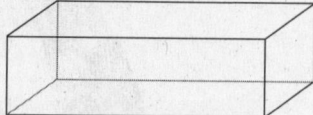 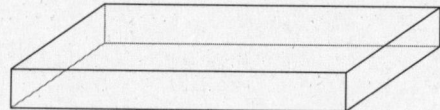

A rectangular solid is simply a box; ETS also sometimes calls it a rectangular *prism*. It has three distinct dimensions: *length, width,* and *height*. The volume of a rectangular solid (the amount of space it contains) is given by this formula:

Volume of a Rectangular Box
$$V = lwh$$

The surface area of a rectangular box is the sum of the areas of all of its sides. A rectangular solid's surface area is given by this formula:

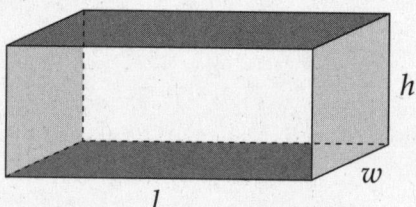

Surface Area of a Rectangular Box
$$SA = 2lw + 2wh + 2lh$$

The volume and surface area of a solid make up the most basic information you can have about that solid (volume is tested more often than surface area). You may also be asked about *lengths* within a rectangular solid—edges and diagonals. The dimensions of the solid give the lengths of its edges, and the diagonal of any *face* of a rectangular solid can be found using the Pythagorean Theorem. There's one more length you may be asked about: the long diagonal that passes from corner to corner through the center of the box. The length of the long diagonal is given by this formula:

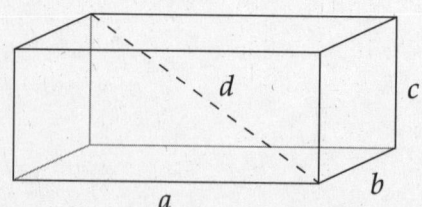

Long Diagonal of a Rectangular Solid
$$a^2 + b^2 + c^2 = d^2$$

This is the Pythagorean Theorem with a third dimension added, and it works just the same way. This formula will work in any rectangular box. The long diagonal is the longest straight line that can be drawn inside any rectangular solid.

Cubes

A cube is a rectangular solid that has the same length in all three dimensions. All six of its faces are squares. This simplifies the formulas for volume, surface area, and the long diagonal:

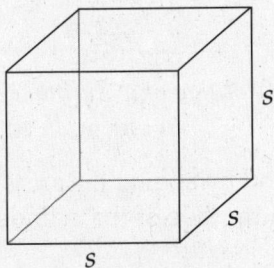

Volume of a Cube
$V = s^3$
Surface Area of a Cube
$SA = 6s^2$

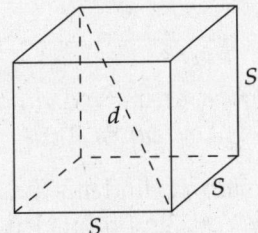

Long Diagonal of a Cube
$d = s\sqrt{3}$

CYLINDERS

A cylinder is a prism with a circular base. It has two important dimensions: its radius and its height. The volume of a cylinder is given by this formula:

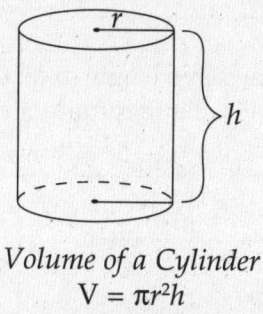

Volume of a Cylinder
$V = \pi r^2 h$

The surface area of a cylinder is found by adding the areas of the two circular bases to the area of the rectangle you'd get if you unrolled the side of the cylinder. That boils down to this formula:

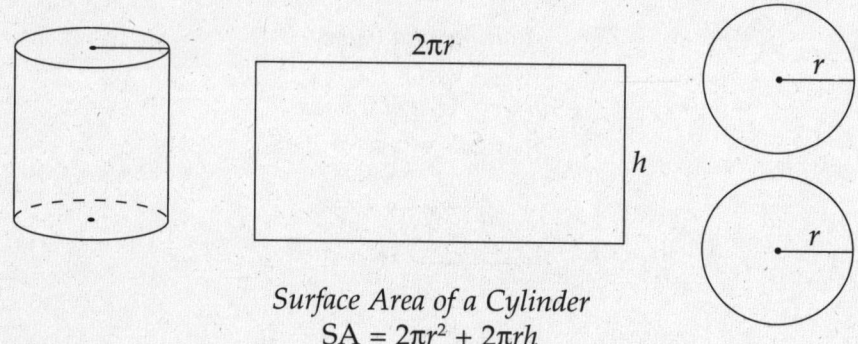

Surface Area of a Cylinder
$SA = 2\pi r^2 + 2\pi rh$

The longest line that can be drawn inside a cylinder is the diagonal of the rectangle formed by the diameter and the height of the cylinder. You can find its length with the Pythagorean Theorem.

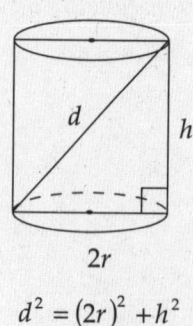

$d^2 = (2r)^2 + h^2$

Cones

If you take a cylinder and shrink one of its circular bases down to a point, then you have a cone. A cone has three significant dimensions, which form a right triangle: Its radius, its height, and the *slant height*, which is the straight-line distance from the tip of the cone to a point on the edge of its base. The formulas for the volume and surface area of a cone are given in the information box at the beginning of both of the Math Subject Tests—but there's a catch. The formula for the volume of a cone is pretty straightforward:

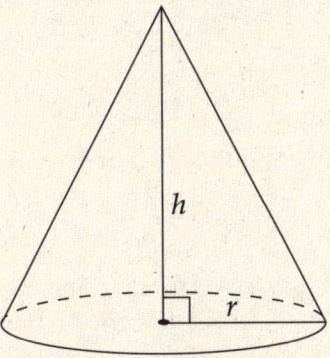

Volume of a Cone

$$V = \frac{1}{3}\pi r^2 h$$

But you have to be careful computing *surface area* for a cone using the formula provided by ETS. The formula at the beginning of the Math Subject Tests is for the *lateral area* of a cone—the area of the sloping sides—not the complete surface area. It doesn't include the circular base. Here's a more useful equation for the surface area of a cone:

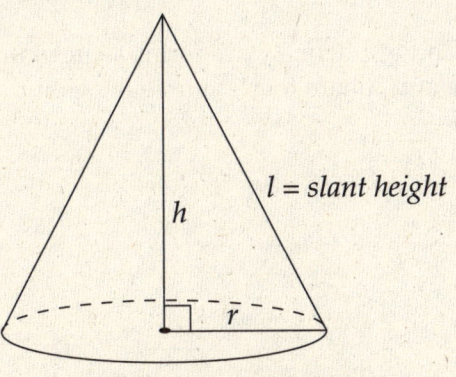

Surface Area of a Cone

$$SA = \pi r l + \pi r^2$$

If you want to calculate only the lateral area of a cone, just use the first half of the above formula—leave the πr^2 off.

Spheres

A sphere is simply a hollow ball. It can be defined as all of the points in space at a fixed distance from a central point. The one important measure in a sphere is its radius. The formulas for the volume and surface area of a sphere are given to you; they appear in the information box at the very beginning of both Math Subject Tests. That means that you don't need to have them memorized, but here they are anyway:

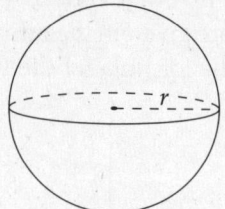

Volume of a Sphere
$$V = \frac{4}{3}\pi r^3$$

Surface Area of a Sphere
$$SA = 4\pi r^2$$

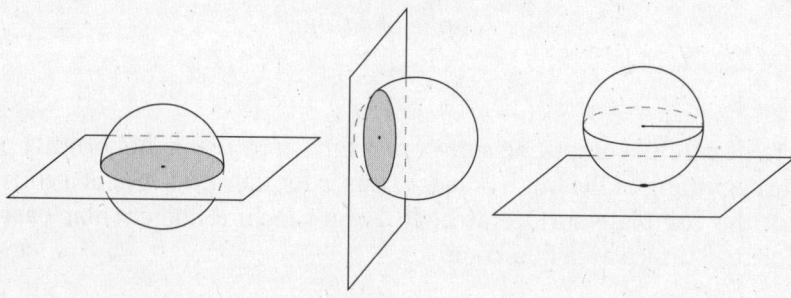

The intersection of a plane and a sphere always forms a circle, unless the plane is *tangent* to the sphere, in which case the plane and sphere touch at only one point.

Pyramids

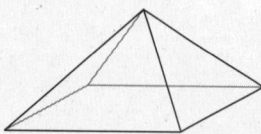 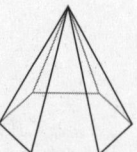 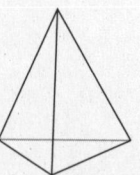

A pyramid is a little like a cone, except that its base is a polygon instead of a circle. Pyramids don't show up often on the Math Subject Tests. When you do run into a pyramid, it will almost always have a rectangular base. Pyramids can be pretty complicated solids, but for the purposes of the Math Subject Tests a pyramid has just two important measures: the area of its base and its height. The height of a pyramid is the length of a line drawn straight down from the pyramid's tip to its base; the height is perpendicular to the base. The volume of a pyramid is given by this formula:

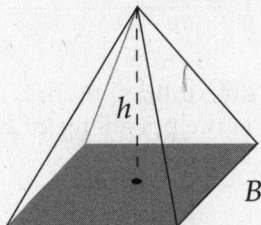

Volume of a Pyramid

$$V = \frac{1}{3}Bh$$

(B = area of base)

It's not really possible to give a general formula for the surface area of a pyramid because there are so many different kinds. At any rate, the information is not generally tested on the Math Subject Tests. If you should be called upon to figure out the surface area of a pyramid, just figure out the area of each face using triangle and quadrilateral rules, and add them up.

TRICKS OF THE TRADE

Here are some of the most common solid-geometry question types you're likely to encounter on the Math Subject Tests. They occur much more often on the Math IIC Test than on the Math IC Test, but can appear on either test.

Triangles in rectangular solids

Many questions about rectangular solids are actually testing triangle rules. Such questions generally ask for the lengths of the diagonals of a box's faces, the long diagonal of a box, or other lengths. These questions are usually solved using the Pythagorean Theorem and the 3-D version of the Pythagorean Theorem that finds a box's long diagonal (see the section on Rectangular Solids).

Drill

Here are some practice questions using triangle rules in rectangular solids. The answers to all drills are found in chapter 12.

32. What is the length of the longest line that can be drawn in a cube of volume 27?

   (A) 3.0
   (B) 4.2
   (C) 4.9
   (D) 5.2
   (E) 9.0

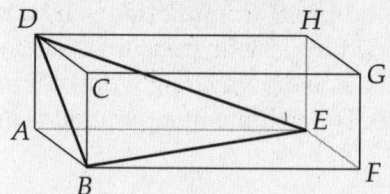

36. In the rectangular solid shown, if $AB = 4$, $BC = 3$, and $BF = 12$, what is the perimeter of triangle EDB?

   (A) 27.33
   (B) 28.40
   (C) 29.20
   (D) 29.50
   (E) 30.02

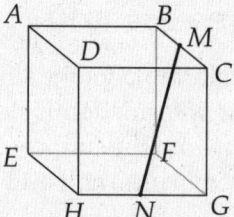

39. In the cube above, M is the midpoint of BC and N is the midpoint of GH. If the cube has a volume of 1, what is the length of MN?

   (A) 1.23
   (B) 1.36
   (C) 1.41
   (D) 1.73
   (E) 1.89

Volume questions

Many solid-geometry questions test your understanding of the relationship between a solid's volume and its other dimensions—sometimes including the solid's area. To solve these questions, just plug the numbers you're given into the solid's volume formula.

Drill

Try the following practice questions. The answers to all drills are found in chapter 12.

17. A cube's volume and surface area are equal. What is the length of an edge of this cube?

   (A) 1
   (B) 2
   (C) 4
   (D) 6
   (E) 9

24. A rectangular solid has a volume of 30, and its edges have integer lengths. What is the greatest possible surface area of this solid?

    (A) 62
    (B) 82
    (C) 86
    (D) 94
    (E) 122

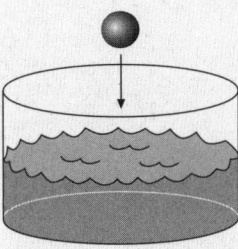

43. A sphere of radius 1 is totally submerged in a cylindrical tank of radius 4, as shown. The water level in the tank rises a distance of h. What is the value of h?

    (A) 0.072
    (B) 0.083
    (C) 0.096
    (D) 0.108
    (E) 0.123

When questions of this type are put together with variables rather than numbers, remember the technique of plugging in. It works as well in geometry as in algebra questions.

17. A cube has a surface area of $6x$. What is the volume of the cube?

    (A) $x^{\frac{2}{3}}$
    (B) $x^{\frac{3}{2}}$
    (C) $6x^2$
    (D) $36x^2$
    (E) x^3

36. A sphere has a radius of r. If this radius is increased by b, then the sphere's surface area is increased by what amount?

    (A) b^2
    (B) $4\pi b^2$
    (C) $8\pi rb + 4\pi b^2$
    (D) $8\pi rb + 2rb + b^2$
    (E) $4\pi r^2 b^2$

SOLID GEOMETRY

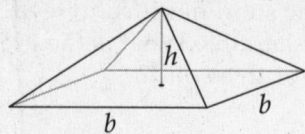

40. If the pyramid shown has a square base with edges of length b, and $b = 2h$, then which of the following is the volume of the pyramid?

    (A) $\dfrac{h^3}{3}$

    (B) $\dfrac{4h^3}{3}$

    (C) $4h^3$

    (D) $8h^2 - h$

    (E) $\dfrac{8h^3 - 4h}{3}$

Inscribed solids

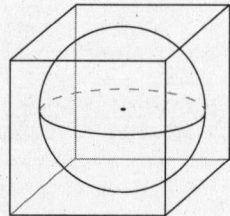 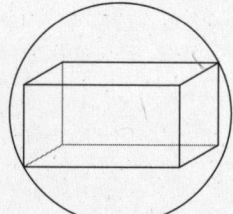 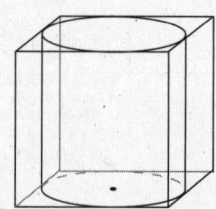

Some questions on the Math Subject Tests will be based on spheres inscribed in cubes, or cubes inscribed in spheres (these are the most popular inscribed shapes). Occasionally you may also see a rectangular solid inscribed in a sphere, or a cylinder inscribed in a rectangular box, etc. The trick to these questions is always figuring out how to get from the dimensions of one of the solids to the dimensions of the other.

Here are a few basic tips that can speed up your work on inscribed-solids questions:

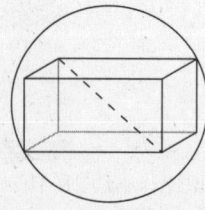

♦ When a cube or rectangular solid is inscribed in a sphere, the long diagonal of that solid is equal to the diameter of the sphere.

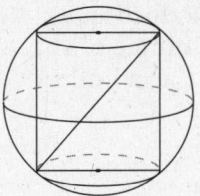

- When a cylinder is inscribed in a sphere, the sphere's diameter is equal to the diagonal of the cylinder's height and diameter.

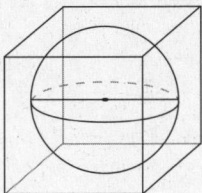

- When a sphere is inscribed in a cube, the diameter of the sphere is equal to the length of the cube's edge.

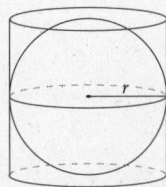

- If a sphere is inscribed in a cylinder, both solids have the same diameter.

Most inscribed-solids questions fall into one of the preceding categories. If you run into a situation not covered by these tips, just look for the way to get from the dimensions of the inner shape to those of the external shape, or vice versa.

Drill

Here are some practice inscribed-solid questions. The answers to all drills are found in chapter 12.

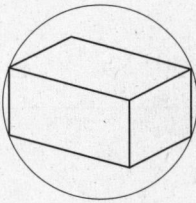

32. A rectangular solid is inscribed in a sphere as shown. If the dimensions of the solid are $3 \times 4 \times 6$, then what is the radius of the sphere?

    (A) 2.49
    (B) 3.91
    (C) 4.16
    (D) 5.62
    (E) 7.81

35. A cylinder is inscribed in a cube with an edge of length 2. What volume of space is enclosed by the cube but not by the cylinder?

(A) 1.41
(B) 1.56
(C) 1.72
(D) 3.81
(E) 4.86

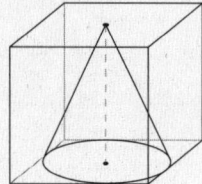

38. A cone is inscribed in a cube of volume 1 in such a way that its base touches 4 edges of the cube. What is the volume of the cone?

(A) 0.21
(B) 0.26
(C) 0.33
(D) 0.42
(E) 0.67

Solids produced by rotation

Three types of solids can be produced by the rotation of simple two-dimensional shapes: spheres, cylinders, and cones. Questions about solids produced by rotation are generally fairly simple; they usually test your ability to visualize the solid generated by the rotation of a flat shape. Sometimes, rotated-solids questions begin with a shape in the coordinate plane—that is, rotated around one of the axes or some other line. Practice will help you figure out the dimensions of the solid from the dimension of the original flat shape.

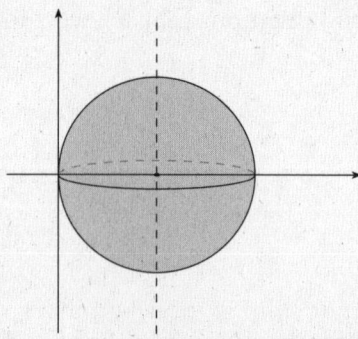

A sphere is produced when a circle is rotated around its diameter. This is an easy situation to work with, as the sphere and the original circle will have the same radius. Find the radius of the circle, and you can figure out anything you want about the sphere.

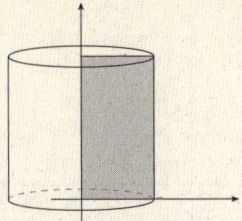

A cylinder is formed by the rotation of a rectangle around a central line *or* one edge.

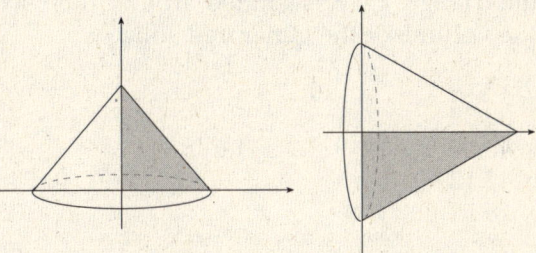

A cone is formed by the rotation of a right triangle around one of its legs, or by the rotation of an isosceles triangle around its line of symmetry.

Drill

Try these rotated-solids questions for practice. The answers to all drills are found in chapter 12.

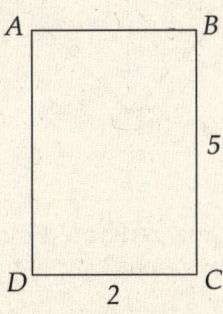

34. What is the volume of the solid generated by rotating rectangle *ABCD* around *AD*?

(A) 15.7
(B) 31.4
(C) 62.8
(D) 72.0
(E) 80.0

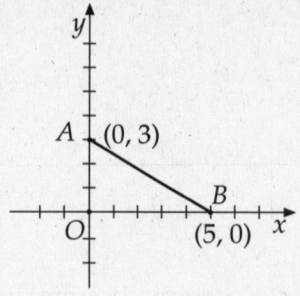

39. If the triangle OAB is rotated around the x-axis, what is the volume of the generated solid?

   (A) 15.70
   (B) 33.33
   (C) 40.00
   (D) 47.12
   (E) 78.54

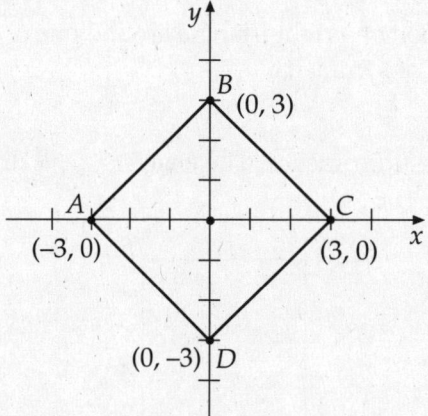

46. What is the volume generated by rotating square $ABCD$ around the y-axis?

   (A) 24.84
   (B) 28.27
   (C) 42.66
   (D) 56.55
   (E) 84.82

Changing dimensions

Some solid-geometry questions will ask you to figure out what happens to the volume of a solid if all of its lengths are increased by a certain factor, or if its area doubles, and so on. To answer questions of this type, just remember a basic rule:

When the lengths of a solid are increased by a certain factor, the surface area of the solid increases by the square of that factor, and the volume increases by the cube of that factor. This rule is only true when the solid's shape doesn't change—its lengths must increase in *every* dimension, not just one. For that reason, cubes and spheres are most often used for this type of question, because their shapes are constant.

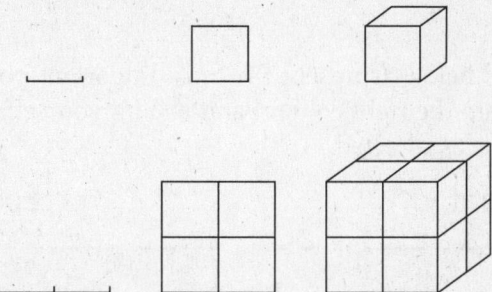

In the illustration above, a length is doubled, which means that the corresponding area is 4 times as great, and the volume is 8 times as great. If the length had been tripled, the area would have increased by a factor of 9, and the volume by a factor of 27.

Drill

Try these practice questions. The answers to all drills are found in chapter 12.

13. If the radius of sphere A is one-third as long as the radius of sphere B, then the volume of sphere A is what fraction of the volume of sphere B?

    (A) $\frac{1}{3}$
    (B) $\frac{1}{4}$
    (C) $\frac{1}{9}$
    (D) $\frac{1}{12}$
    (E) $\frac{1}{27}$

18. A rectangular solid with length l, width w, and height h has a volume of 24. What is the volume of a rectangular solid with length $\frac{l}{2}$, width $\frac{w}{2}$, and height $\frac{h}{2}$?

    (A) 18
    (B) 12
    (C) 6
    (D) 3
    (E) 2

21. If the surface area of a cube is increased by a factor of 2.25, then its volume is increased by what factor?

    (A) 1.72
    (B) 3.38
    (C) 4.50
    (D) 5.06
    (E) 5.64

FLASHCARDS

The rules and formulas listed below represent the most important points to study in this chapter. Memorize them by covering up the right column and testing yourself on the left column, or, better still, by making real flashcards as directed.

Front of Card	Back of Card
What is the formula for the volume of a rectangular solid?	$V = lwh$
What is the formula for the surface area of a rectangular solid?	$SA = 2lw + 2wh + 2lh$
What is the formula for the length of the long diagonal of a rectangular box?	$a^2 + b^2 + c^2 = d^2$
What is the formula for the volume of a cube?	$V = s^3$
What is the formula for the surface area of a cube?	$SA = 6s^2$

Front of Card | **Back of Card**

What is the formula for the long diagonal of a cube?	$d = s\sqrt{3}$
What is the formula for the volume of a cylinder?	$V = \pi r^2 h$
What is the formula for the surface area of a cylinder?	$SA = 2\pi r^2 + 2\pi rh$

Coordinate Geometry

COORDINATE GEOMETRY ON THE SUBJECT TESTS

About 12 percent of the questions on each Math Subject Test will concern graphs on the coordinate plane. Most coordinate-geometry questions on the Math IC Test are about lines, slopes, and distances. On the Math IIC Test, you're more likely to see parabolas, ellipses, and more complicated curves. Simple circles and parabolas can appear on either test. The techniques in this chapter will prepare you for all major coordinate-geometry question types.

DEFINITIONS

Here are some geometry terms that appear on the Math Subject Tests. Make sure you're familiar with them. If the meaning of any of these vocabulary words keeps slipping your mind, add that word to your flashcards.

Coordinate Plane: A system of two perpendicular axes used to describe the position of a point using a pair of coordinates—also called the *rectangular coordinate system*, or the *Cartesian plane*.

Slope: For a straight line, the ratio of vertical change to horizontal change.

x-axis: The horizontal axis of the coordinate plane.

y-axis: The vertical axis of the coordinate plane.

Origin: The intersection of the *x* and *y* axes, with coordinates (0, 0).

x-intercept: The *x*-coordinate at which a line or other function intersects the *x*-axis.

y-intercept: The *y*-coordinate at which a line or other function intersects the *y*-axis.

THE COORDINATE PLANE

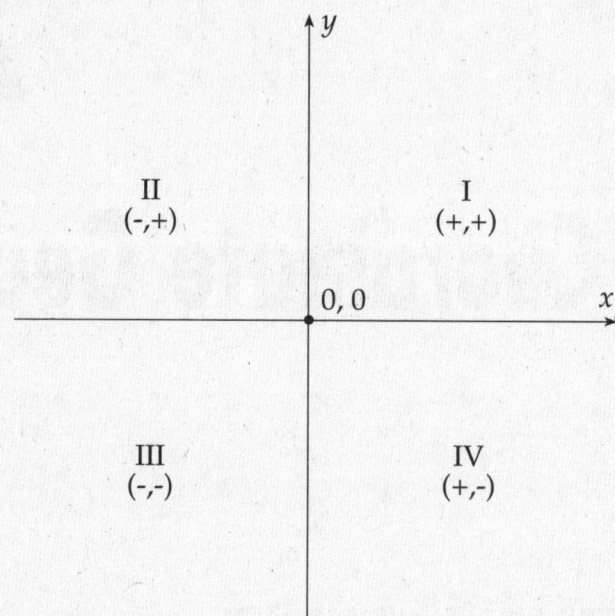

The coordinate plane is a perfectly flat surface that extends an infinite distance in two dimensions. Oh, and it doesn't exist. It's an abstract idea, a way of seeing mathematical relationships visually. The plane is divided into four regions by two perpendicular axes called the *x*- and *y*-axes; these axes are like rulers that measure horizontal distances (the *x*-axis) and vertical distances (the *y*-axis). Each axis has a positive direction and a negative direction; up and right are positive, down and left are negative. The four regions created by the axes are known as *quadrants*. The quadrants are numbered from I to IV, starting on the upper right and moving counterclockwise.

The location of every point on the coordinate plane can be expressed by a pair of *coordinates* that show the point's position with relation to the axes. The *x*-coordinate is always given first, followed by the *y*-coordinate, like so: (2,3). This is called a coordinate pair—it is read as "two over, three up." These coordinates reflect the distance on each axis from the *origin*, or intersection of the axes.

Drill

On the coordinate plane below, match each coordinate pair to the corresponding point on the graph, and identify the quadrant in which the point is located. The answers to all drills are found in chapter 12.

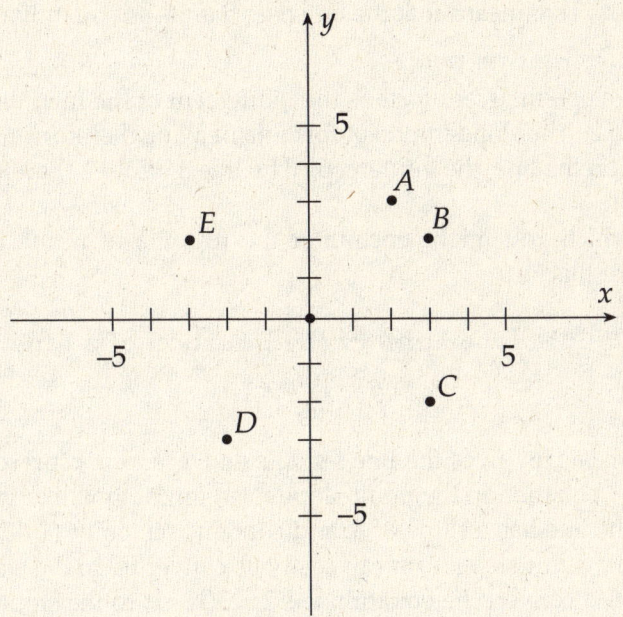

1. (–3, 2)   Point _____, quadrant _____
2. (2, 3)   Point _____, quadrant _____
3. (3, –2)   Point _____, quadrant _____
4. (–2, –3)   Point _____, quadrant _____
5. (3, 2)   Point _____, quadrant _____

THE EQUATION OF A LINE

Most of the coordinate-geometry questions on the Math IC test will deal with the equations and graphs of lines. Lines will also be tested on the Math IIC test, but will generally be outnumbered on the Math IIC test by more complicated functions.

The equation of a line can show up on the test in two forms. The more common form is called the *slope-intercept* formula, and it looks like this:

Slope-Intercept Form of the Equation of a Line
$$y = mx + b$$

In this form, *m* and *b* are constants. An equation in this form might look like this: $y = 5x - 4$. The constant *m* is the slope of the line—the ratio of vertical change to horizontal change. To understand what slope means graphically, just convert the slope to a fraction, if it isn't a fraction already. In this case, 5 can be written as $\frac{5}{1}$. That means that the line rises 5 units as you move 1 unit from left to right. Slope is often referred to as *rise* over *run*.

The constant *b* in the slope-intercept form is the *y*-intercept of the line, the *y*-coordinate at which the line intersects the *y*-axis. The slope-intercept formula of a line therefore gives you the slope of the line and a specific point on the line, the *y*-intercept. The line $y = 5x - 4$ therefore has a slope of 5 and contains the point (0, –4).

The other form in which you might encounter the formula of a line is called the *point-slope* formula, which looks like this:

Point-Slope Form of the Equation of a Line
$$y - y_1 = m(x - x_1)$$

In this formula, *m* once again gives the line's slope, and x_1 and y_1 represent the coordinates of a point on the line. A line's equation in the point-slope form might look like this: $y + 2 = 3(x - 1)$. The equation tells you that the line has a slope of 3, and contains the point (1, –2).

A line's equation in the point-slope form can easily be converted to the slope-intercept form. Just isolate *y*. Here's how you'd convert the equation $y + 2 = 3(x - 1)$ to the slope-intercept form:

$$y + 2 = 3(x - 1)$$
$$y + 2 = 3x - 3$$
$$y = 3x - 5$$

The line therefore contains the point (0, –5).

In general, whenever you're given the equation of a line in any form, you can find the *y*-intercept by making $x = 0$ and finding the value of *y*. In the same way, you can find the *x*-intercept by making $y = 0$ and solving for the value of *x*. The *x*- and *y*-intercepts are often the easiest points on a line to find. If you need to identify the graph of a linear equation, and the slope of the line isn't enough to narrow your choices down to one, finding the *x*- and *y*-intercepts will help.

To graph a line, simply plug a couple of *x*-values into the equation of the line, and plot the coordinates that result. The *y*-intercept is generally the easiest point to plot. Often, the *y*-intercept and the slope are enough to graph a line accurately enough, or to identify the graph of a line.

Drill

Try the following practice questions. The answers to all drills are found in chapter 12.

7. If a line of slope 0.6 contains the point (3, 1), then it must also contain which of the following points?

   (A) (–2, –2)
   (B) (–1, –4)
   (C) (0, 0)
   (D) (2, –1)
   (E) (3, 4)

10. The line $y - 1 = 5(x - 1)$ contains the point $(0, n)$. What is the value of n?

    (A) 0
    (B) −1
    (C) −2
    (D) −3
    (E) −4

11. What is the slope of the line whose equation is $2y - 13 = -6x - 5$?

    (A) −5
    (B) −3
    (C) −2
    (D) 0
    (E) 3

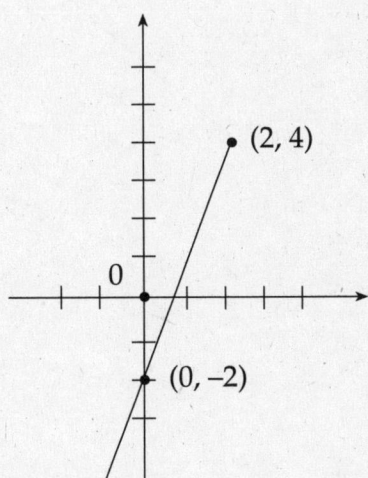

19. If the line $y = mx + b$ is graphed above, then which of the following statements is true?

    (A) $m < b$
    (B) $m = b$
    (C) $2m = 3b$
    (D) $2m + 3b = 0$
    (E) $m = \dfrac{2b}{3}$

23. Which of the following could be the graph of
    $2(y+1) = -6(x-2)$?

(A)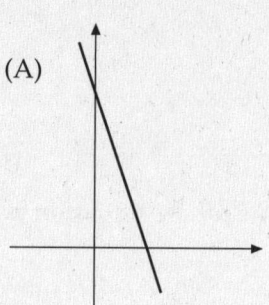

(D)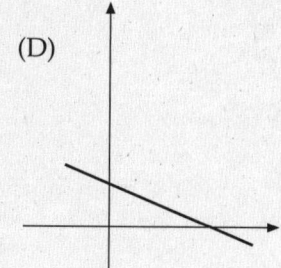

(B)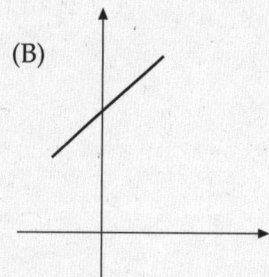

(E)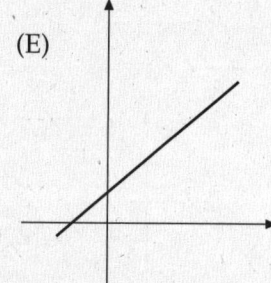

(C)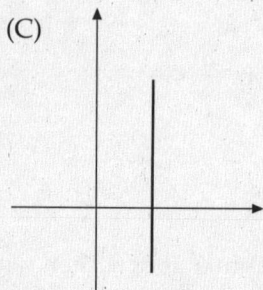

More About Slope

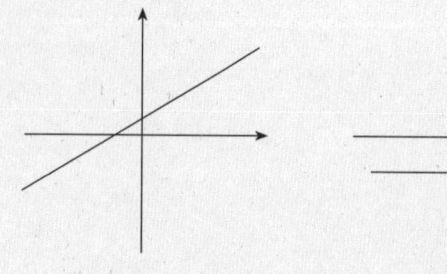 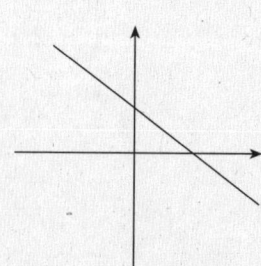

Figure 1 Figure 2 Figure 3

Often, slope is all you need to match the equation of a line to its graph. To begin with, it's easy to distinguish positive slopes from negative slopes. A line with positive slope is shown in figure 1, above; it goes uphill from left to right. A line with zero slope is shown in figure 2; it's horizontal, and neither rises nor falls. A line with negative slope is shown in figure 3; it goes downhill from left to right.

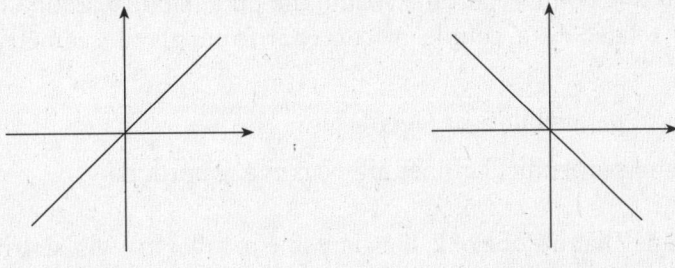

Figure 4　　　　　　　　Figure 5

A line with a slope of 1 rises at a 45° angle, as shown in figure 4. A line with a slope of −1 falls at a 45° angle, as shown in figure 5.

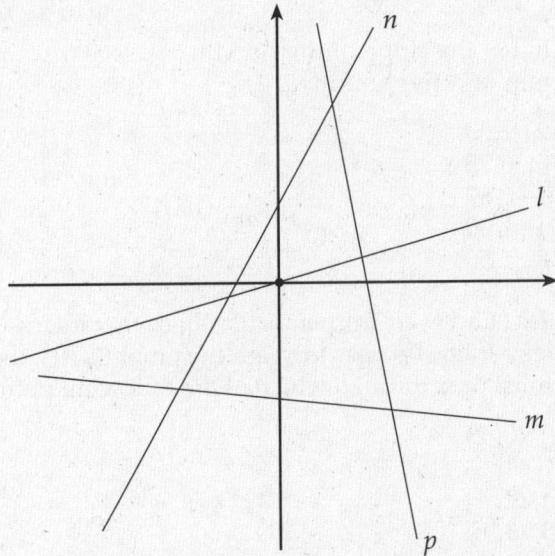

Because a line with a slope of 1 or −1 forms a 45° angle with either axis, you can figure out even more about a line's slope by comparing that line's slope to a 45° angle. Lines that are closer to horizontal have fractional slopes. Lines that are closer to vertical have slopes greater than 1 or less than −1. On the graph above, for example, line *l* has a positive fractional slope. Line *m* has a negative fractional slope. Line *n* has a positive slope greater than 1. Line *p* has a negative slope less than −1. Estimating slope can be a valuable time-saver.

Of course, if you have the equation of a line, you can use it to figure out the line's slope exactly. But what if you're only given the coordinates of a couple of points on a line? Because slope is the ratio of rise (change in *y*) to run (change in *x*), the coordinates of two points on a line provide you enough information to figure out a line's slope. Just use this formula, which assumes that you know the coordinates of two points, (x_1, y_1) and (x_2, y_2):

Slope Formula

$$m = \frac{y_2 - y_1}{x_2 - x_1}$$

Using the slope formula, you can figure out the slope of any line given only two points on that line—which means that you can figure out the complete equation of the line. Just find the line's slope and plug the slope and one point's coordinates into the point-slope equation of a line.

Slope can also be related to a couple of concepts from plane geometry—parallel lines and perpendicular lines:

- The slopes of parallel lines are identical.
- The slopes of perpendicular lines are negative reciprocals.

That means that if line l has a slope of 2, then any line parallel to l will also have a slope of 2. Any line perpendicular to l will have a slope of $-\frac{1}{2}$.

Drill

The answers to all drills are found in chapter 12.

4. What is the slope of the line that passes through the origin and the point (−3, 2)?

   (A) −1.50
   (B) −0.75
   (C) −0.67
   (D) 1.00
   (E) 1.50

17. Lines l and m are perpendicular lines that intersect at the origin. If line l passes through the point (2, −1), then line m must pass through which of the following points?

    (A) (0, 2)
    (B) (1, 3)
    (C) (2, 1)
    (D) (3, 6)
    (E) (4, 0)

47. Line f and line g are perpendicular lines with slopes of x and y, respectively. If $xy \neq 0$, which of the following are possible values of $x-y$?

    I. 0.8
    II. 2.0
    III. 5.2

    (A) I only
    (B) I and II only
    (C) I and III only
    (D) II and III only
    (E) I, II, and III

LINE SEGMENTS

A line by definition goes on forever—it has infinite length. Coordinate-geometry questions may also ask about line *segments*, however. Any coordinate-geometry question asking for the distance between two points is a line-segment question. Any question that draws or describes a rectangle, triangle, or other polygon in the coordinate plane may also involve line-segment formulas. The most commonly requested line-segment formula gives the length of a line segment, given the coordinates of the segment's endpoints, (x_1, y_1) and (x_2, y_2):

The Distance Formula

$$d = \sqrt{(x_2 - x_1)^2 + (y_2 - y_1)^2}$$

This formula may look complicated, but if you look at it carefully, you'll see that it's just a variation on the Pythagorean Theorem. Square both sides of the distance formula, and you get $d^2 = (x_2 - x_1)^2 + (y_2 - y_1)^2$, which is the same as $a^2 + b^2 = c^2$. Finding horizontal and vertical distances in the coordinate plane is easy; you can just count them. It's diagonal distances that require the distance formula. On the graph below, for example, the length of BC cannot be counted directly. You have to use the distance formula.

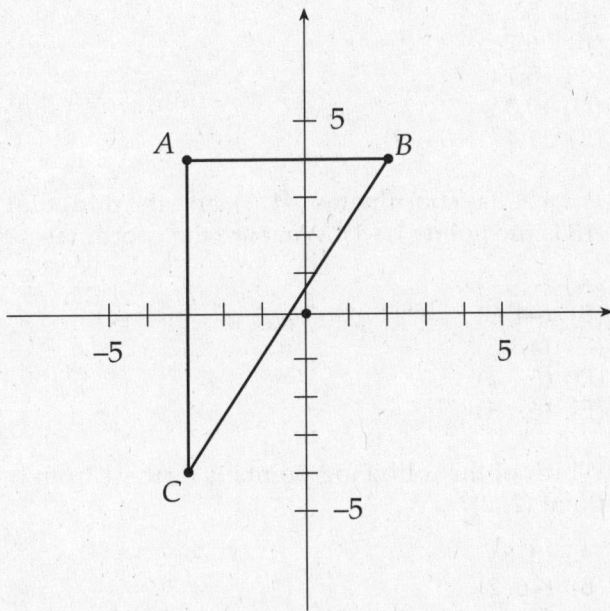

The coordinates of B are (2, 4). The coordinates of C are (–3, –4). If you plug these coordinates into the distance formula, you get:

$$d = \sqrt{(2-(-3))^2 + (4-(-4))^2}$$
$$d = \sqrt{(5)^2 + (8)^2}$$
$$d = \sqrt{25 + 64}$$
$$d = \sqrt{89}$$
$$d = 9.434$$

Notice that you would get the same answer by counting the vertical distance between B and C (8) and the horizontal distance between B and C (5), and using the Pythagorean Theorem to find the diagonal distance.

The other important line-segment formula is used to find the coordinates of the middle point of a line segment with endpoints (x_1, y_1) and (x_2, y_2):

Coordinates of the Midpoint of a Line Segment

$$M = \left(\frac{x_1 + x_2}{2}, \frac{y_1 + y_2}{2} \right)$$

The midpoint and distance formulas used together can answer any line-segment question.

Drill

The answers to all drills are found in chapter 12.

12. What is the distance between the origin and the point (–5, 9)?

    (A) 5.9
    (B) 6.7
    (C) 8.1
    (D) 10.3
    (E) 11.4

19. Point A has coordinates (–4, 3), and the midpoint of AB is the point (1, –1). What are the coordinates of B?

    (A) (–3, 4)
    (B) (–4, 5)
    (C) (4, –5)
    (D) (5, –4)
    (E) (6, –5)

27. Which of the following points is farthest from the point (2, 2)?

    (A) (8, 8)
    (B) (–6, 2)
    (C) (4, –6)
    (D) (–5, –3)
    (E) (9, 4)

LINEAR INEQUALITIES

A linear inequality looks just like a linear equation, except that an inequality sign replaces the equal sign. They are graphed just as lines are graphed, except that the "greater than" or "less than" is represented by shading above or below the line. If the inequality is a "greater than or equal to" or "less than or equal to," then the line itself is included, and is drawn as a solid line. If the inequality is a "greater than" or "less than," then the line itself is not included; the line is drawn as a dotted line, and only the shaded region is included in the inequality.

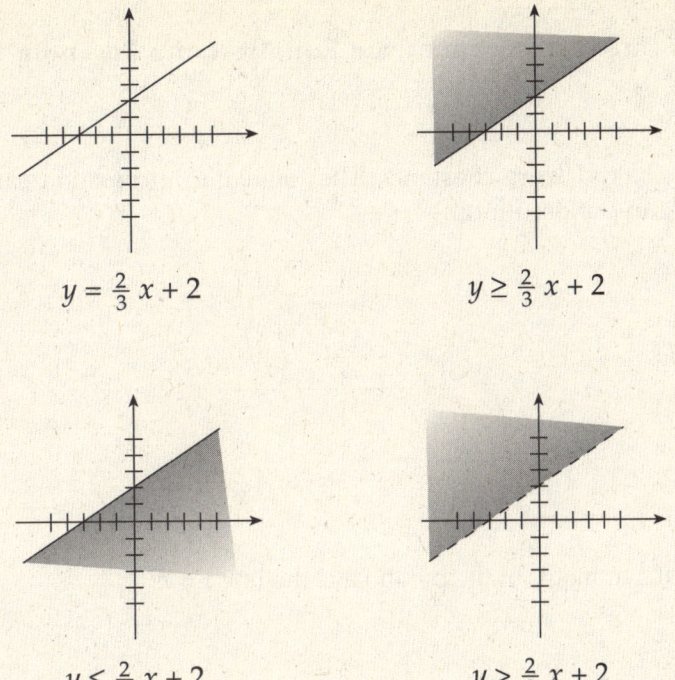

General Equations

While lines are the focus of most coordinate geometry questions, you may also be required to work with the graphs of other shapes in the coordinate plane. In the next few pages, you'll find the general forms of the equations of a number of shapes, and listings of the special information each equation contains.

When ETS asks a coordinate-geometry question about nonlinear shapes, the questions are generally very simple. If you remember the basic equations in this chapter and the shapes of their graphs, you should have little trouble. Questions on this material generally test your understanding of the information contained in the standard forms of these equations.

The parabola

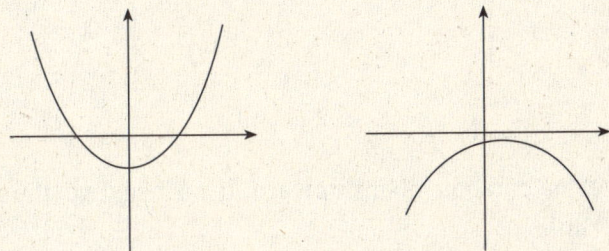

A parabola takes the form of a single curve opening either upward or downward, becoming increasingly steep as you move away from the center of the curve. Parabolas are the graphs of *quadratic* functions, which were discussed in chapter 4. The equation of a parabola can come in two forms. Here is the one that will make you happiest:

Standard Form of the Equation of a Parabola

$$y = a(x - h)^2 + k$$

In this formula, a, h, and k are constants. The following information can be gotten from the equation of a parabola in standard form:

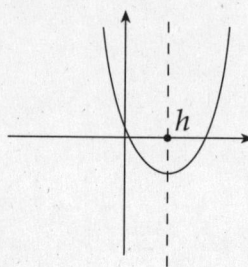

- The axis of symmetry of the parabola is the line $x = h$.

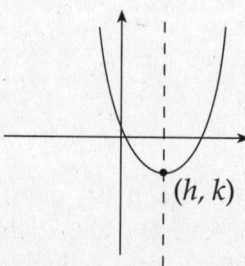

- The vertex of the parabola is the point (h, k).

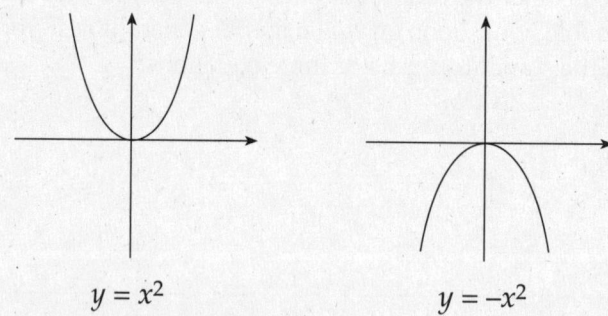

- If a is positive, the parabola opens upward. If a is negative, the parabola opens downward.

The other form that the equation of a parabola can take is the form you saw in chapter 4:

General Form of the Equation of a Parabola
$$y = ax^2 + bx + c$$

In this formula, a, b, and c are constants. The following information can be gotten from the equation of a parabola in general form:

- The axis of symmetry of the parabola is the line $x = -\frac{b}{2a}$.

- The x–coordinate of the parabola's vertex is $-\frac{b}{2a}$. The y–coordinate of the vertex is whatever you get when you plug $-\frac{b}{2a}$ into the equation as x.

- The y-intercept of the parabola is the point $(0, c)$.

- If a is positive, the parabola opens upward. If a is negative, the parabola opens downward.

Drill
The answers to all drills are found in chapter 12.

34. What is the minimum value of $f(x)$ if $f(x) = x^2 - 6x + 8$?

    (A) −3
    (B) −2
    (C) −1
    (D) 0
    (E) 2

37. What are the coordinates of the vertex of the parabola defined by the equation $y = \frac{1}{2}x^2 + x + \frac{5}{2}$?

    (A) (−2, 4)
    (B) (−1, 2)
    (C) (1, 2)
    (D) (2, 4)
    (E) (2, −4)

38. At which of the following x-values does the parabola defined by $y=(x-3)^2 - 4$ cross the x-axis?

    (A) −3
    (B) 3
    (C) 4
    (D) 5
    (E) 9

The circle

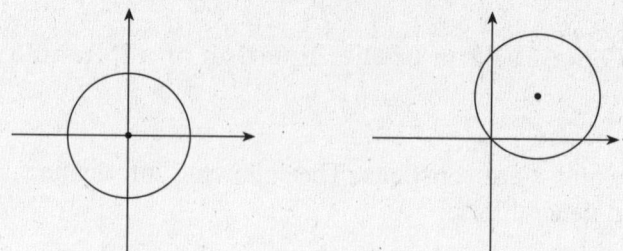

The circle is that round shape you know and love. It's also probably ETS's favorite nonlinear shape in the coordinate plane, particularly on the Math IC Test. Here's the formula for a circle:

Standard Form of the Equation of a Circle

$$(x - h)^2 + (y - k)^2 = r^2$$

In this formula, h, k, and r are constants. The following information can be gotten from the equation of a circle in standard form:

- The center of the circle is the point (h, k).
- The length of the circle's radius is r.

And that's all there is to know about a circle; once you know its radius and the position of its center, you can sketch the circle yourself or identify its graph easily. It's also a simple matter to estimate the radius and center coordinates of a circle whose graph is given, and make a good guess at the equation of that circle. One last note: If the circle's center is the origin, then $(h, k) = (0, 0)$. This greatly simplifies the equation of the circle:

Equation of a Circle with Center at Origin

$$x^2 + y^2 = r^2$$

Drill

The answers to all drills are found in chapter 12.

30. Which of the following points does NOT lie on the circle whose equation is $(x - 2)^2 + (y - 4)^2 = 9$?

    (A) (−1, 4)
    (B) (−1, −1)
    (C) (2, 1)
    (D) (2, 7)
    (E) (5, 4)

33. If the equation $x^2 + 6x + y^2 - 4y = 12$ defines a circle with center P, which of the following is P?

    (A) (−3, 2)
    (B) (−2, −4)
    (C) (2, 6)
    (D) (3, −2)
    (E) (3, 4)

34. Points S and T lie on the circle $x^2 + y^2 = 16$. If S and T have identical y-coordinates but distinct x-coordinates, then which of the following is the distance between S and T?

    (A) 4.0
    (B) 5.6
    (C) 8.0
    (D) 11.3
    (E) It cannot be determined from the information given.

The ellipse

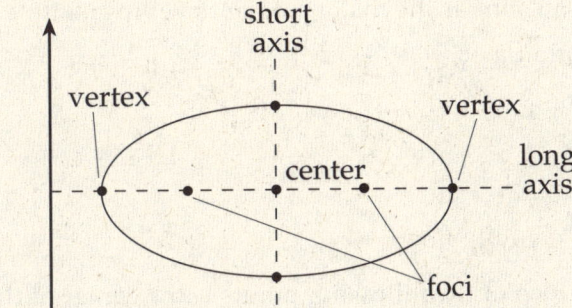

The equation of an ellipse looks similar to the equation of a circle, but an ellipse is actually a much more complex shape. You don't need to worry about the ellipse if you're taking the Math IC Test; it appears exclusively on the Math IIC Test.

An ellipse has a center, like a circle, but, since it's squashed a little flatter than a circle, it has no constant radius. Instead, an ellipse has two *vertices* at the ends of its long axis; and two *foci*, points within the ellipse. The foci of an ellipse (just one is called a focus) are important to the definition of an ellipse. The distances from the two foci to a point on the ellipse always add up to the same number—for every point on the ellipse. This is the formula for an ellipse:

General Equation of an Ellipse

$$\frac{(x-h)^2}{a^2} + \frac{(y-k)^2}{b^2} = 1$$

In this formula, a, b, h, and k are constants. The following information can be gotten from the equation of an ellipse in standard form:

- The center of an ellipse is the point (h, k).

An ellipse can be longer either horizontally or vertically. If the constant under the $(x - h)^2$ term is larger than the constant under the $(y - k)^2$ term, then the major axis of the ellipse is horizontal. If the constant under the $(y - k)^2$ term is bigger, then the major axis is vertical. Like a circle, the equation for an ellipse becomes simpler when it's centered at the origin, and $(h, k) = (0, 0)$:

Equation of an Ellipse with Center at Origin

$$\frac{x^2}{a^2} + \frac{y^2}{b^2} = 1$$

Most of the ellipses that show up on the Math IIC Test are in this simplified form; they are centered at the origin.

Drill

The answers to all practice questions are in chapter 12.

15. How long is the major axis of the ellipse with a formula of $\frac{x^2}{16} + \frac{y^2}{25} = 1$?

    (A) 1
    (B) 4
    (C) 5
    (D) 8
    (E) 10

40. Which of the following points is the center of the ellipse whose formula is $\frac{(x+5)^2}{9} + \frac{(y-3)^2}{4} = 1$?

    (A) $\left(\frac{25}{9}, -\frac{9}{4}\right)$
    (B) $\left(-\frac{5}{9}, \frac{3}{4}\right)$
    (C) $(-5, 3)$
    (D) $(25, -9)$
    (E) $(9, 16)$

43. What is the center of the ellipse defined by the equation $9x^2 - 18x + 4y^2 + 16y = 11$?

    (A) $(-3, 2)$
    (B) $(1, -2)$
    (C) $(4, 9)$
    (D) $(9, 4)$
    (E) $(9, 16)$

The hyperbola

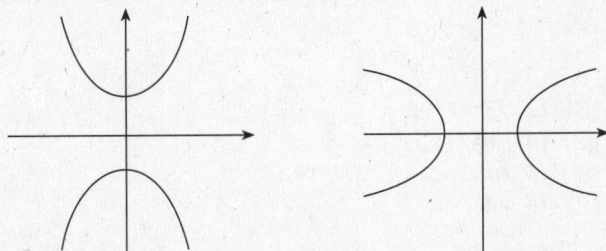

A hyperbola is basically an oval turned inside out. Like the ellipse, the hyperbola only shows up on the Math IIC test, and it doesn't show up frequently. The equation of a hyperbola differs from the equation of an ellipse only by a sign.

General Equation of a Hyperbola

$$\frac{(x-h)^2}{a^2} - \frac{(y-k)^2}{b^2} = 1$$

In this formula, a, b, h, and k are constants. The following information can be gotten from the equation of a hyperbola in standard form:

- The hyperbola's center is the point (h, k).

Like an ellipse, a hyperbola can be oriented either horizontally or vertically. If the y-term is negative, then the curves open out to the right and left. If the x-term is negative, then the curves open up and down. Like an ellipse, a hyperbolas equation becomes simpler when it is centered at the origin, and $(h, k) = (0, 0)$:

Equation of a Hyperbola with Center at Origin

$$\frac{x^2}{a^2} - \frac{y^2}{b^2} = 1$$

Most of the hyperbolas that show up on the Math Subject Tests are in this simplified form; they are centered at the origin.

Drill

38. The hyperbola $\frac{(x+4)^2}{9} - \frac{(y+5)^2}{4} = 1$ has its center at which of the following points?

    (A) $(-9, -4)$
    (B) $(-4, -5)$
    (C) $(4, 5)$
    (D) $(9, -4)$
    (E) $(16, 25)$

43. If P is the center of the hyperbola whose equation is $4x^2 - 24x - y^2 - 6y = -23$, then which of the following is P?

    (A) $(3, -3)$
    (B) $(2, 3)$
    (C) $(4, -6)$
    (D) $(4, 6)$
    (E) $(6, -4)$

Triaxial Coordinates: Thinking 3-D

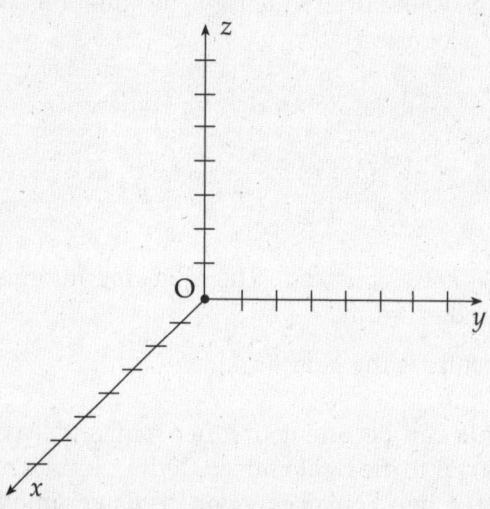

On the Math IIC Test (or the difficult third of the Math IC), you may run into a twist on the coordinate plane: A coordinate *space*. A third dimension can be added to the coordinate plane by introducing a third axis (conventionally called the z-axis) that passes through the origin at right angles to both the x-axis and the y-axis. While the x- and y-axes define the location of a point in a plane, the x-, y-, and z-axes define the location of a point in a three-dimensional space.

Such a system of three axes is called a three-dimensional coordinate system, a triaxial coordinate system, or a coordinate space. Sometimes, it's not called anything at all; ETS will simply show you a diagram of a three-dimensional graph like the one above, or a set of triple coordinates, and expect you to understand what you're looking at. The coordinates of a point in three dimensions are given in this form: (x, y, z). The point $(3, 4, 5)$ is located 3 units along the x-axis, 4 units along the y-axis, and 5 units along the z-axis. Always check the labels on the axes if you're given a diagram, because there's no firm convention about which axis is pictured in which position.

Just as the Pythagorean Theorem is used to calculate distances in the coordinate plane, distances in a coordinate space can be calculated with the souped-up version of the Pythagorean Theorem used for the long diagonal of a rectangular solid:

Distance in a Three-Dimensional Space
$$d^2 = x^2 + y^2 + z^2$$

Calculating a distance in three-dimensional coordinates can be visualized as finding the long diagonal of a rectangular box:

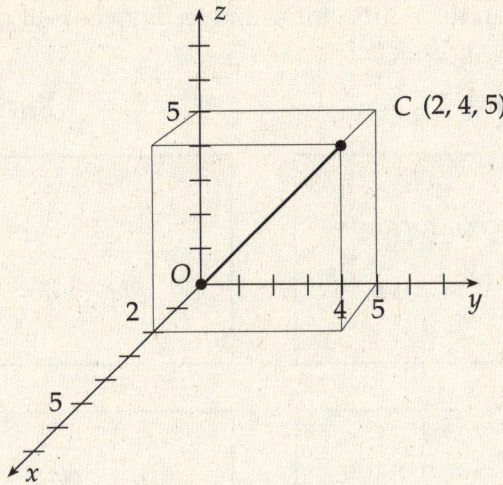

The distance between the point (2, 4, 5) and the origin (0, 0, 0) is the same as the length of the long diagonal of a rectangular solid with dimensions 2 × 4 × 5.

Calculating the distance between a couple of points in three dimensions is only slightly more complicated when neither point is the origin. You'll still use the formula for the long diagonal of a box. For the dimensions of the box, you'll use the *difference* between the corresponding coordinates of the two points. For example, in order to calculate the distance between the point (3, 4, 5) and the point (6, 2, 10), you would find the differences of the coordinates: the x-coordinates differ by 3, the y-coordinates by 2, and the z-coordinates by 5. The distance between the two points will be the same as the long diagonal of a box with dimensions 3 × 2 × 5.

Most of the three-dimensional coordinate questions on the Math Subject Tests require you to calculate a distance between two points in a 3-D coordinate system.

Drill

Try the following practice questions. The answers to all drills are found in chapter 12.

29. What is the distance between the origin and the point (5, 6, 7)?

    (A) 4.24
    (B) 7.25
    (C) 10.49
    (D) 14.49
    (E) 18.00

34. Sphere O has a radius of 6, and its center is the origin. Which of the following points is NOT inside the sphere?

    (A) (−3, 5, 1)
    (B) (−4, −4, 3)
    (C) (5, −2, 2)
    (D) (4, 1, −4)
    (E) (2, −4, −3)

FLASHCARDS

The rules and formulas listed below represent the most important points to study in this chapter. Memorize them by covering up the right column and testing yourself on the left column, or, better still, by making real flashcards as directed.

Front of Card	Back of Card
What is the slope-intercept form of the equation of a line?	$y = mx + b$ m = slope; b = y-intercept
What is the point-slope form of the equation of a line?	$y - y_1 = m(x - x_1)$ m = slope (x_1, y_1) is a point on the line
What is the slope formula?	$m = \dfrac{y_2 - y_1}{x_2 - x_1}$
What is the distance formula?	$d = \sqrt{(x_2 - x_1)^2 + (y_2 - y_1)^2}$
What is the midpoint formula?	$M = \left(\dfrac{x_2 - x_1}{2}, \dfrac{y_2 - y_1}{2} \right)$

Front of Card	Back of Card
What is the standard equation of a parabola?	$y = a(x - h)^2 + k$ (h, k) is the vertex of the parabola
What is the standard form of the equation of a circle?	$(x - h)^2 + (y - k)^2 = r^2$ (h, k) is the center of the circle r is the circle's radius
What is the standard form of the equation of an ellipse?	$\dfrac{(x - h)^2}{a^2} + \dfrac{(y - k)^2}{b^2} = 1$ (h, k) is the center of the ellipse
What is the standard form of the equation of a hyperbola?	$\dfrac{(x - h)^2}{a^2} - \dfrac{(y - k)^2}{b^2} = 1$ (h, k) is the center of the hyperbola

Trigonometry

TRIGONOMETRY ON THE SUBJECT TESTS

The rules of trigonometry tested on the Math IC test are much more limited than those tested on the Math IIC test. Trigonometry on the Math IC test is confined to right triangles and the most basic relationships between the sine, cosine, and tangent functions. If you're taking the Math IC test, that's the only material from this chapter you need to know. If you plan to take the Math IIC test, then this entire chapter is your domain; rule it wisely.

DEFINITIONS

Here are some trigonometric terms that appear on the Math Subject Tests. Make sure you're familiar with them. If the meaning of any of these vocabulary words keeps slipping your mind, add that word to your flashcards.

Acute Angle: An angle whose measure in degrees is between 0 and 90.

Obtuse Angle: An angle whose measure in degrees is between 90 and 180.

Radian: A unit of angle measure—1 radian is the measure of a central angle in a circle that intercepts an arc equal in length to the circle's radius.

arc-: Prefix added to trigonometric functions, meaning *inverse*.

THE BASIC FUNCTIONS

The basis of trigonometry is the relationship between the parts of a right triangle. When you know the measure of one of the acute angles in a right triangle, you have complete information about the shape of that triangle. For example, if you know that a right triangle contains a 20° angle, then you know all three angles; the triangle must also have a 90° angle, and because there are 180° in a triangle, the third angle must measure 70°. You don't know the *lengths* in the triangle, but you know its *shape*, its proportions.

A right triangle that contains a 20° angle can have only one shape, though it can be any size. The same is true for a triangle containing any other acute angle. That's the fundamental idea of trigonometry: Once you know the measure of an acute angle in a right triangle, you know that triangle's proportions.

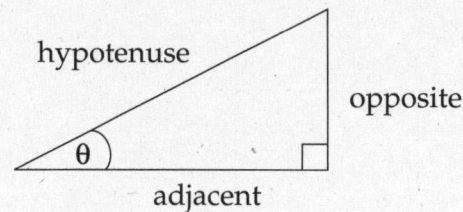

The three basic functions in trigonometry—the sine, cosine, and tangent—are ways of expressing proportions in a right triangle. A trigonometric function of any angle comes from the proportions of a right triangle containing that angle. For any given angle, there is only one possible set of proportions.

Sine

The sine of an angle is the ratio of the opposite side to the hypotenuse. The sine function of an angle θ is abbreviated this way: $\sin \theta$.

Cosine

The cosine of an angle is the ratio of the adjacent side to the hypotenuse. The cosine function of an angle θ is abbreviated this way: $\cos \theta$.

Tangent

The tangent of an angle is the ratio of the opposite side to the adjacent side. The tangent function of an angle θ is abbreviated this way: $\tan \theta$.

You can remember the meanings of these functions by memorizing the word *SOHCAHTOA*:

SOHCAHTOA

$$\sin = \frac{opposite}{hypotenuse} \qquad \cos = \frac{adjacent}{hypotenuse} \qquad \tan = \frac{opposite}{adjacent}$$

These three functions form the basis of everything else in trigonometry. All of the more complicated functions and rules in trigonometry can be derived from the information contained in SOHCAHTOA.

WHAT YOUR CALCULATOR CAN DO FOR YOU

Tables of sine, cosine, and tangent values are programmed into your calculator—that's what the "sin," "cos," and "tan" keys do:

- If you enter an angle-measure into your calculator and hit one of the three trigonometric function keys, your calculator will give you the function (sine, cosine, or tangent) of that angle. This operation is written like this:

    $\sin 30° = 0.5 \qquad \cos 30° = 0.866 \qquad \tan 30° = 0.577$

- Your calculator can also take a trig-function value and tell you what angle would produce that value. Just enter the sine, cosine, or tangent of an angle, then press the "INV" or "2ND" key and the appropriate trig-function key, and your calculator will give you the measure of that angle. This is called taking an *inverse function*, and it's written like this:

    $\sin^{-1}(0.5) = 30° \qquad \cos^{-1}(0.866) = 30° \qquad \tan^{-1}(0.577) = 30°$

    OR like this:

    $\arcsin(0.5) = 30° \qquad \arccos(0.866) = 30° \qquad \arctan(0.577) = 30°$

The expressions "$\sin^{-1}(0.5)$" and "$\arcsin(0.5)$" have the same meaning. Both mean "the angle whose sine is 0.5." While ordinary trig functions take angle measures and give the values of their functions, inverse trig functions take function values and produce the corresponding angle measures; they work in reverse.

FINDING TRIG FUNCTIONS IN RIGHT TRIANGLES

Drill

On the Math IC Test, the three basic trigonometric functions always occur in right triangles—particularly the Pythagorean triplets from chapter 5. Use the definitions of the sine, cosine, and tangent to fill in the requested quantities in the following triangles. The answers to all drills are found in chapter 12.

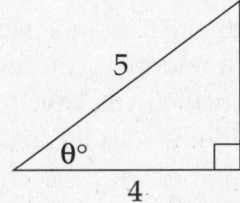

1. $\sin \theta =$ _____

    $\cos \theta =$ _____

    $\tan \theta =$ _____

2. $\sin \theta =$ _____

    $\cos \theta =$ _____

    $\tan \theta =$ _____

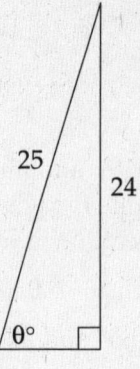
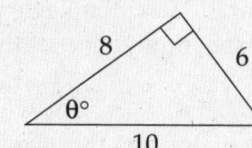

3. sin θ = _____   4. sin θ = _____

    cos θ = _____   cos θ = _____

    tan θ = _____   tan θ = _____

COMPLETING TRIANGLES

The preceding examples have all involved figuring out the values of trigonometric functions from lengths in a right triangle. Slightly more difficult trigonometry questions may require you to go the other way: to figure out lengths or measures of angles using trigonometry. For example:

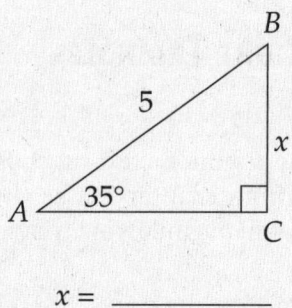

x = _____

In triangle *ABC*, you know only two quantities: The length of *AB* and the measure of ∠*A*. This question, unlike previous examples, doesn't give you enough information to use the Pythagorean Theorem. What you need is an equation that relates the information you have (*AB* and ∠*A*) to the information you don't have (*x*). Use the SOHCAHTOA definitions to set up an equation. Solve that equation, and you find the value of the unknown:

$$\sin = \frac{opposite}{hypotenuse}$$

$$\sin 35° = \frac{x}{5}$$

$$5(\sin 35°) = x$$

$$5(0.5736) = x$$

$$2.8679 = x$$

Side *BC* of triangle *ABC* therefore has a length of 2.87.

You can use a similar technique to find the measure of an acute angle in a right triangle. For example:

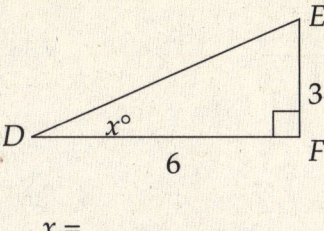

x = _____

In triangle *DEF*, you know two quantities—the lengths of *EF* and *FD*. Once again, in order to find the value of x you must create an equation relating *EF* and *FD* to ∠*D*. You do this by finding the SOHCAHTOA definition that relates all three quantities:

$$\tan = \frac{opposite}{adjacent}$$

$$\tan x = \frac{EF}{FD}$$

$$\tan x = \frac{3}{6}$$

$$\tan x = 0.5$$

To solve for x, take the *inverse tangent* of both sides of the equation. On the left side, that just gives you x. To take the inverse tangent of the right side, punch 0.5 into your calculator, press the inverse key and then the tangent key. The result is the angle whose tangent is 0.5:

$$\tan^{-1}(\tan x) = \tan^{-1}(0.5)$$
$$x = 26.57°$$

The measure of ∠*D* is therefore 26.57°

Drill

Use the techniques you've just reviewed to complete the following triangles. The answers to all drills are found in chapter 12.

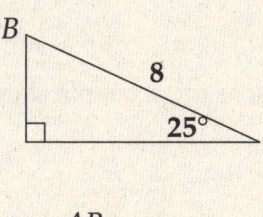

1. AB ∞ _____

   CA ∞ _____

   ∠B ∞ _____

2. EF ∞ _____

   FD ∞ _____

   ∠D ∞ _____

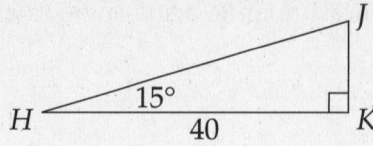
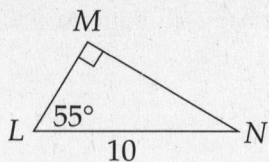

3. HJ = _____   4. LM = _____

   JK = _____   MN = _____

   ∠J = _____   ∠N = _____

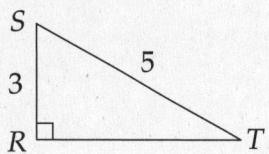
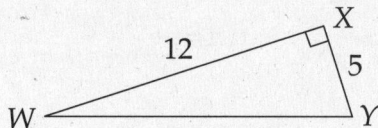

5. TR = _____   6. YW = _____

   ∠S = _____   ∠W = _____

   ∠T = _____   ∠Y = _____

TRIGONOMETRIC IDENTITIES

Some Subject Test questions will ask you to do algebra with trigonometric functions. These questions usually involve using the SOHCAHTOA definitions of sine, cosine, and tangent. Often, the way to simplify equations that are mostly made up of trigonometric functions is to express the functions this way:

$$\sin = \frac{O}{H} \qquad \cos = \frac{A}{H} \qquad \tan = \frac{O}{A}$$

Writing trig functions this way can simplify trig equations, as this example shows:

$$\frac{\sin x}{\cos x} =$$

$$\frac{O}{H} \div \frac{A}{H} =$$

$$\frac{O}{A} = \tan x$$

150 ◆ CRACKING THE SAT II: MATH SUBJECT TESTS

Working with trig functions this way lets you simplify expressions. The equation above is actually a commonly used *trigonometric identity*. You should memorize this, as it can often be used to simplify equations itself:

$$\frac{\sin x}{\cos x} = \tan x$$

Here's the breakdown of another frequently used trigonometric identity:

$$\sin^2 \theta + \cos^2 \theta =$$

$$(\sin \theta)(\sin \theta) + (\cos \theta)(\cos \theta) =$$

$$\left(\frac{O}{H}\right)\left(\frac{O}{H}\right) + \left(\frac{A}{H}\right)\left(\frac{A}{H}\right) =$$

$$\frac{O^2}{H^2} + \frac{A^2}{H^2} =$$

$$\frac{O^2 + A^2}{H^2} = 1$$

That last step may seem a little baffling, but it's really simple. This equation is based on a right triangle, in which O and A are legs of the triangle, and H is the hypotenuse. Consequently you know that $O^2 + A^2 = H^2$. That's just the Pythagorean Theorem. That's what lets you do the last step, in which $\frac{O^2 + A^2}{H^2} = 1$. This completes the second commonly used identity that you should memorize:

$$\sin^2 \theta + \cos^2 \theta = 1$$

In addition to memorizing these two identities, you should practice working algebraically with trig functions in general. Some questions may require you to use the SOHCAHTOA definitions of the trig functions; others may require you to use the two identities you've just reviewed. Take a look at these examples:

35. If $\sin x = 0.707$, then what is the value of $(\sin x) \cdot (\cos x) \cdot (\tan x)$?

    (A) 1.0
    (B) 0.707
    (C) 0.5
    (D) 0.4
    (E) 0.207

This is a tricky question. To solve it, simplify that complicated trigonometric expression. Writing in the SOHCAHTOA definitions works just fine, but in this case it's even faster to use one of those identities:

$$(\sin x) \cdot (\cos x) \cdot (\tan x) =$$
$$(\sin x) \cdot (\cos x) \cdot \left(\frac{\sin x}{\cos x}\right) =$$
$$(\sin x) \cdot (\sin x) =$$
$$\sin^2 x =$$

Now it's a simpler matter to answer the question. If $\sin x = 0.707$, then $\sin^2 x = 0.5$. The answer is (C). Take a look at this one:

36. If $\sin a = 0.4$, and $1 - \cos^2 a = x$, then what is the value of x?

    (A) 0.8
    (B) 0.6
    (C) 0.44
    (D) 0.24
    (E) 0.16

Here again, the trick to the question is simplifying the complicated trig expression. Using the second trig identity, you can quickly take these steps:

$$1 - \cos^2 a = x$$

$$\sin^2 a = x$$

$$(0.4)^2 = x$$

$$x = 0.16$$

And that's the answer. (E) is correct.

Using the SOHCAHTOA definitions and the two trigonometric identities reviewed in this section, simplify trigonometric expressions to answer the following sample questions:

Drill

Try the following practice questions. The answers to all drills are found in chapter 12.

1. $(\sin x)(\cos x)(\tan x) =$

    (A) $\cos^2 x$
    (B) $\sin^2 x$
    (C) $\tan^2 x$
    (D) 1
    (E) $2\sin x$

2. $(1-\sin x)(1 + \sin x) =$

   (A) $\cos x$
   (B) $\sin x$
   (C) $\tan x$
   (D) $\cos^2 x$
   (E) $\sin^2 x$

3. $\dfrac{\tan x \cos x}{\sin x} =$

   (A) $\dfrac{1}{\tan x}$
   (B) $\dfrac{1}{\cos x}$
   (C) 1
   (D) $\cos^2 x$
   (E) $\tan x$

4. $\dfrac{1}{\cos x} - (\sin x)(\tan x) =$

   (A) $\cos x$
   (B) $\sin x$
   (C) $\tan x$
   (D) $\cos^2 x$
   (E) $\sin^2 x$

5. $\dfrac{\tan x - \sin x \cos x}{\tan x} =$

   (A) $1-\cos x$
   (B) $1-\sin x$
   (C) $\tan x + 1$
   (D) $\cos^2 x$
   (E) $\sin^2 x$

The Other Trig Functions

On the Math IIC Test, you may run into the *other* three trigonometric functions: the cosecant, secant, and cotangent. These functions are abbreviated as csc θ, sec θ, and cot θ, respectively, and they are simply the reciprocals of the three basic trigonometric functions you've already reviewed. Here's how they relate:

$$\csc\theta = \frac{1}{\sin\theta} \qquad \sec\theta = \frac{1}{\cos\theta} \qquad \cot\theta = \frac{1}{\tan\theta}$$

You can also express these functions in terms of the sides of a right triangle—just by flipping over the SOHCAHTOA definitions of the three basic functions:

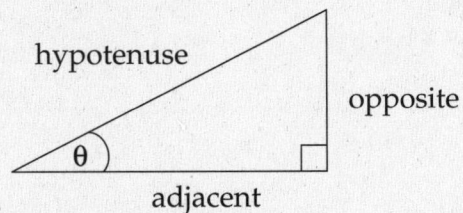

$$\text{cosecant} = \frac{\text{hypotenuse}}{\text{opposite}} \qquad \text{secant} = \frac{\text{hypotenuse}}{\text{adjacent}} \qquad \text{cotangent} = \frac{\text{adjacent}}{\text{opposite}}$$

These three functions generally show up in algebra-style questions, which require you to simplify complex expressions containing trig functions. The goal is usually to get an expression into the simplest form possible, one that contains no fractions. Such questions are like algebra-style questions involving the three basic trig functions; the only difference is that the addition of three more functions increases the number of possible forms an expression can take. For example:

$$(\cos x)(\cot x) + (\sin^2 x \csc x) =$$
$$(\cos x)(\cos x \div \sin x) + (\sin^2 x)(1 \div \sin x) =$$
$$\cos^2 x \div \sin x + \sin^2 x \div \sin x =$$
$$\cos^2 x + \sin^2 x \div \sin x =$$
$$1 \div \sin x =$$
$$\csc x$$

The entire expression $(\cos x)(\cot x)+(\sin^2 x \csc x)$ is therefore equivalent to a single trig function, the cosecant of x. That's generally the way algebraic trigonometry questions work on the Math IIC Test.

Drill

Simplify each of the following expressions to a single trigonometric function. Keep an eye out for the trigonometric identities reviewed on page 150—they'll still come in handy. The answers to all drills are found in chapter 12.

19. $\sec^2 x - 1 =$

    (A) $\sin x \cos x$
    (B) $\sec^2 x$
    (C) $\cos^2 x$
    (D) $\sin^2 x$
    (E) $\tan^2 x$

23. $\dfrac{1}{\sin x \cot x} =$

    (A) $\cos x$
    (B) $\sin x$
    (C) $\tan x$
    (D) $\sec x$
    (E) $\csc x$

24. $\sin x + (\cos x)(\cot x)$

    (A) $\csc x$
    (B) $\sec x$
    (C) $\cot x$
    (D) $\tan x$
    (E) $\sin x$

GRAPHING TRIGONOMETRIC FUNCTIONS

There are two common ways to represent trigonometric functions graphically: on the *unit circle*, or on the coordinate plane (you'll get a good look at both methods in the coming pages). Both of these graphing approaches are ways of showing the repetitive nature of trigonometric functions. All of the trig functions (sine, cosine, and the rest) are called *periodic* functions. That simply means that they cycle repeatedly through the same values.

For example, if you picked a certain angle and its sine, cosine, and tangent, and then slowly changed the measure of that angle, you'd see the sine, cosine, and tangent change as well. But after a while, you would have increased the angle by 360°—in other words, you would come full circle, back to the angle you started with. The new angle, equivalent to the old one, would have the same sine, cosine, and tangent as the original. As you continued to increase the angle's measure, the sine, cosine, and tangent would cycle through the same values all over again. All trigonometric functions repeat themselves every 360 degrees.

For that reason, angles of 0° and 360° are mathematically equivalent. So are angles of 40° and 400°, or 360° and 720°. Any two angle measures separated by 360 degrees are equivalent. In the next few sections, you'll see how that's reflected in the graphs of trigonometric functions.

THE UNIT CIRCLE

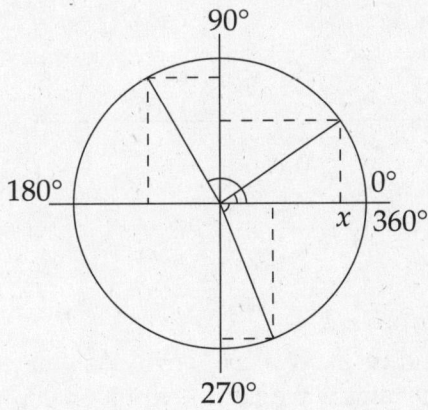

This is the unit circle. It looks a little like the coordinate plane; in fact, it *is* the coordinate plane, or at least a piece of it. The circle is called the *unit circle* because it has a radius of 1 (a single unit). This is convenient because it makes trigonometric values easy to figure out. The radius touching any point on the unit circle is the hypotenuse of a right triangle; the horizontal leg of the triangle is the cosine; the vertical leg is the sine. It works out this way because sine = opposite ÷ hypotenuse, and cosine = adjacent ÷ hypotenuse; and here the hypotenuse is 1, so the sine is simply the length of the opposite side, and the cosine simply the length of the adjacent.

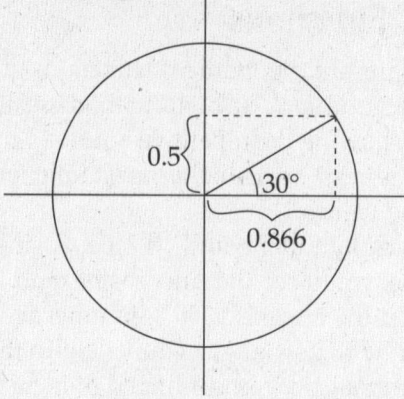

Suppose you wanted to show the sine and cosine of a 30° angle. That angle would appear on the unit circle as a radius drawn at a 30° angle to the positive x-axis (above). The x-coordinate of the point where the radius intercepts the circle is 0.866; that's the value of cos 30°; the y-coordinate of that point is 0.5, which is the value of sin 30°.

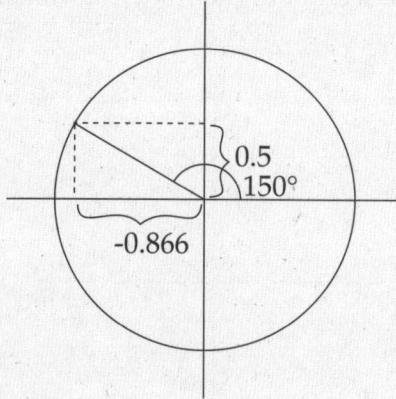

Now take a look at the sine and cosine of a 150° angle. As you can see, it looks just like the 30° angle, flipped over the x-axis. Its y-value is the same—sin 150° = 0.5—but its x-value is now *negative*. The cosine of 150° is –0.866.

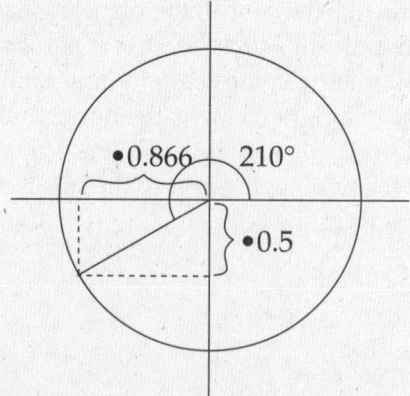

Here, you see the sine and cosine of a 210° angle. Once again, this looks just like the 30° angle, but this time flipped over the x- and y-axes. The sine of 210° is –0.5; the cosine of 210° is –0.866.

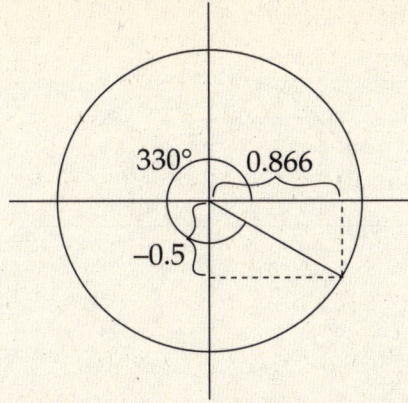

This is the sine and cosine of a 330° angle. Like the previous angles, the 330° angle has a sine and cosine equivalent in magnitude to those of the 30° angle. In the case of the 330° angle, the sine is negative and the cosine positive; so sin 330° = –0.5 and cos 330° = 0.866. Notice that a 330° angle is equivalent to an angle of –30°.

Following these angles around the unit circle gives us some useful information about the sine and cosine functions:

- Sine is positive between 0° and 180° and negative between 180° and 360°. At 0°, 180°, and 360°, sine is zero.

- Cosine is positive between 0° and 90° and between 270° and 360°. (You could also say that cosine is positive between –90° and 90°.) Cosine is negative between 90° and 270°. At 90° and 270°, cosine is zero.

When these angles are sketched on the unit circle, sine is positive in quadrants I and II, and cosine is positive in quadrants I and IV. There's another important piece of information you can get from the unit circle. The biggest value that can be produced by a sine or cosine function is 1; the smallest value that can be produced by a sine or cosine function is –1.

Following the tangent function around the unit circle also yields useful information:

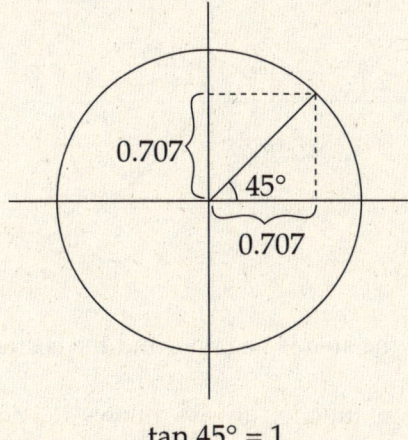

tan 45° = 1

The sine of 45° is $\frac{\sqrt{2}}{2}$ or 0.707, and the cosine of 45° is also $\frac{\sqrt{2}}{2}$ or 0.707. Since the tangent is the ratio of the sine to the cosine, that means that the tangent of 45° is 1.

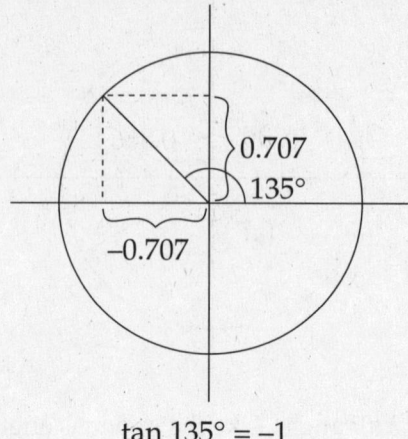

tan 135° = –1

The tangent of 135° is –1; here the sine is positive but the cosine is negative.

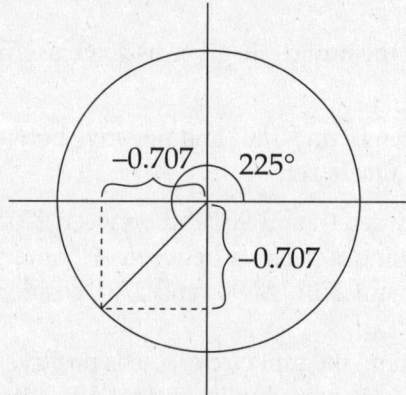

The tangent of 225° is 1; here the sine and cosine are both negative.

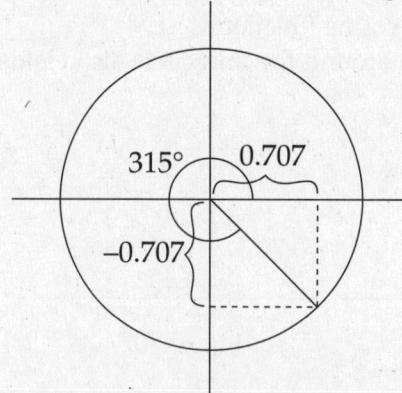

The tangent of 315° is –1; here the sine is negative and the cosine is positive.

This is the pattern that the tangent function always follows. It's positive in quadrants I and III, and negative in quadrants II and IV:

- Tangent is positive between 0° and 90°, and between 180° and 270°.
- Tangent is negative between 90° and 180°, and between 270° and 360°.

The unit circle is extremely useful for identifying equivalent angles (like 270° and −90°), and also for seeing other correspondences between angles, like the similarity between the 45° angle and the 135° angle, which are mirror images of one another on the unit circle.

Drill

Make simple sketches of the unit circle to answer the following questions about angle equivalencies. The answers to all drills are found in chapter 12.

18. If sin 135° = sin x, then x could equal

    (A) −225°
    (B) −45°
    (C) 225°
    (D) 315°
    (E) 360°

21. If cos 60° = cos n, then n could be

    (A) 30°
    (B) 120°
    (C) 240°
    (D) 300°
    (E) 360°

26. If sin 30° = cos t, then t could be

    (A) −30°
    (B) 60°
    (C) 90°
    (D) 120°
    (E) 240°

30. If tan 45° = tan x, then which of the following could be x?

    (A) −45°
    (B) 135°
    (C) 225°
    (D) 315°
    (E) 360°

36. If $0° \leq \theta \leq 360°$ and $(\sin\theta)(\cos\theta) < 0$, which of the following gives the possible values of θ?

    (A) $0° \leq \theta \leq 180°$
    (B) $0° \leq \theta \leq 180°$ or $270° \leq \theta \leq 360°$
    (C) $0° < \theta < 90°$ or $180° < \theta < 270°$
    (D) $90° < \theta < 180°$ or $270° < \theta < 360°$
    (E) $0° < \theta < 180°$ or $270° < \theta < 360°$

DEGREES AND RADIANS

On the Math IIC Test, you may run into an alternate means of measuring angles. This alternate system measures angles in *radians*, rather than degrees. One degree is defined as $\frac{1}{360}$ of a full circle; one radian, on the other hand, is the measure of an angle that intercepts an arc exactly as long as the circle's radius. Since the circumference of a circle is 2π times the radius, the circumference is about 6.28 times as long as the radius, and there are about 6.28 radians in a full circle.

A number like 6.28 isn't easy to work with. For that reason, angle measurements in radians are usually given in multiples or fractions of π. For example, there are exactly 2π radians in a full circle. There are π radians in a semicircle. There are $\frac{\pi}{2}$ radians in a right angle. Because 2π radians and 360 degrees both describe a full circle, you can relate degrees and radians with the following proportion:

$$\frac{\text{degrees}}{360} = \frac{\text{radians}}{2\pi}$$

To convert degrees to radians, just plug the number of degrees into the proportion and solve for radians; the same technique works in reverse for converting radians to degrees. This is what the unit circle looks like in radians, compared to the unit circle in degrees:

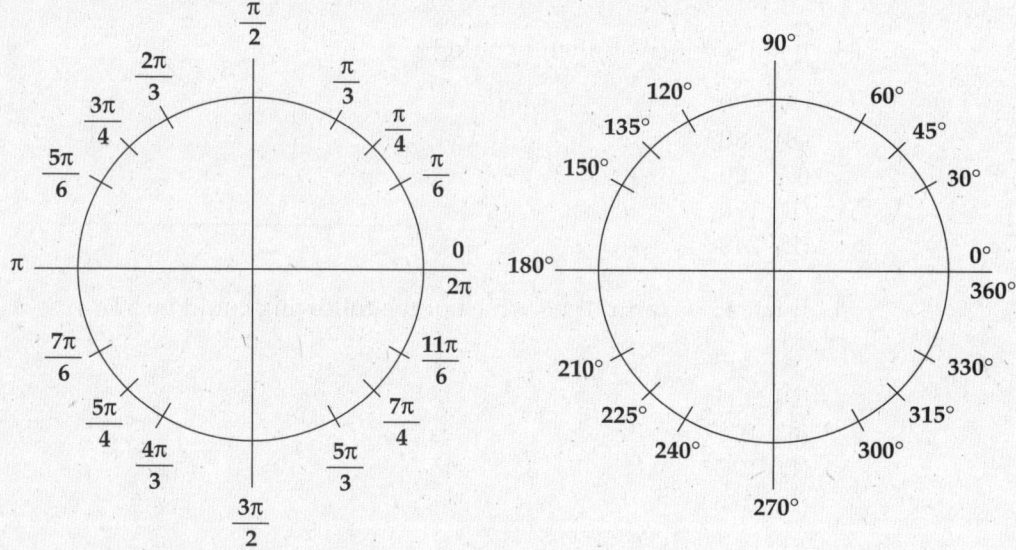

 **Drill**

By referring to these unit circles and using the proportion given above, fill in the following chart of radian-degree equivalencies. The answers to all drills are found in chapter 12.

Degrees	Radians
30°	
45°	
	$\frac{\pi}{3}$
	$\frac{\pi}{2}$
120°	
	$\frac{3\pi}{4}$
150°	
	π
	$\frac{5\pi}{4}$
240°	
	$\frac{3\pi}{2}$
300°	
315°	
330°	$\frac{11\pi}{6}$
	2π

A scientific calculator can calculate trigonometric functions of angles entered in radians, as well; however, it is necessary to shift the calculator from degree-mode into radian-mode. Consult your calculator's operating manual and make sure you know how to do this.

TRIGONOMETRIC GRAPHS ON THE COORDINATE PLANE

In a unit-circle diagram, the x-axis and y-axis represent the horizontal and vertical components of an angle, just as they do on the coordinate plane. The angle itself is represented by the angle of a certain radius to the positive x-axis. Any trigonometric function can be reflected on a single unit-circle diagram.

When a single trigonometric function is graphed, however, the axes take on different meanings. The x-axis represents the value of the angle; this axis is usually marked in radians. The y-axis represents a specific trigonometric function of that angle. For example, here is the coordinate-plane graph of the sine function:

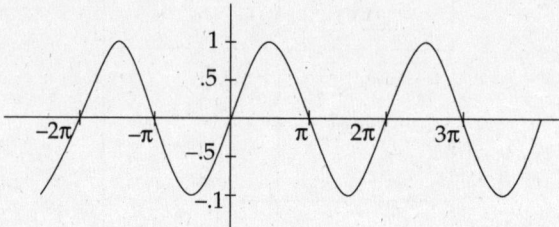

Compare this graph to the unit circle on page 153. A quick comparison will show you that both graphs present the same information. At an angle of zero the sine is zero; at a quarter circle ($\frac{\pi}{2}$ radians or 90 degrees), the sine is 1; and so on.

Remember that trigonometric functions are called *periodic* functions, meaning that they repeat. The distance a function travels before it repeats is called the *period* of the function. As you can see from the graph, the period of the sine function is 2π radians, or 360°.

Here is the graph of the *cosine* function:

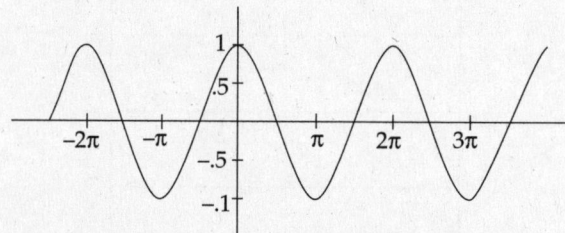

Notice that the cosine curve is identical to the sine curve, only shifted to the left by $\frac{\pi}{2}$ radians, or 90 degrees. The cosine function also has a period of 2π radians.

Finally, here is the graph of the tangent function:

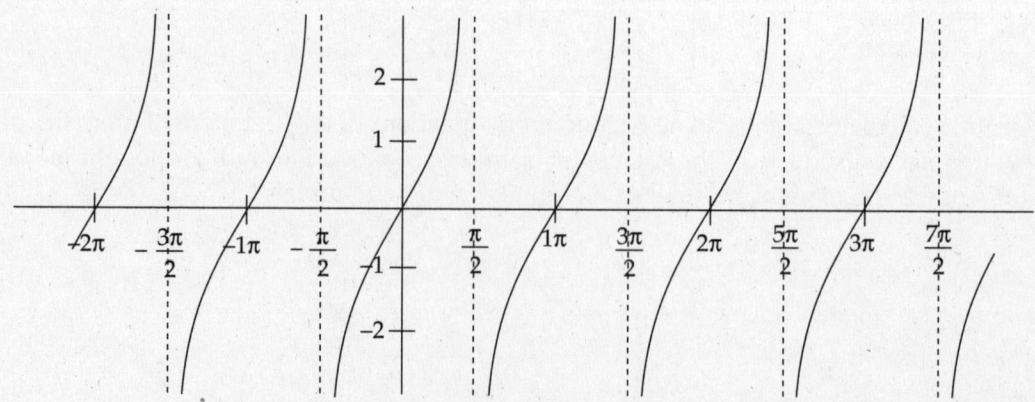

This function, obviously, is very different from the others. Firstly, the tangent function has no upper or lower limit, unlike the sine and cosine functions, which produce values no higher than 1 or lower than –1. Secondly, the tangent function has *asymptotes*. These are values on the *x*-axis at which the tangent function does not exist; they are represented by vertical dotted lines.

It's easy to see why the tangent function's graph has asymptotes, if you recall the definition of the tangent:

$$\tan\theta = \frac{\sin\theta}{\cos\theta}$$

A fraction is undefined whenever its denominator equals zero. At any value where the cosine function equals zero, therefore, the tangent function is undefined—it doesn't exist. As you can see by comparing the cosine and tangent graphs, the tangent has an asymptote wherever the cosine function equals zero.

It's important to be able to recognize the graphs of the three basic trigonometric functions. You'll find more information about these functions and their graphs in the following chapter on functions.

Trigonometry in Non-Right Triangles

The rules of trigonometry are based on the right triangle, as you've seen in the preceding sections. Right triangles are *not*, however, the only places you can use trigonometric functions. There are a couple of powerful rules relating angles and lengths that you can use in *any* triangle. These are rules that only come up on the Math IIC test, and there are only two basic laws you need to know: The Law of Sines and the Law of Cosines.

The Law of Sines
The Law of Sines can be used to complete the dimensions of a triangle about which you have partial information. This is what the law says:

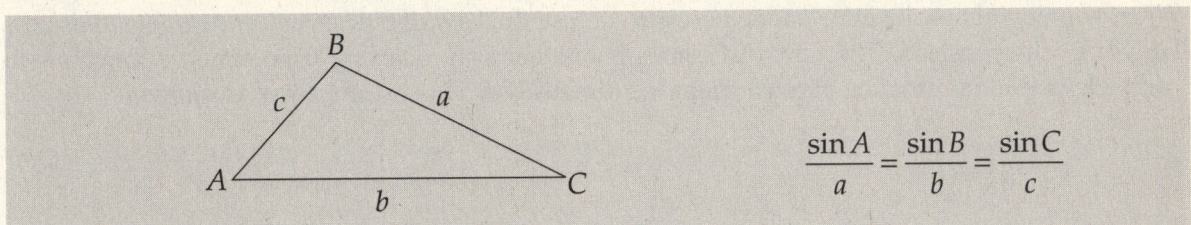

In English, this law means that the sine of each angle in a triangle is related to the length of the opposite side by a constant proportion. Once you figure out the proportion relating the sine of one angle to the opposite side, you know the proportion for every angle. Take a look at this example:

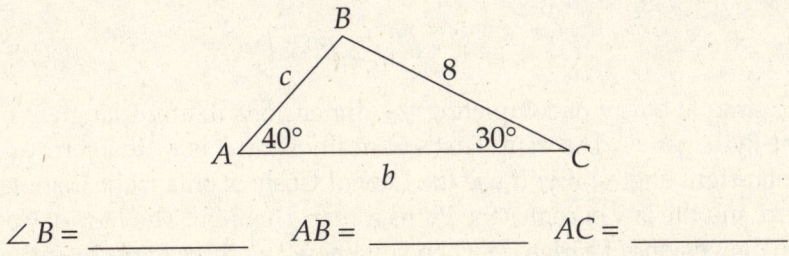

∠B = _____   AB = _____   AC = _____

In this triangle, you know only two angles and one side. Immediately, you can fill in the third angle, knowing that there are 180° in a triangle. Then, you can fill in the missing sides using the Law of Sines. Write out the proportions of the Law of Sines, filling in the values you know:

$$\frac{\sin 40°}{8} = \frac{\sin 110°}{b} = \frac{\sin 30°}{c}$$

$$\frac{0.643}{8} = \frac{0.940}{b} = \frac{0.5}{c}$$

$$0.0803 = \frac{0.940}{b} = \frac{0.5}{c}$$

At this point, you can set up two individual proportions and solve them individually for b and c, respectively:

$$0.0803 = \frac{0.940}{b} \qquad 0.0803 = \frac{0.5}{c}$$

$$b = \frac{0.940}{0.0803} \qquad c = \frac{0.5}{0.0803}$$

$$b = 11.70 \qquad c = 6.23$$

So the length of AC is 11.70, and the length of AB is 6.23. You now know every dimension of triangle ABC. The Law of Sines can be used in any triangle if you know:

- two sides and one of their opposite angles
- two angles and any side

The Law of Cosines

When you don't have the information necessary to use the Law of Sines, you may be able to use the Law of Cosines instead. The Law of Cosines is another way of using trigonometric functions to complete partial information about a triangle's dimensions. This is the Law of Cosines:

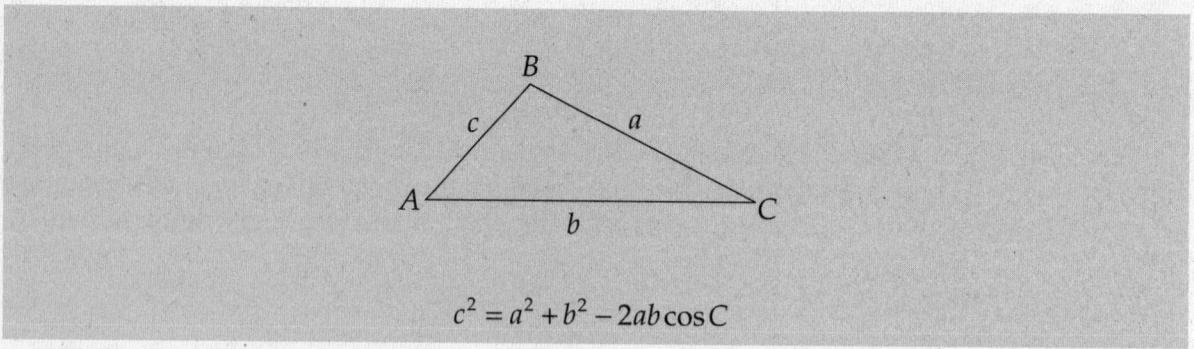

$$c^2 = a^2 + b^2 - 2ab \cos C$$

The Law of Cosines is a way of completing the dimensions of any triangle. You'll notice that it looks a bit like the Pythagorean Theorem. That's basically what it is, with a term added to the end to compensate for non-right angles; if you use the Law of Cosines on a right triangle, the "$2ab \cos C$" term becomes zero, and the law becomes the Pythagorean Theorem. The Law of Cosines can be used to fill in unknown dimensions of a triangle when you know any three of the quantities in the formula. Take a look at this example:

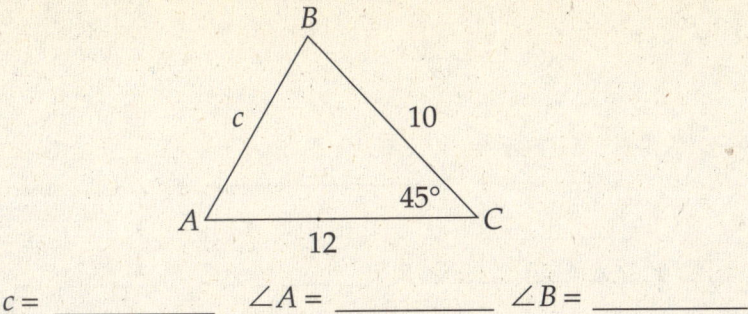

$c =$ _____ $\angle A =$ _____ $\angle B =$ _____

In this triangle, you know only two sides and an angle—the angle between the known sides. That is, you know a, b, and C. In order to find the length of the third side, c, just fill the values you know into the Law of Cosines, and solve:

$$c^2 = a^2 + b^2 - 2ab\cos C$$

$$c^2 = (10)^2 + (12)^2 - 2(10)(12)\cos 45°$$

$$c^2 = 100 + 144 - 240(0.707)$$

$$c^2 = 244 - 169.7$$

$$c^2 = 74.3$$

$$c = 8.62$$

The length of AB is therefore 8.62. Now that you know the lengths of all three sides, just use the Law of Sines to find the values of the unknown angles, or rearrange the Law of Cosines to put the other unknown angles in the C position, and solve to find the measures of the unknown angles. The Law of Cosines can be used in any triangle if you know:

- all three sides
- two sides and the angle between them

 **Drill**

In the following practice exercises, use these two trigonometric laws to complete the dimensions of these non-right triangles. The answers to all drills are found in chapter 12.

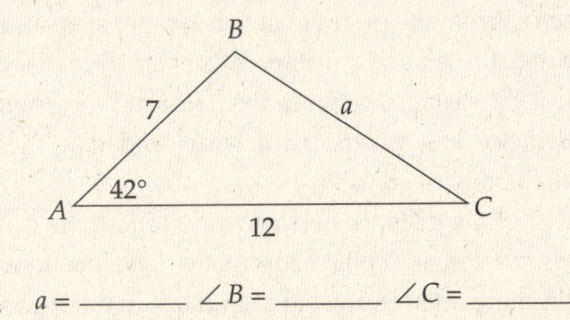

1. $a =$ _____ $\angle B =$ _____ $\angle C =$ _____

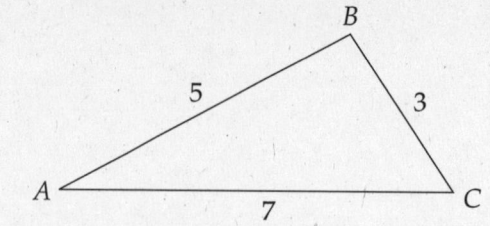

2. ∠A = _____ ∠B = _____ ∠C = _____

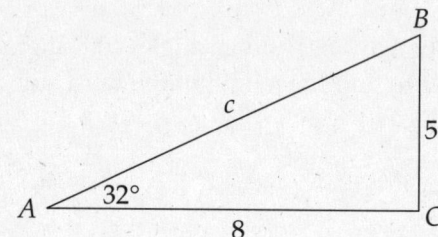

3. c = _____ ∠B = _____ ∠C = _____

 ## Polar Coordinates

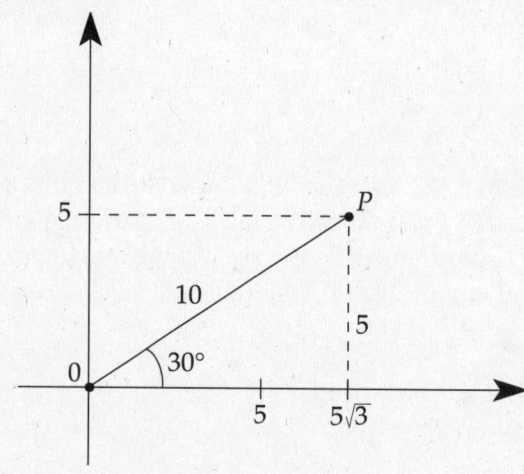

Polar coordinates are another way of describing the position of a point in the coordinate plane. In the figure above, the position of point P can be described in two ways. In standard rectangular coordinates, you would count across from the origin to get an x-coordinate and up from the origin to get a y-coordinate. (As we've noted, these x and y distances can be regarded as legs of a right triangle—the hypotenuse of the triangle is the distance between the point and the origin.) Rectangular coordinates consist of a horizontal distance and a vertical distance, and take the form (x,y). In rectangular coordinates, point P would be described as $(5\sqrt{3}, 5)$.

Polar coordinates consist of the distance between a point and the origin, and the angle between that segment and the positive x-axis. Polar coordinates take the form (r, θ). This angle can be expressed in degrees, but is more often expressed in radians. In polar coordinates, therefore, P could be described as (10, 30°) or $\left(10, \dfrac{\pi}{6}\right)$.

As you saw in the unit circle, there's more than one way to express any angle. For any angle, there is an infinite number of equivalent angles that can be produced by adding 360° (or 2π, if you're working in radians) any number of times. Therefore, there is an infinite number of equivalent polar coordinates for any point. Point P, at (10, 30°), can also be expressed as (10, 390°), or $\left(10, \dfrac{13\pi}{6}\right)$. You could go on producing equivalent expressions as long as you wanted by adding or subtracting 360° (or 2π).

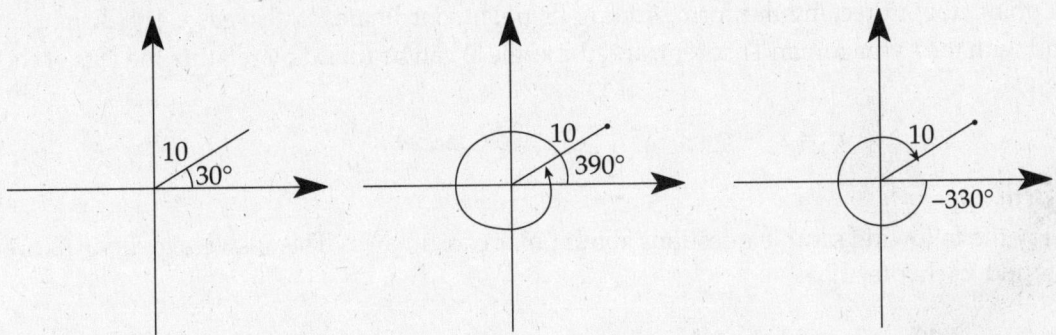

There's still another way to produce equivalent polar coordinates. The distance from the origin—the r in (r, θ)—can be negative. This means that once you've found the angle at which the hypotenuse must extend, a negative distance extends in the *opposite* direction, 180° away from the angle. Therefore, you can also create equivalent coordinates by changing the angle by 180° and flipping the sign on the distance. The point $P(10, 30°)$ or $\left(10, \dfrac{\pi}{6}\right)$ could also be expressed as $(-10, 210°)$ or $\left(-10, \dfrac{7\pi}{6}\right)$. Other equivalent coordinates can be generated by pairing equivalent angles with these negative distances.

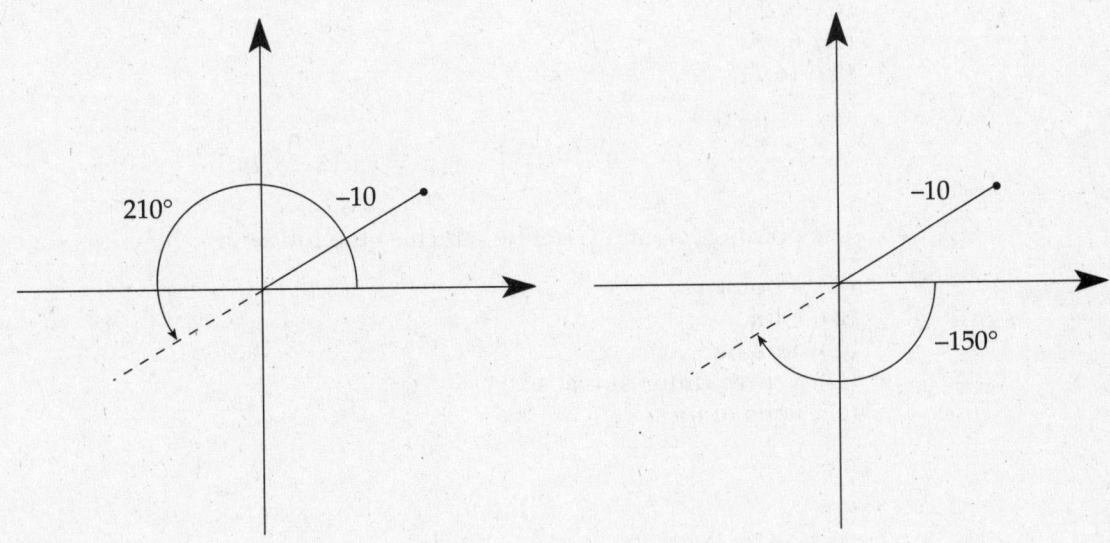

Converting coordinates

Converting rectangular coordinates to polar coordinates and vice versa is simple. You just use the trigonometry techniques reviewed in this chapter:

Given a point (r, θ) in polar form, you can find its rectangular coordinates this way:

$$x = r \cos \theta \qquad\qquad y = r \sin \theta$$

Given a point (x, y) in rectangular form, finding its polar coordinates is also easy. The distance r can be found with the Pythagorean Theorem, and the angle θ can be found by relating the legs of the right triangle:

$$r^2 = x^2 + y^2 \qquad\qquad \tan \theta = \frac{x}{y}$$

Drill

Try the following sample questions about polar coordinates. The answers to all drills are found in chapter 12.

39. Which of the following rectangular coordinate pairs is equivalent to the polar coordinates $\left(6, \frac{\pi}{3}\right)$?

    (A) (0.5, 1.7)
    (B) (2.6, 5.2)
    (C) (3.0, 5.2)
    (D) (4.2, 4.8)
    (E) (5.2, 15.6)

42. The point $\left(7, \frac{3\pi}{4}\right)$ is how far from the x-axis?

    (A) 3.67
    (B) 4.95
    (C) 5.25
    (D) 6.68
    (E) 16.71

$$A = \left(6, \frac{\pi}{3}\right) \qquad B = \left(6, \frac{5\pi}{3}\right) \qquad C = (3, 2\pi)$$

45. The points A, B, and C define which of the following?

    (A) a point
    (B) a line
    (C) a plane
    (D) a three-dimensional space
    (E) none of these

FLASHCARDS

The rules and formulas listed below represent the most important points to study in this chapter. Memorize them by covering up the right column and testing yourself on the left column, or, better still, by making real flashcards as directed.

Front of Card	Back of Card
What does SOHCAHTOA stand for?	$\sin = \dfrac{opposite}{hypotenuse}$ $\cos = \dfrac{adjacent}{hypotenuse}$ $\tan = \dfrac{opposite}{adjacent}$
$\tan x$ can also be expressed as	$\dfrac{\sin x}{\cos x}$
$\sin^2 x + \cos^2 x =$	1
$\csc x$ can also be expressed as	$\dfrac{1}{\sin x}$
$\sec x$ can also be expressed as	$\dfrac{1}{\cos x}$

TRIGONOMETRY ◆ 169

Front of Card	Back of Card
cot x can also be expressed as 	$\dfrac{1}{\tan x}$
What is the Law of Sines?	$\dfrac{\sin A}{a} = \dfrac{\sin B}{b} = \dfrac{\sin C}{c}$
What is the Law of Cosines?	$c^2 = a^2 + b^2 - 2ab \cos C$

9
Functions

FUNCTIONS ON THE SUBJECT TESTS

In chapter 4, simple algebraic functions were introduced. These were functions of the style you may remember from the SAT I; they used a symbol of some kind to represent a series of algebraic operations. Here are a couple of examples:

$$¥x = (x + 3)^2 - x^2 \qquad \|\underline{a}\| = |7a - 12|$$

Questions that involve functions in this form are generally testing your algebra skills, everything from PEMDAS to FOIL. Functions sometimes do show up in this form on the Math Subject Tests—especially the Math IC Test—but you're more likely to encounter functions in mathematical notation, like this:

$$f(x) = (x + 3)^2 - x^2 \qquad f(a) = |7a - 12|$$

Functions in this notation are treated just like functions represented symbolically. You'll still need to remember PEMDAS and other rules of algebra. When dealing with functions in the $f(x)$ form, however, you can expect to need a few other rules as well—rules relating to the properties of functions and their graphs.

DEFINITIONS

Here are some terms concerning algebraic functions that appear on the Math Subject Tests. Make sure you're familiar with them. If the meaning of any of these vocabulary words keeps slipping your mind, add that word to your flashcards.

Domain: The set of values that may be put into a function.

Range: The set of values that can be produced by a function.

Even Function: A function for which $f(x) = f(-x)$—even functions are symmetrical across the y-axis.

Odd Function: A function for which $-f(x) = f(-x)$—odd functions have radial symmetry, which means that they are the same when rotated 180° around the origin.

Root: Values in a function's domain at which the function equals zero—a root is also called a *zero* of a function.

Degree: The exponent of an algebraic term—the degree of a polynomial is the highest degree of any term in the polynomial.

Asymptote: A line on the graph of a function that indicates values at which that function does not exist.

Period: In periodic functions, the distance traveled by the function before it repeats itself.

Frequency: The number of times a graph repeats itself in a given distance; the reciprocal of the function's period.

Amplitude: In a periodic function, the distance that the graph rises above a central value.

BASIC FUNCTIONS

A function is a type of relation between two sets of numbers called the *domain* and *range* of the function. Specifically, a function is a relation in which every element in the domain corresponds to only one element in the range; for every x in the function, there is only one possible $f(x)$.

The most basic function questions test only your understanding of functions and the algebra required to work with them. Here are some examples of algebraic functions:

$$f(x) = |x^2 - 16| \qquad g(x) = \frac{1}{4}(x-2)^3$$

$$t(a) = a(a-6) + 8 \qquad p(q) = \frac{3-q}{q}$$

Notice that it's not the *letters* of the variables that matter, but only their positions. The domain of the function is represented by the *independent variable*—usually x, which is *put into* the function. The range of the function is represented by the *dependent variable*, $f(x)$—which *comes out* of the function. A function could be written as $g(h)$, in which case h would be the independent variable (you would treat it like x), and $g(h)$ would be the dependent variable (you would treat it like $f(x)$).

There are a few unusual function types that you should be prepared for. It is possible, for example, for elements in the domain to consist of more than one value, like this:

$$f(a,b) = \frac{a^2 + b^2}{ab} \qquad g(x,y) = (x+2)^2 - (y-2)^2$$

In each of these functions, an element in the domain is a *pair* of values. Functions of this kind are fairly rare on the Math Subject Tests, but you may run into one. Although they're unusual, they're not difficult. Simply treat them like ordinary functions: to calculate the value of $f(3, 4)$, for example, simply take the values 3 and 4 and plug them into the definition of the function in the positions of a and b, respectively (you get $\frac{25}{12}$).

You may also have to work with a *split function*. A split function is one that has different definitions, depending on some condition that is part of the function. Here are a couple of examples of split functions:

$$y(x) = \begin{cases} x^2, & x > 0 \\ 1, & x = 0 \\ -x^2, & x < 0 \end{cases} \qquad f(x) = \begin{cases} 5x, & x \text{ is odd} \\ 4x, & x \text{ is even} \end{cases}$$

Functions of this type are fairly self-explanatory. It's just necessary to check the conditions of the function before plugging values in, to make sure you're using the right function definition.

When questions ask you to work with algebraic functions, you'll be required to do one of two things: plug numbers into a function and get a numerical answer, or plug variables into a function and get an algebraic answer. For example, given the function $g(x) = (x + 2)^2$, you could run into two types of questions:

3. If $g(x) = (x + 2)^2$, what is the value of $g(4)$?

   (A)   8
   (B)   12
   (C)   16
   (D)   36
   (E)   64

Answering this question is a simple matter of plugging 4 into the function, and simplifying $(4 + 2)^2$ to get 36. Here, on the other hand, is an algebraic version of the same question:

18. If $g(x) = (x + 2)^2$, what is the value of $g(x + 2)$?

    (A) $x^2 + 4$
    (B) $x^2 + 6$
    (C) $x^2 + 4x + 4$
    (D) $x^2 + 4x + 6$
    (E) $x^2 + 8x + 16$

To solve this question, it's necessary to plug the quantity $(x + 2)$ into the function in the x-position and do the algebra:

$$g(x) = (x+2)^2$$
$$g(x+2) = ((x+2)+2)^2$$
$$g(x+2) = (x+4)^2$$
$$g(x+2) = x^2 + 8x + 16$$

If you do the algebra correctly, you'll find that (E) is the correct answer.

Drill

Practice working with functions in the following practice questions. The answers to all drills are found in chapter 12.

14. If $f(x) = x^2 - x^3$, then $f(-1) =$

   (A) -2
   (B) -1
   (C) 0
   (D) 1
   (E) 2

17. If $f(z) = \sqrt{z^2 + 8z}$, then how much does $f(z)$ increase as z goes from 7 to 8?

   (A) 0.64
   (B) 1.07
   (C) 2.96
   (D) 3.84
   (E) 5.75

26. If $g(t) = t^3 + t^2 - 9t - 9$, then $g(3) =$

   (A) -9
   (B) 0
   (C) 9
   (D) 27
   (E) 81

29. If $f(x,y) = \dfrac{xy}{x+y}$, which of the following is equal to $f(3,-6)$?

   (A) -48
   (B) -6
   (C) 3
   (D) 6
   (E) 18

30. If $h(x) = x^2 + x - 2$, and $h(n) = 10$, then n could be which of the following?

    (A) −4
    (B) −3
    (C) −1
    (D) 1
    (E) 2

33. The function f is given by $f(x) = x \cdot [x]$, where $[x]$ is defined to be the greatest factor of x that does not equal x. What is $f(75)$?

    (A) 25
    (B) 225
    (C) 625
    (D) 1125
    (E) 1875

$$g(x) = \begin{cases} 2|x| & \text{if } x \leq 0 \\ -|x| & \text{if } x > 0 \end{cases}$$

34. What is the value of $g(-y)$ if $y = 3$?

    (A) −6.0
    (B) −3.0
    (C) −1.5
    (D) 1.5
    (E) 6.0

Compound Functions

A compound function is a combination of two or more functions, in sequence. It's essentially a function of a function—you take the product of the first function and put it into the second function. For example:

$$f(x) = x^2 + 10x + 3 \qquad g(x) = \frac{1}{\sqrt{x+22}}$$

$$g(f(x)) =$$

The expression $g(f(x))$ is a compound function made up of the functions $f(x)$ and $g(x)$. As with any algebraic expression with parentheses, you start with the innermost part. To find $g(f(x))$ for any x, calculate the value of $f(x)$, and plug that value into $g(x)$. The result is the product of $g(f(x))$. Like questions based on simple algebraic functions, compound-function questions come in two flavors: questions that require you to plug numbers into compound functions and do the arithmetic, and questions that require you to plug terms with variables into compound functions and find an algebraic answer. For example:

$$f(x) = x^2 + 10x + 3$$

$$g(x) = \frac{1}{\sqrt{x+22}}$$

34. What is the value of $g(f(-4))$?

    (A) 0.15
    (B) 1.00
    (C) 2.75
    (D) 3.00
    (E) 6.56

To find the value of $g(f(-4))$, just plug –4 into $f(x)$; you should find that $f(-4) = -21$. Then, plug –21 into $g(x)$. You should find that $g(-21) = 1$. The correct answer is (B).

The more complicated type of compound-function question asks you to find the algebraic expression of a compound function. Essentially, that means you'll be combining the definitions of two functions. For example:

$$f(x) = x^2 + 10x + 3$$

$$g(x) = \frac{1}{\sqrt{x+22}}$$

36. Which of the following is $g(f(x))$?

    (A) $\dfrac{1}{x-5}$

    (B) $\dfrac{1}{x+5}$

    (C) $\sqrt{x^2 + 10x + 3}$

    (D) $\dfrac{1}{x^2 + 10x + 3}$

    (E) $\dfrac{1}{(x+5)^2}$

To find the algebraic expression of a compound function, simply plug the definition of one function into the definition of the other. In this case, $f(x)$ is the inner function—the one that's applied first—so you will plug $f(x)$ into $g(x)$:

$$g(x) = \frac{1}{\sqrt{x+22}}$$

$$g(f(x)) = \frac{1}{\sqrt{(x^2 + 10x + 3) + 22}}$$

$$g(f(x)) = \frac{1}{\sqrt{x^2 + 10x + 25}}$$

$$g(f(x)) = \frac{1}{\sqrt{(x+5)^2}}$$

$$g(f(x)) = \frac{1}{x+5}$$

When you plug the definition of *f*(*x*) into *g*(*x*) and simplify, you find that the whole expression can be boiled down to something very simple, and answer (B) is correct.

Drill

Practice working with compound functions in the following practice questions. The answers to all drills are found in chapter 12.

17. If $f(x) = 3x$ and $g(x) = x + 4$, what is the difference between $f(g(x))$ and $g(f(x))$?

    (A) 0
    (B) 2
    (C) 4
    (D) 8
    (E) 12

24. If $f(x) = |x| - 5$ and $g(x) = x^3 - 5$, what is $f(g(-2))$?

    (A) –18
    (B) –5
    (C) 0
    (D) 3
    (E) 8

$$f(x) = x^2 + 10x + 25$$
$$g(x) = \sqrt{x} + 4$$

32. Which of the following is $g(f(x))$?

    (A) $x - 1$
    (B) $x + 1$
    (C) $x + 7$
    (D) $x + 9$
    (E) $x^2 - 2x - 1$

$$f(x) = \sqrt{x}$$
$$g(x) = x^3 - 2$$

36. What is the positive difference between $f(g(3))$ and $g(f(3))$?

    (A) 0.7
    (B) 0.9
    (C) 1.8
    (D) 3.4
    (E) 6.8

INVERSE FUNCTIONS

Inverse functions are opposites—functions that undo each other. Here's a simple example:

$$f(x) = 5x \qquad\qquad f^{-1}(x) = \frac{x}{5}$$

Here, the function $f(x)$ multiplies x by 5. Its inverse, symbolized by $f^{-1}(x)$, divides x by 5. Any number put through one of these functions and then the other would come back to where it started. Here's a slightly more complex pair of inverse functions:

$$f(x) = 5x + 2 \qquad\qquad f^{-1}(x) = \frac{x-2}{5}$$

Here, the function $f(x)$ multiplies x by 5 and then adds 2. The inverse function $f^{-1}(x)$ does the opposite steps *in reverse order*, subtracting 2 and then dividing by 5. Let's add one more step:

$$f(x) = \frac{5x+2}{4} \qquad\qquad f^{-1}(x) = \frac{4x-2}{5}$$

Now, the function $f(x)$ multiplies x by 5, adds 2, and then divides by 4. The inverse function $f^{-1}(x)$ once again does the reverse; it multiplies by 4, subtracts 2, and then divides by 5. An inverse function always works this way; it does the opposite of each operation in the original function, in reverse order.

Compound functions and inverse functions are often used together in questions on the Math Subject Tests. It's the characteristic of inverse functions that they have opposite effects—they undo each other. For that reason, whenever you see the statement $f(g(x)) = x$, you know that the functions $f(x)$ and $g(x)$ are inverse functions. When a value x is put through one function and then the other, it returns to its original value. That means that whatever changes $f(x)$ makes are undone by $g(x)$. The statement $f(g(x)) = x$ means that $g(x) = f^{-1}(x)$.

The typical inverse-function question gives you the definition of a function and asks you to identify the function's inverse:

40. If $f(x) = \frac{x}{4} + 3$ and $f(g(x)) = x$, which of the following is $g(x)$?

    (A) $x - \frac{3}{4}$
    (B) $x - 12$
    (C) $4x - 3$
    (D) $4x - 12$
    (E) $4(x + 12)$

In this question, the statement $f(g(x)) = x$ tells you that $f(x)$ and $g(x)$ are inverse functions. Finding $g(x)$, then, amounts to finding the inverse of $f(x)$. You could do this by picking out the function that does the opposite of the operations in $f(x)$, in reverse order; but there's an easier way. By definition, inverse functions undo each other. In practice, this means that if you plug an easy number into $f(x)$ and get a result, the inverse function will be the function that turns that result back into your original number.

For example, given the function $f(x)$, you might decide to plug in 8, a number that makes the math easy:

$$f(x) = \frac{x}{4} + 3$$

$$f(8) = \frac{8}{4} + 3$$

$$f(8) = 2 + 3$$

$$f(8) = 5$$

You find that $f(x)$ turns 8 into 5. The inverse function $g(x)$ will be the one that does the reverse—that is, turns 5 into 8. To find $g(x)$, plug 5 into each of the answer choices. The answer choice that gives you 8 will be the correct answer. In this case, the correct answer is (D).

Drill

Practice your inverse-function techniques on these practice questions. The answers to all drills are found in chapter 12.

22. If $f(x) = \dfrac{4x - 5}{2}$ and $f(g(x)) = x$, what is $g(x)$?

   (A) $2x + \dfrac{5}{4}$
   (B) $\dfrac{2x + 5}{4}$
   (C) $x + \dfrac{5}{2}$
   (D) $\dfrac{x}{4} + \dfrac{2}{5}$
   (E) $\dfrac{5x + 2}{4}$

33. If $f(x) = 4x^2 - 12x + 9$ for $x \geq 0$, what is $f^{-1}(9)$?

   (A) 1
   (B) 3
   (C) 5
   (D) 12
   (E) 16

35. If $f(3) = 9$, then $f^{-1}(4) =$

   (A) −2
   (B) 0
   (C) 2
   (D) 16
   (E) It cannot be determined from the information given.

FUNCTIONS ◆ 179

Domain and Range

Some function questions will ask you to make statements about the domain and range of functions. With a few simple rules, it's easy to figure out what limits there are on the domain or range of a function.

Domain

The domain of a function is the set of values that may be put into a function without violating any laws of math. When you're dealing with a function in the $f(x)$ form, the domain includes all of the allowable values of x. Sometimes a function question will limit the function's domain in some way, like this:

> For all integers n, $f(n) = (n-2)\pi$. What is the value of $f(7)$?

In this function, the independent variable n is limited; n can only be an integer. The domains of most functions, however, are *not* obviously limited. Generally, you can put whatever number you want into a function; the domain of the average function is *all real numbers*. Only certain functions have domains that are mathematically limited. To figure out the limits of a function's domain, you need to use a few basic rules. Here are the laws that can limit a function's domain:

- It's impossible for a fraction to have a denominator of zero. Any values that would make the bottom of a fraction equal to zero must be excluded from the domain of that function.

- It's impossible take the square root of a negative number. Any values that would make a number under a square root sign negative must be excluded from the domain of that function.

- It's also impossible to take any other even-numbered root of a negative number ($\sqrt[4]{}$, $\sqrt[6]{}$, etc.). No value in the domain can make the function include an even-numbered root of a negative number.

Whenever a function contains a fraction, a square root, or another even-numbered root, it's possible that the function will have a limited domain. Look for any values that would make denominators zero, or even-numbered roots negative. Those values must be eliminated from the domain. Take a look at these examples:

$$f(x) = \frac{x+5}{x}$$

In this function, there is a variable in the denominator of a fraction. This denominator must not equal zero, so the domain of $f(x)$ is $\{x \neq 0\}$.

$$g(x) = \frac{x}{x+5}$$

Once again, this function has a variable in the denominator of a fraction. In this case, the value of x that would make the denominator equal zero is –5. Therefore, the domain of $g(x)$ is $\{x \neq -5\}$.

$$t(a) = 4\sqrt{a}$$

This function has a variable under a square root sign. The quantity under a square root sign must not be negative, so the domain of $t(a)$ is $\{a \geq 0\}$.

$$s(a) = 3\sqrt{10-a}$$

Here again, you have a function with a variable under a square root. This time, the values that would make the expression negative are values greater than 10; all of these values must be eliminated from the function's domain. The domain of $s(a)$ is therefore $\{a \leq 10\}$.

A function can involve both fractions and square roots; always pay careful attention to any part of a function that could place some limitation on the function's domain. It's also possible to run into a function where it's not easy to see what values violate the denominator rule or the square root rule. Generally, factoring is the easiest way to make these relationships clearer. For example:

$$f(x) = \frac{1}{x^3 + 2x^2 - 8x}$$

Here, you've got variables in the denominator. You know this is something to watch out for, but it's not obvious what values might make the denominator equal zero. To make it clearer, factor the denominator:

$$f(x) = \frac{1}{x(x^2 + 2x - 8)}$$

$$f(x) = \frac{1}{x(x+4)(x-2)}$$

Now, things are much clearer. Whenever quantities are being multiplied, the entire product will equal zero if any one piece equals zero. Any value that makes the denominator equal zero must be eliminated from the function's domain. In this case, the values 0, –4, and 2 all make the denominator zero. The domain of $f(x)$ is $\{x \neq -4, 0, 2\}$. Take a look at one more example:

$$g(x) = \sqrt{x^2 + 4x - 5}$$

Once again, you've got an obvious warning sign: variables under a radical. Any values of x that make the expression under the radical negative must be eliminated from the domain. But what values are those? Are there any? To make it clear, factor the expression:

$$g(x) = \sqrt{(x+5)(x-1)}$$

The product of two expressions can be negative only when one of the expressions is negative and the other positive. If both expressions are positive, their product is positive. If both expressions are negative, their product is still positive. So the domain of $g(x)$ must contain only values that make $(x + 5)$ and $(x - 1)$ both negative or both positive. With a little experimentation, you'll find that both expressions are negative when $x < -5$, and both expressions are positive when $x > 1$. The domain of $g(x)$ is therefore $\{x < -5 \text{ or } x > 1\}$.

Domain notation

The domain of a function is generally described using the variable x. This is sometimes the case even when the variable x does not appear in the function. A function $f(x)$, whose domain includes only values greater than 0 and less than 24, could be described in the following ways:

The domain of $f(x)$ is $\{0 < x < 24\}$.

The domain of f is the set $\{x : 0 < x < 24\}$

A function in the form $f(x)$ can be referred to either as $f(x)$ or simply as f.

Range

The range of a function is the set of possible values that can be *produced* by the function. When you're dealing with a function in the $f(x)$ form, the range consists of all the allowable values of $f(x)$. The rage of a function, like the domain, is limited by a few laws of mathematics. Several of these laws are the same laws that limit the domain. Here are the major rules that limit a function's range:

- An even exponent produces only nonnegative numbers. Any term raised to an even exponent must be positive or zero.

- The square root of a quantity represents only the positive root. Like even powers, a square root can't be negative. The same is true for other even-numbered roots ($\sqrt[4]{}$, $\sqrt[6]{}$, etc.).

- Absolute values produce only nonnegative values.

These three operations—even exponents, even roots, and absolute values—can produce only nonnegative values. Consider these three functions:

$$f(x) = x^4 \qquad f(x) = \sqrt{x} \qquad f(x) = |x|$$

These functions all have the same range: $\{f(x) \geq 0\}$. These are the three major mathematical operations that often limit the ranges of functions. They can operate in unusual ways. The fact that a term in a function must be nonnegative can affect the entire function in different ways. Take a look at the following examples:

$$f(x) = -x^4 \qquad f(x) = -\sqrt{x} \qquad f(x) = -|x|$$

Each of these functions once again contains a nonnegative operation, but in each case the sign is now flipped by a negative sign. The range of each function is now $\{f(x) \leq 0\}$. In addition to being flipped by negative signs, ranges can also be slid upward or downward by addition and subtraction. Take a look at these examples:

$$f(x) = x^4 - 5 \qquad f(x) = \sqrt{x} - 5 \qquad f(x) = |x| - 5$$

Each of these functions contains a nonnegative operation that is then decreased by 5. The range of each function is consequently also decreased by 5, becoming $\{f(x) \geq -5\}$. Notice the pattern: a nonnegative operation has a range of $\{f(x) \geq 0\}$; when the sign of the nonnegative operation is flipped, the sign of the range also flips; when a quantity is added to the operation, the same quantity is added to the range. These changes can also be made in combination:

$$g(x) = \frac{-x^2 + 6}{2}$$

In this function, the sign of the nonnegative operation is flipped, 6 is added, and the whole thing is divided by 2. As a result, the range of $g(x)$ is $\{g(x) \leq 3\}$. The range of x^2, which is $\{y : y \geq 0\}$, has its sign flipped, is increased by 6, and is then divided by 2.

Range notation

Ranges can be represented in several ways. If the function $f(x)$ can produce values between –10 and 10, then a description of its range could look like any of the following:

The range of $f(x)$ is given by $\{f: -10 < f(x) < 10\}$.

The range of $f(x)$ is $\{-10 < f(x) < 10\}$.

The range of $f(x)$ is the set $\{y: -10 < y < 10\}$.

Because a function's range is represented on the y-axis when the function is graphed, the range is sometimes described using the variable y, even when y doesn't appear in the function.

Plugging in on range questions

On the Math Subject Tests, any range question you see will be multiple-choice. If the answer to a range question is not immediately clear, you can get to the right answer by the Process of Elimination. Just keep plugging in numbers until you've eliminated all but one of the answer choices. Look at this example:

25. If $f(x) = |-x^2 - 8|$ for all real numbers x, then which of the following sets is the range of f?

    (A) $\{y : y \geq -8\}$
    (B) $\{y : y > 0\}$
    (C) $\{y : y \geq 0\}$
    (D) $\{y : y \leq 8\}$
    (E) $\{y : y \geq 8\}$

The best way to find the range of $f(x)$ is to start with the range of x^2 (which must be greater than or equal to zero) and change it step-by-step, following the operations in the function. In this case, you would take the range $\{f(x) \geq 0\}$, flip the sign, subtract eight, and then make it positive, ending up with $\{f(x) \geq 8\}$. If you are confused by this process or uncertain what steps to take, you can always fall back on plugging in. Just plug in values for x that will make the math easy, and eliminate answer choices based on the results you get.

For example, if you plugged in $x = 3$, you would find that $f(3) = 17$. That's enough to eliminate answer choice (D). If you plugged in $x = 0$, you'd find that $f(0) = 8$. Plugging in other numbers would soon show you that it's impossible to get a value smaller than 8. Answer choice (E) wins again.

FUNCTIONS WITHIN INTERVALS: DOMAIN MEETS RANGE

A question that introduces a function will sometimes ask about that function only within a certain interval. This *interval* is a set of values for the variable in the x position—the independent variable. This interval is an artificial limitation on the function's domain. An interval can be written in two ways:

If $f(x) = 4x - 5$ for $0 \le x \le 10$, then what is the range of f?

If $f(x) = 4x - 5$ for [0, 10] then which of the following sets equals the range of f?

These two questions present the same information and ask the same question. The second version simply uses a different notation to describe the interval, or domain, in which $f(x)$ is being looked at. You've got to be alert when domains or ranges are given in this notation, because it's easy to mistake intervals in this form for coordinate pairs, which is confusing.

The example given above also demonstrates the most common form of function-interval question, in which you're given a domain for the function and asked for the range. Whenever the function has no exponents, finding the range is easy: just plug the upper and lower bound of the *domain* into the function. The results will be the upper and lower bounds of the range. In the example above, the function's range is the set $\{y : -5 \le y \le 35\}$.

When a function is more complicated and contains exponents and fractions, finding the functions of the domain's upper and lower bounds is just the beginning. Then use the range techniques from the previous section to see whether there are other limits on the function's range.

Drill

Practice your domain and range techniques on the following practice questions. The answers to all drills are found in chapter 12.

24. If $f(x) = \dfrac{1}{x^3 - x^2 - 6x}$, then which of the following sets is the domain of f?

    (A) $\{x : x \ne -2, 0, 3\}$
    (B) $\{x : x \ne 0\}$
    (C) $\{x : x > -2\}$
    (D) $\{x : x > 0\}$
    (E) $\{x : x \ge 3\}$

27. If $g(x) = \sqrt{x^2 - 4x - 12}$, then the domain of g is given by which of the following?

    (A) $\{x : x \ge -2\}$
    (B) $\{x : x \ne 3, 4\}$
    (C) $\{x : -2 \le x \le 6\}$
    (D) $\{x : -2 < x < 6\}$
    (E) $\{x : x \le -2 \text{ or } x \ge 6\}$

30. If $t(a) = \dfrac{a^2 + 5}{3}$, then which of the following sets is the range of t?

   (A) $\{y: \ y \neq 0\}$
   (B) $\{y: \ y \geq 0\}$
   (C) $\{y: \ y \geq 0.60\}$
   (D) $\{y: \ y \geq 1.67\}$
   (E) $\{y: \ y \geq 2.24\}$

34. If $f(x) = 4x + 3$ for $-1 \leq x \leq 4$, then which of the following gives the range of f?

   (A) $\{y: \ -4 \leq y \leq 7\}$
   (B) $\{y: \ -4 \leq y \leq 19\}$
   (C) $\{y: \ -1 \leq y \leq 7\}$
   (D) $\{y: \ -1 \leq y \leq 19\}$
   (E) $\{y: \ 1 \leq y \leq 19\}$

GRAPHING FUNCTIONS

All of the function techniques covered in this chapter so far have dealt with the algebra involved in doing functions. Most of the function questions on each Math Subject Test will be algebra questions like the ones you've seen so far. However, there's another class of function question that appears on the Subject Tests: Graphical questions.

Graphical-function questions require you to relate an algebraic function to the graph of that function in some way. Here are some of the tasks you might be required to do on a graphical-function question:

- Match a function's graph with the function's domain or range.
- Match the graph of a function with the function's algebraic definition.
- Decide whether statements about a function are true or false, based on its graph.
- Answer questions about the symmetry of a function's graph.

None of these tasks is very difficult, as long as you're prepared for them. The next few pages will tell you everything you need to know.

IDENTIFYING GRAPHS OF FUNCTIONS

When a function $f(x)$ is graphed, the x-axis represents the values of x. The y-axis represents the values of $f(x)$. When you look at the coordinates (x, y) of any point on the function's graph, x represents a point in the function's domain, and y represents the function of that point.

The most useful tool for identifying the graph of a function is the *vertical-line test*. Remember, a function is a relation of a domain and a range, in which each value in the domain matches up with only one value in the range. Simply put, there's only one $f(x)$ for each x. Graphically, that means that any vertical line drawn through the x-axis can intersect a function only once. If you can intersect a graph more than once with a vertical line, it isn't a function. Here's the vertical-line test in action:

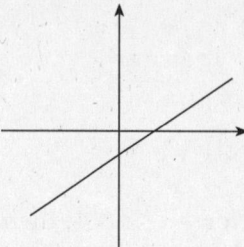

This is a function, because no vertical line can intersect it more than once. All straight lines are functions, with only one exception: A vertical line is not a function, because another vertical line would intersect it at an infinite number of points.

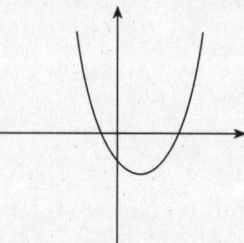

This is also a function. Any parabola that opens up or down is a function.

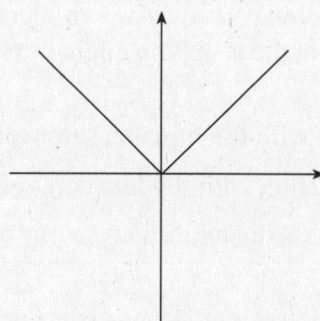

This is the graph of $f(x) = |x|$, and it's a function as well.

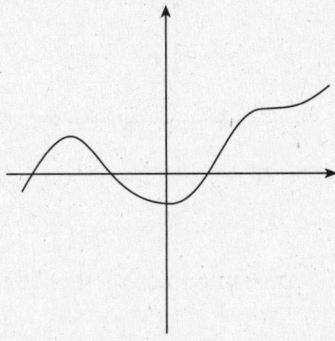

This complicated curve also passes the vertical-line test for functions.

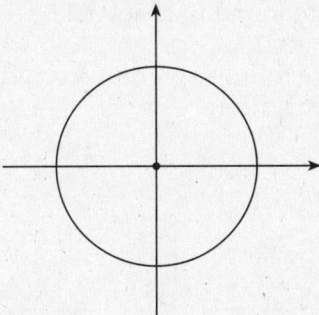

This is *not* a function; there are may places where a vertical line can intersect a circle twice.

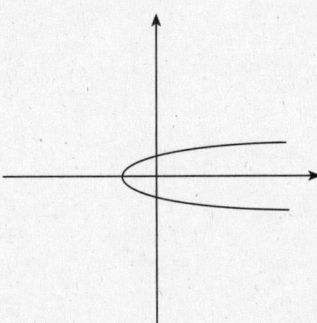

This isn't a function either. Although this graph is parabolic in shape, it fails the vertical-line test.

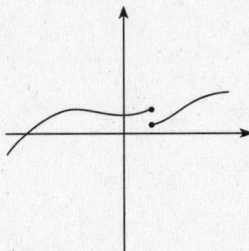

Nope. It's close, but there's one point where a vertical line can intersect this graph twice—it can't be a function.

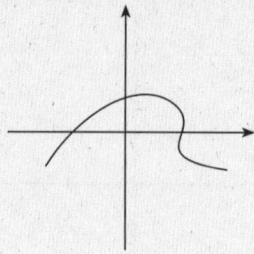

This curve is also not a function. It's possible to cross this curve more than once with one vertical line.

Drill

Use the vertical-line test to distinguish functions from non-functions in the following practice questions. The answers to all drills are found in chapter 12.

9. Which of the following graphs could NOT be the graph of a function?

(A)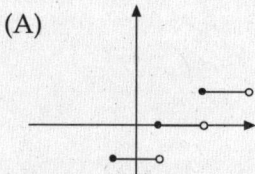

(B)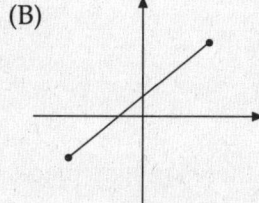

(C)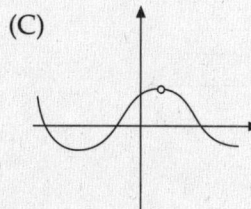

(D)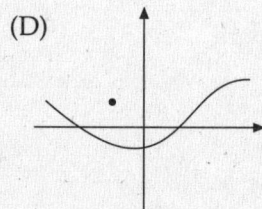

(E)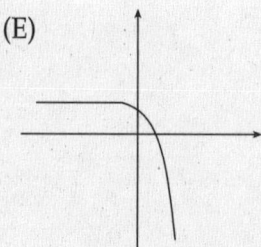

15. Which of the following could NOT be the graph of a function?

(A)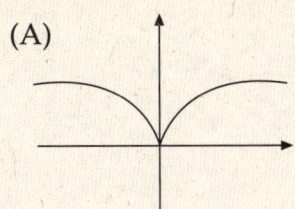

(B)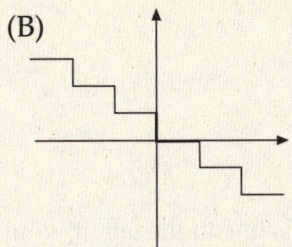

(C)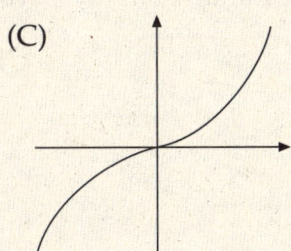

(D)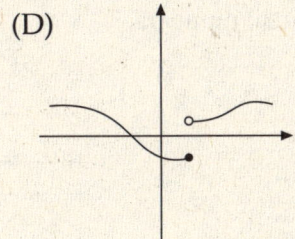

(E)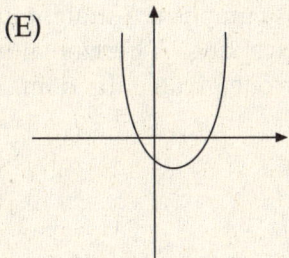

Range and Domain in Graphs

The graph of a function gives important information about the function itself. You can generally state a function's domain and range accurately just by looking at its graph. Even when the graph doesn't give you enough information to state them exactly, it will often let you eliminate incorrect answers about the range and domain. Take a look at the following graphs of functions and the information they provide:

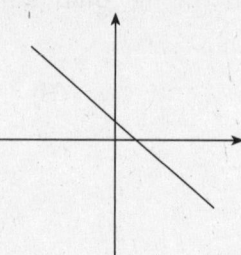

If you followed this line to the left, it would continue to rise forever. Likewise, if you followed it to the right it would continue to fall. The range of this line (the set of *y*-values it occupies) goes on forever; the range is said to be "all real numbers." Because the line also continues to the left and right forever, there are no *x*-values that the line does not pass through. The domain of this function, like its range, is the set of all real numbers.

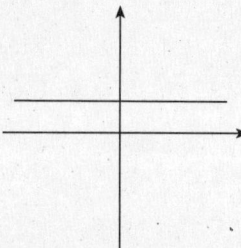

The same thing is true of all linear functions (whose graphs are straight lines); their ranges and domains include all real numbers. There's only one exception. A horizontal line extends forever to the left and right (through all *x*-values) but has only one *y*-value. Its domain is therefore all real numbers, while its range contains only one value.

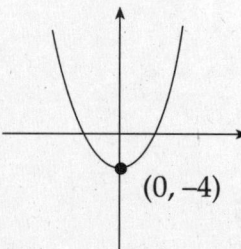

The domain of this function is the set of all real numbers, because parabolas continue widening forever. Its range however, *is* limited. The parabola extends upward forever, but never descends lower along the *y*-axis than $y = -4$. The range of this function is therefore $\{y: y \geq -4\}$.

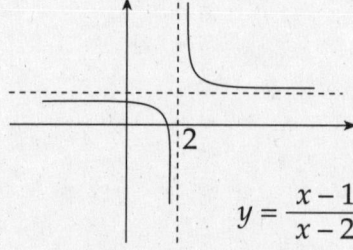

This function has two *asymptotes*. Asymptotes are lines that the function approaches but never reaches; they mark values in the domain or range at which the function does not exist or is undefined.

The asymptotes on this graph mean that it's impossible for x to equal 2, and it's impossible for y to equal 1. The domain of $f(x)$ is therefore $\{x : x \neq 2\}$, and the range is $\{y: y \neq 1\}$.

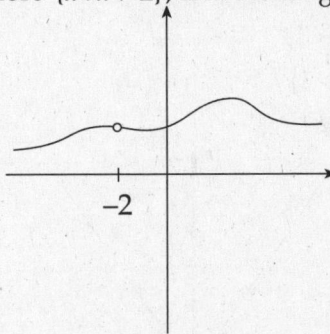

The hole in this function's graph means that there's an x-value missing at that point. The domain of any function whose graph sports a little hole like this one must exclude the corresponding x value. The domain of this function, for example, would simply be $\{x: x \neq -2\}$.

To estimate range and domain based on a function's graph, just use common sense and remember these rules:

- If something about a function's shape will prevent it from continuing forever up and down, then that function has a limited range.

- If the function has a horizontal asymptote at a certain y-value, then that value is excluded from the function's range.

- If anything about a function's shape will prevent it from continuing forever to the left and right, then that function has a limited domain.

- If a function has a vertical asymptote or hole at a certain x-value, then that value is excluded from the function's domain.

Drill

Test your understanding of range and domain with the following practice questions. The answers to all drills are found in chapter 12.

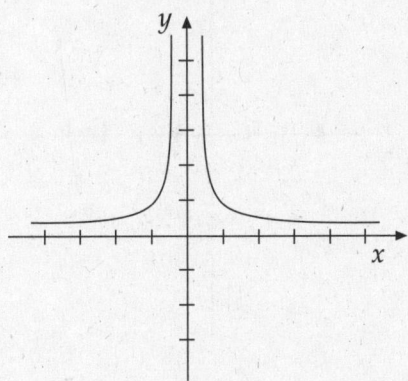

17. If the graph of $f(x)$ is shown above, which of the following could be the domain of f?

(A) $\{x: x \neq 0\}$
(B) $\{x: x > 0\}$
(C) $\{x: x \geq 0\}$
(D) $\{x: x > 1\}$
(E) $\{x: x \geq 1\}$

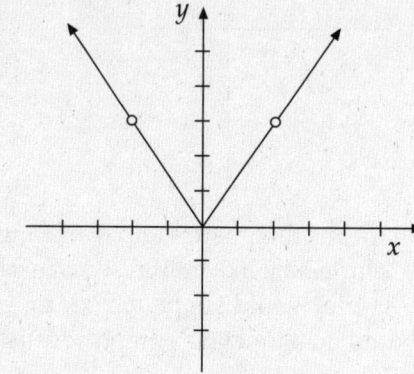

24. Which of the following could be the domain of the function graphed above?

    (A) $\{x: x \neq 2\}$
    (B) $\{x: -2 < x < 2\}$
    (C) $\{x: x < -2 \text{ or } x > 2\}$
    (D) $\{x: |x| \neq 2\}$
    (E) $\{x: |x| > 2\}$

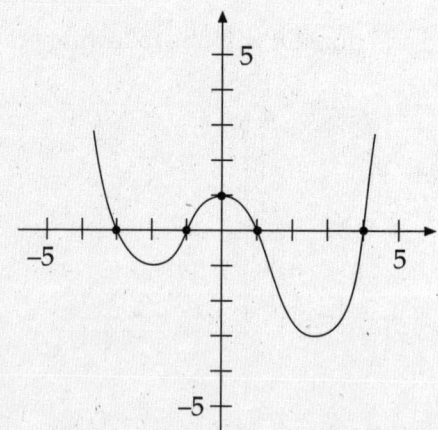

28. If $g(x)$ is graphed above, which of the following sets could be the range of $g(x)$?

   (A) $\{y: y \leq -1\}$
   (B) $\{y: y \geq -1\}$
   (C) $\{y: y \geq -3\}$
   (D) $\{y: -3 \leq y \leq -1\}$
   (E) $\{y: y \leq -3 \text{ or } y \geq 1\}$

ROOTS OF FUNCTIONS IN GRAPHS

The roots of a function are the values that make the function equal to zero. To find the roots of a function $f(x)$ algebraically, you simply set $f(x)$ equal to zero and solve for x. The solutions are the roots of the function.

Graphically, the roots of a function are the values of x at which the graph crosses the x-axis. That makes them easy to spot on a graph. If you are asked to match a function to its graph, it's often helpful to find the roots of the function using algebra; then it's a simple matter to compare the function's roots to the x-intercepts on the graph. Take a look at this function:

$$f(x) = x^3 + 3x^2 - 4x$$

If you factor it to find its roots, you get this:

$$f(x) = x(x + 4)(x - 1)$$

The roots of $f(x)$ are therefore $x = -4, 0,$ and 1. You can expect the graph of $f(x)$ to cross the x-axis at those three x-values.

Drill

Try the following practice questions by working with the roots of functions. The answers to all drills are found in chapter 12.

16. Which of the following is a zero of $f(x) = 2x^2 - 7x + 5$?

   (A) 1.09
   (B) 1.33
   (C) 1.75
   (D) 2.50
   (E) 2.75

19. The function $g(x) = x^3 + x^2 - 6x$ has how many distinct roots?

   (A) 1
   (B) 2
   (C) 3
   (D) 4
   (E) It cannot be determined from the information given.

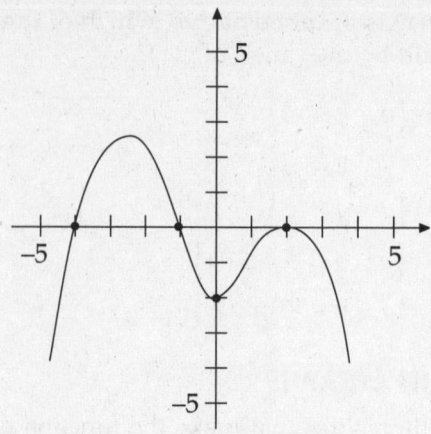

25. If the graph of f(x) is shown above, which of the following sets could be the roots of f(x)?

(A) $\{x = -2, 0, 2\}$
(B) $\{x = -4, -1, 0\}$
(C) $\{x = -1, 2\}$
(D) $\{x = -4, -1, 2\}$
(E) $\{x = -4, -1\}$

Symmetry in Functions

Some Math Subject Test questions will ask about the symmetry of graphs. There are just a few kinds of symmetry that come up often on the test.

Symmetry across the y-axis

A function is symmetrical across the y-axis when the graph seems to have a mirror image reflected in the y-axis. If the paper were folded along the y-axis, the left and right halves of the graph would meet perfectly. Functions with symmetry across the y-axis are sometimes called *even functions*. This is because only functions with even exponents can have this kind of symmetry.

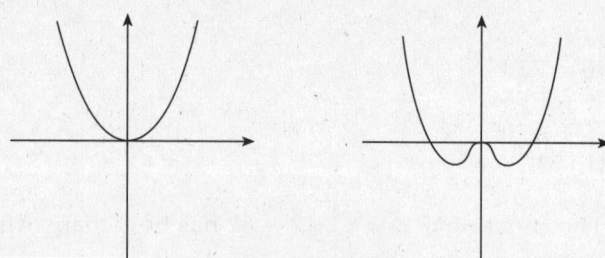

Even Functions

This is the algebraic definition of symmetry across the *y*-axis:

> **A function is symmetrical across the *y*-axis when**
>
> $f(x) = f(-x)$

This means that the negative and positive versions of any *x*-value produce the same *y*-value.

Origin symmetry

A function has radial symmetry when rotating the graph 180° produces a shape identical to the original. Functions with origin symmetry are sometimes called *odd functions*, because only functions with odd exponents can have this kind of symmetry.

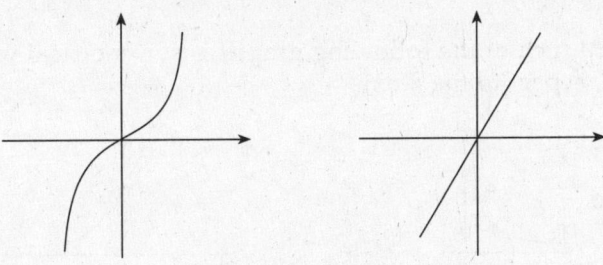

Odd Functions

This is the algebraic definition of origin symmetry:

> **A function has origin symmetry when**
>
> $f(x) = -f(-x)$

Symmetry across the *x*-axis

Some equations will produce graphs that are symmetrical across the *x*-axis. These equations can't be functions, however, because each *x*-value would then have to have two corresponding *y*-values. A graph that is symmetrical across the *x*-axis automatically fails the vertical-line test:

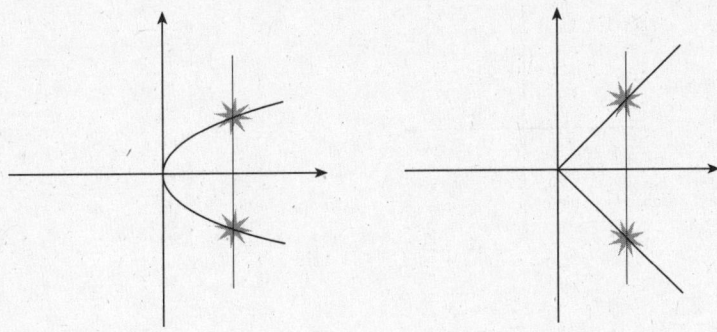

Because it doesn't apply to functions, *x*-axis symmetry comes up much less often than the previous two kinds do. Still, just in case, here's the algebraic definition of *x*-axis symmetry:

A relation is symmetrical across the *x*-axis when

$f(x)$ and $-f(x)$ exist for every x

Questions asking about symmetry generally test basic comprehension of one or both of these definitions. It's also important to understand the connection between these algebraic definitions and the appearance of graphs with different kinds of symmetry.

Drill

Try these practice questions. The answers to all drills are found in chapter 12.

6. Which of the following graphs is symmetrical with respect to the *x*-axis?

(A)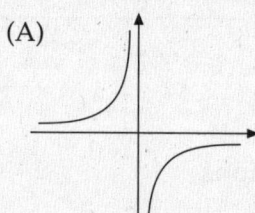

(B)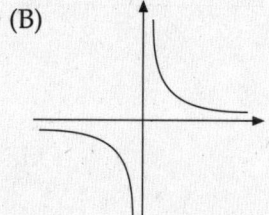

(C)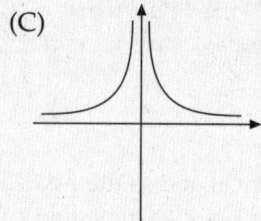

(D)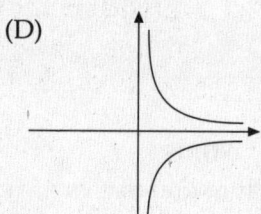

(E)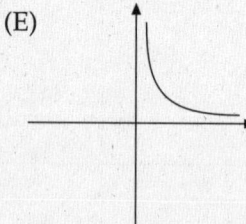

17. If an even function is one for which $f(x)$ and $f(-x)$ are equal, then which of the following is an even function?

   (A) $g(x) = 5x + 2$

   (B) $g(x) = x$

   (C) $g(x) = \dfrac{x}{2}$

   (D) $g(x) = x^3$

   (E) $g(x) = -|x|$

22. The function $f(x)$ has radial symmetry. If $f(2) = -3$, then $f(-2) =$

   (A) −6
   (B) −5
   (C) −3
   (D) 3
   (E) 6

DEGREES OF FUNCTIONS

IIC ONLY

The degree of a function is the degree of the largest exponent in that function. For example, the function $p(x) = x^3 - 4x^2 + 7x - 12$ is a third-degree function. This means that $p(x)$ has three roots. These roots can be distinct or identical. For example, look at these two sixth-degree functions:

$$f(x) = x^6$$
$$g(x) = (x-1)(x-2)(x-3)(x-4)(x-5)(x-6)$$

The function $f(x)$ has six roots, but they're all the same: $f(x) = 0$ when $x = 0$, which makes the function equal $0 \cdot 0 \cdot 0 \cdot 0 \cdot 0 \cdot 0$. Basically, the function has six roots of zero—it has only one *distinct* root. This is an important distinction, because the graph of a function with six roots of zero is very different from the shape of a function with three roots of zero.

The function $g(x)$ has six distinct roots: $f(x) = 0$ when $x = 1, 2, 3, 4, 5,$ or 6. A sixth-degree function can have anywhere from 1 to 6 distinct roots. For example, a function might have four roots of 2, a root of 3, and a root of 4, for a total of three distinct roots. The equation of this function would look like this:

$$f(x) = (x-2)^4(x-3)(x-4)$$

This is still a sixth-degree function, and it has six roots. That's the algebraic meaning of the degree of a function: It equals the number of roots that the function has.

The degree of a function tells you a great deal about the shape of the function's graph. Take a look at the following graphs:

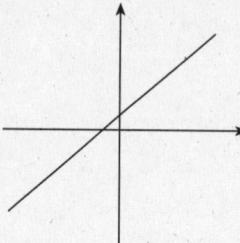

This is the graph of a first-degree function. All first-degree functions are linear functions, whose graphs are straight lines. A first-degree function has no extreme values—that is, it has no point which is higher or lower than all of the others.

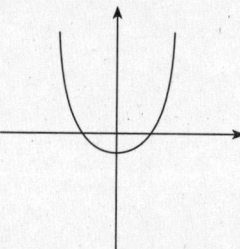

A second-degree function is usually a parabola. The function graphed above must be at least a second-degree function. A second-degree function has one extreme value, a maximum or minimum. This function's extreme value is a minimum.

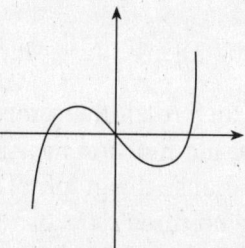

A third-degree function can have as many as two local extreme values. The function graphed above, which has a local maximum and a local minimum, must be at least a third-degree function.

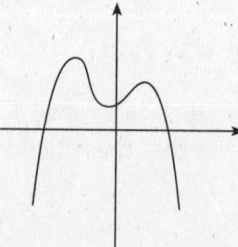

A fourth-degree function can have as many as three local extreme values. The function above has three extreme values, two local maxima and a local minimum between them. It must be at least a fourth-degree function.

By now, you should see the pattern. A fourth-degree function can have a maximum of three extreme values in its graph; a fifth-degree function can have a maximum of four extreme values in its graph. This pattern goes on forever. An nth-degree function has a maximum of n distinct roots and a maximum of $(n-1)$ extreme values in its graph. These two rules are the basis of a number of Math Subject Test questions. Take a look at the following practice questions.

Drill

The answers to all practice questions are located in chapter 12.

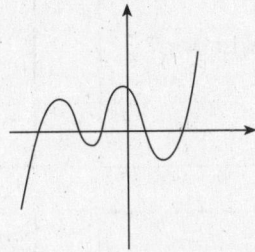

31. If the graph above is a portion of the graph of $f(x)$, then which of the following could be $f(x)$?

   (A) $ax + b$
   (B) $ax^2 + bx + c$
   (C) $ax^3 + bx^2 + cx + d$
   (D) $ax^4 + bx^3 + cx^2 + dx + e$
   (E) $ax^5 + bx^4 + cx^3 + dx^2 + ex + f$

35. If $g(x)$ is a fourth-degree function, then which of the following could be the definition of $g(x)$?

   (A) $g(x) = (x - 3)(x + 5)$
   (B) $g(x) = x(x + 1)^2$
   (C) $g(x) = (x - 6)(x + 1)(x - 5)$
   (D) $g(x) = x(x + 8)(x - 1)^2$
   (E) $g(x) = (x - 2)^3 (x + 4)(x - 3)$

Reviewing Functions Further

It's impossible to cover every aspect of functions that may turn up on the Math Subject Tests; this is one of the most varied question categories on the tests. To be thoroughly prepared for function questions on the Math Subject Tests—particularly on the Math IIC test—you should read this chapter carefully and then take a cruise through your precalculus textbook.

FLASHCARDS

The rules and formulas listed below represent the most important points to study in this chapter. Memorize them by covering up the right column and testing yourself on the left column, or, better still, by making real flashcards as directed.

Front of Card	Back of Card
What is the domain of a function?	The set of x-values that can be put into the function.
What is the range of a function?	The set of y-values that a function can produce.
What are the roots of a function?	The values that make the function equal zero.
What is the algebraic definition of symmetry across the y-axis?	$f(x) = f(-x)$
What is the algebraic definition of radial (or origin) symmetry?	$f(x) = -f(-x)$

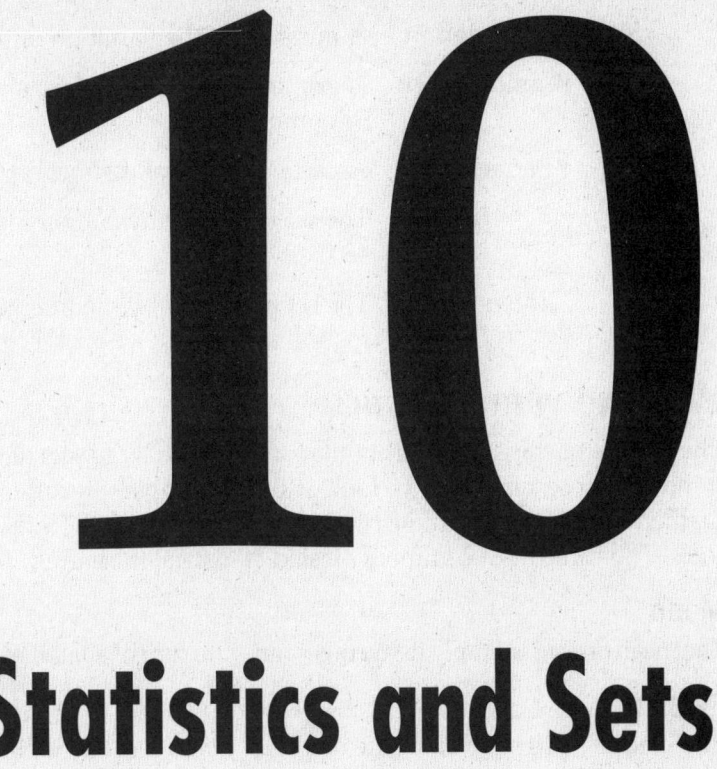

Statistics and Sets

STATISTICS ON THE SUBJECT TESTS

Math Subject Test questions about statistics and sets deal with the arrangements and combinations of large sets, probability, overlapping groups, and statistical measures like mean, median, and mode. On each Math Subject Test, only about one question in twenty will involve statistics and sets, so spend time on this chapter only after you've mastered the more essential material in earlier chapters.

DEFINITIONS

Here are some terms dealing with sets and statistics that appear on the Math Subject Tests. Make sure you're familiar with them. If the meaning of any of these vocabulary words keeps slipping your mind, add that word to your flashcards.

- **Mean:** An average—also called an arithmetic mean.
- **Median:** The middle value in a set when they are arranged in order—when there is no single middle value, the average of the two middle values.
- **Mode:** The value that occurs most often in a set.
- **Range:** The difference between the largest and smallest values in a set.

Standard Deviation: A measure of the variation of the values in a set.

Combination: A grouping of distinct objects in which order is not important.

Permutation: An arrangement of distinct objects in a definite order.

Union: The set produced by combining two or more distinct sets.

Intersection: The set produced by the intersection or overlap of two or more distinct sets.

WORKING WITH STATISTICS

The science of statistics is all about working with large sets of numbers and trying to see patterns and trends in those numbers. To look at those numbers in different ways, statisticians use a variety of mathematical tools. And, just to keep you guessing, ETS tests your knowledge of several of these tools. The three most commonly tested statistical measures are the mean, the median, and the mode.

Mean

The mean of a set is simply its average value, the sum of all its elements divided by the number of elements.

Median

The median is the middle value of a set. To find a set's median, you must first put all of its elements in order. If the set has an odd number of elements, then there will be one value in the exact middle, which is the median value. If the set has an even number of elements, then there will be *two* middle values; the median value is the average of these two middle values.

Mode

The mode of a set is simply the value that occurs most often in that set.

Many statistics questions require you to work with all three of these measures. The calculations involved are usually not very difficult, however, the real challenge of these questions is simply understanding these terms and knowing how to use them. Similarly, there are two more statistical terms that you may be required to know for certain questions: Range and standard deviation.

Range

The range of a set is the difference between the set's highest and lowest values.

Standard deviation

(IIC ONLY)

The standard deviation of a set is a measure of the set's variation from its mean. A set composed of ten identical values (having a range of 0) could have the same mean as a set with widely scattered values; the first set would have a much smaller standard deviation than the second.

Standard deviation comes up very infrequently on the Math Subject Tests. Computing a standard deviation is an annoying process; a question asking about it is more likely to test your comprehension of the measure than your ability to calculate it. If you have to calculate a standard deviation, follow these five steps:

- ◆ Find the mean of the set.
- ◆ Find the difference between each value in the set and the mean of the set; write these numbers down.
- ◆ Square all of these differences.

- Find the average of the resulting squares.
- Take the square root of this average.

The resulting number is the standard deviation. As you can see, it's annoying to calculate. It's far more important to understand what it measures. Remember, the standard deviation is a measure of how far the typical value in a set is from the set's arithmetic mean. The bigger the standard deviation, the more widely dispersed the values are. The smaller the standard deviation, the more closely grouped the values in a set are around the mean.

Drill

Try the following practice questions using these statistical definitions. The answers to all drills are found in chapter 12.

21. Set A contains only the factors of 21 that are greater than 1, and set B contains only the factors of 48 that are greater than 1. If set C is the union of sets A and B, what is the mode of set C?

    (A) 2
    (B) 3
    (C) 8
    (D) 12
    (E) Set C has no mode.

25. Set M contains ten elements whose sum is zero. Which of the following statements must be true?

    I. The mean of M is zero.
    II. The median of M is zero.
    III. The mode of M is zero.

    (A) None
    (B) I only
    (C) I and II only
    (D) II and III only
    (E) I, II, and III

Probability

Probability is a mathematical expression of the likelihood of an event. The basis of probability is simple. The likelihood of any event is discussed in terms of all of the possible outcomes. To express the probability of a given event, x, you would count the number of possible outcomes, count the number of outcomes that give you what you want, and arrange them in a fraction, like this:

$$\text{Probability of } x = \frac{\text{number of outcomes that are } x}{\text{total number of possible outcomes}}$$

Every probability is a fraction. The biggest a probability can get is 1; a probability of 1 indicates total certainty.

Figuring out the probability of any single event is usually simple. When you flip a coin, there are only two possible outcomes, heads and tails; the probability of getting heads is therefore 1 out of 2, or $\frac{1}{2}$. When you roll a die, there are six possible outcomes, 1 through 6; the odds of getting a 6 is therefore $\frac{1}{6}$. The odds of getting an even result when rolling a die are $\frac{1}{2}$ since there are three even results in six possible outcomes. Here's a typical example of a simple probability question:

11. A bag contains 7 blue marbles and 14 marbles that are not blue. If one marble is drawn at random from the bag, what is the probability that the marble is blue?

    (A) $\frac{1}{7}$

    (B) $\frac{1}{3}$

    (C) $\frac{1}{2}$

    (D) $\frac{2}{3}$

    (E) $\frac{3}{7}$

Here, there are 21 marbles in the bag, 7 of which are blue. The probability that a marble chosen at random would be blue is therefore $\frac{7}{21}$, or $\frac{1}{3}$. The correct answer is (B).

Probability of multiple events:

Some advanced probability questions require you to calculate the probability of more than one event. Here's a typical example:

23. If a fair coin is flipped three times, what is the probability that the result will be tails exactly twice?

    (A) $\frac{1}{8}$

    (B) $\frac{1}{5}$

    (C) $\frac{3}{8}$

    (D) $\frac{5}{8}$

    (E) $\frac{2}{3}$

When the number of possibilities involved is small enough, the easiest and safest way to do a probability question like this is to write out all of the possibilities and count the ones that give you what you want. Here are all the possible outcomes of flipping a coin three times:

heads, heads, heads	tails, tails, tails
heads, heads, tails	tails, tails, heads
heads, tails, heads	tails, heads, tails
heads, tails, tails	tails, heads, heads

As you can see by counting, only three of the eight possible outcomes produce tails exactly twice. The chance of getting exactly two tails is therefore $\frac{3}{8}$. The correct answer is (C).

Sometimes, however, you'll be asked to calculate probabilities for multiple events when there are too many outcomes to write out easily. Consider, for example, this variation on an earlier question:

41. A bag contains 7 blue marbles and 14 marbles that are not blue. What is the probability that the first three marbles drawn at random from this bag will be blue?

    (A) $\frac{1}{3}$

    (B) $\frac{1}{9}$

    (C) $\frac{1}{21}$

    (D) $\frac{1}{38}$

    (E) $\frac{1}{46}$

Three random drawings from a bag of 21 objects produce a huge number of possible outcomes. It's not practical to write them all out. To calculate the likelihood of three events combined, you need to take advantage of a basic rule of probability:

> The probability of multiple events occurring together is the product of the probabilities of the events occurring individually.

In order to calculate the probability of a series of events, calculate the odds of each event happening separately and multiply them together. This is especially important in processes like drawings, because each event affects the odds of following events. This is how you'd calculate the probability of those three marble drawings:

The first drawing is just like the simple question you did earlier; there are 7 blue marbles out of 21 total—a probability of $\frac{1}{3}$.

For the second drawing, the numbers are different. There are now 6 blue marbles out of a total of 20, making the probability of drawing another blue marble $\frac{6}{20}$, or $\frac{3}{10}$.

For the third drawing, there are now 5 blue marbles remaining out of a total of 19. The odds of getting a blue marble this time are $\frac{5}{19}$.

To calculate the odds of getting blue marbles on the first three random drawings, just multiply these numbers together:

$$\frac{1}{3} \times \frac{3}{10} \times \frac{5}{19} = \frac{1}{38}$$

The odds of getting three blue marbles is therefore $\frac{1}{38}$, and the answer is (D). This can also be expressed decimally, as 0.026. ETS often asks for answers in decimal form on the Math Subject Tests, just to make sure you haven't forgotten how to push the little buttons on your calculator. Just bear with them.

Drill

Try the following practice questions about probability. The answers to all drills are found in chapter 12.

13. If the probability that it will rain is $\frac{5}{12}$, then what is the probability that it will *not* rain?

    (A) $\frac{7}{12}$

    (B) $\frac{5}{7}$

    (C) $\frac{12}{7}$

    (D) $\frac{12}{5}$

    (E) It cannot be determined from the information given.

16. In an experiment, it is found that the probability that a released bee will land on a painted target is $\frac{2}{5}$. It is also found that when a bee lands on the target, the probability that it will attempt to sting it is $\frac{1}{3}$. In this experiment, what is the probability that a released bee will land on the target and attempt to sting it?

    (A) $\frac{2}{15}$

    (B) $\frac{1}{5}$

    (C) $\frac{2}{5}$

    (D) $\frac{1}{3}$

    (E) $\frac{6}{5}$

	Daily Cookie Production	Number Burned
Monday	256	34
Tuesday	232	39
Wednesday	253	41

20. The chart above shows the cookie production at MunchCo for three days. What is the probability that a cookie made on one of these three days will be burned?

    (A) $\dfrac{1}{26}$
    (B) $\dfrac{2}{13}$
    (C) $\dfrac{1}{7}$
    (D) $\dfrac{3}{13}$
    (E) It cannot be determined from the information given.

24. If two six-sided dice are rolled, each having faces numbered 1–6, what is the probability that the product of the two numbers will be odd?

    (A) $\dfrac{1}{6}$
    (B) $\dfrac{1}{4}$
    (C) $\dfrac{1}{3}$
    (D) $\dfrac{1}{2}$
    (E) $\dfrac{7}{12}$

44. In a basketball-shooting contest, if the probability that Heather will make a basket on any given attempt is $\dfrac{4}{5}$, then what is the probability that she will make at least one basket in three attempts?

    (A) $\dfrac{12}{125}$
    (B) $\dfrac{64}{125}$
    (C) $\dfrac{124}{125}$
    (D) 1
    (E) $\dfrac{12}{5}$

Permutations and Combinations

Questions about permutations and combinations are fairly rare on the Math Subject Tests, and more common on the Math IIC Test than the Math IC Test. As is the case with many of the odds and ends of precalculus, questions about permutations and combinations are rarely mathematically difficult; they just test your understanding of the concepts and ability to work with them. Both permutations and combinations are simply ways of counting sets.

Simple permutations

A permutation is an arrangement of objects of a definite order. The simplest sort of permutations question might ask you how many different arrangements are possible for 6 different chairs in a row, or how many different 4-letter arrangements of the letters in the word FUEL are possible. Both of these simple questions can be answered with the same technique.

Just draw a row of boxes corresponding to the positions you have to fill. In the case of the chairs, there are six positions, one for each chair. You would make a sketch like this:

☐ ☐ ☐ ☐ ☐ ☐

Then, in each box, write the number of objects available to be put into that box. Keep in mind that objects put into previous boxes are no longer available. For the chair-arranging example, there would be 6 chairs available for the first box; only 5 left for the second box; 4 for the third, and so on until only one chair remained to be put into the last position. Finally, just multiply the numbers in the boxes together, and the product will be the number of possible arrangements, or permutations:

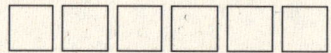 = 720

There are 720 possible permutations of a set of 6 chairs. This number can also be written as 6!. That's not a display of enthusiasm—the exclamation point means *factorial*. The number is read "six factorial," and it means $6 \cdot 5 \cdot 4 \cdot 3 \cdot 2 \cdot 1$, which equals 720. A factorial is simply the product of a series of integers counting down to 1 from the specified number. For example, the number 70! means $70 \cdot 69 \cdot 68 \ldots 3 \cdot 2 \cdot 1$.

The number of possible arrangements of any set with n members is simply $n!$. In this way, the number of possible arrangements of the letters in FUEL is 4!, because there are 4 letters in the set. That means $4 \cdot 3 \cdot 2 \cdot 1$ arrangements, or 24. If you sketched 4 boxes for the 4 letter positions and filled in the appropriate numbers, that's exactly what you'd get.

Advanced permutations

Permutations get a little trickier when you work with smaller arrangements. For example, what if you were asked how many 2-letter arrangements could be made from the letters in FUEL? It's just a modification of the original counting procedure. Sketch 2 boxes for the 2 positions. Then fill in the number of letters available for each position. As before, there are 4 letters available for the first space, and 3 for the second; the only difference is that you're done after two spaces:

[4][3] = 12

As you did before, multiply the numbers in the boxes together to get the total number of arrangements. You should find there are 12 possible 2-letter arrangements from the letters in FUEL.

That's all there is to permutations. The box-counting procedure is the safest way to approach them. Just sketch the number of positions available, and fill in the number of objects available for each position, from first to last—then multiply those numbers together.

On to combinations

Combinations differ from permutations in just one way: In combinations, order doesn't matter. A permutations question might ask you to form different numbers from a set of digits. Order would certainly matter in that case, because 135 is very different from 513. Similarly, a question about seating arrangements would be a permutations question, because the word "arrangements" tells you that order is important.

Combinations questions, on the other hand, deal with groupings in which order *isn't* important. Combinations questions often deal with the selection of committees: Josh–Lisa–Andy isn't any different from Andy–Lisa–Josh, as far as committees go. In the same way, a question about the number of different 3-topping pizzas you could make from a 10-topping list would be a combinations question, because the order in which the toppings are put on is irrelevant.

Combination and permutation questions can be very similar in appearance. Always ask yourself carefully whether sequence is important in a certain question before you proceed.

Calculating combinations

Calculating combinations is surprisingly easy. All you have to do is throw out duplicate answers that count as separate permutations, but not as separate combinations. For example, let's make a full-fledged combination question out of that pizza example.

pepperoni	sausage
meatballs	anchovies
green peppers	onion
mushrooms	garlic
tomato	broccoli

36. If a pizza must have 3 toppings chosen from the list above, and no topping may be used more than once on a given pizza, how many different kinds of pizza can be made?

    (A) 720
    (B) 360
    (C) 120
    (D) 90
    (E) 30

To calculate the number of possible combinations, start by figuring out the number of possible *permutations*:

$$\boxed{10}\,\boxed{9}\,\boxed{8} = 720$$

That tells you that there are 720 possible 3-topping permutations that can be made from a list of 10 toppings. You're not done yet, though. Because this is a list of *permutations*, it contains many arrangements that duplicate the same group of elements in different orders. For example, those 720 permutations would include these:

pepperoni, mushrooms, onion   mushrooms, onion, pepperoni
pepperoni, onion, mushrooms   onion, pepperoni, mushrooms
mushrooms, pepperoni, onion   onion, mushrooms, pepperoni

All six of these listings are different permutations of the same group. In fact, for every 3-topping combination, there will be 6 different permutations. You've got to divide 720 by 6 to get the true number of combinations, which is 120. The correct answer is (C).

So, how do you know what number to divide permutations by to get combinations? It's simple. For the 3-position question above, we divided by 6, which is 3!. That's all there is to it. To calculate a number of possible combinations, calculate the possible permutations first, and divide that number by the number of positions, factorial. Take a look at one more:

29. How many different 4-person teams can be made from a roster of 9 players?

    (A) 3,024
    (B) 1,512
    (C) 378
    (D) 254
    (E) 126

This is definitely a combination question. Start by sketching 4 boxes for the 4 team positions:

Then fill in the number of possible contestants for each position, and multiply them together. This gives you the number of possible *permutations*:

 = 3,024

Finally, divide this number by 4!, for the 4 positions you're working with. This gets rid of different permutations of identical groups. You divide 3,024 by 24 and get the number of possible combinations, 126. The correct answer is (E).

Drill

Try the following practice questions about permutations and combinations. The answers to all drills are found in chapter 12.

27. How many different 4-student committees can be chosen from a panel of 12 students?

    (A) 236
    (B) 495
    (C) 1,980
    (D) 11,880
    (E) 20,736

32. In how many different orders may 6 books be placed on a shelf?

    (A) 36
    (B) 216
    (C) 480
    (D) 720
    (E) 46,656

45. How many committees of 4 females and 3 males may be assembled from a pool of 17 females and 12 males?

   (A) 523,600
   (B) 1,560,780
   (C) 1.26×10^7
   (D) 7.54×10^7
   (E) 7.87×10^9

GROUP QUESTIONS

Group questions are a very specific type of counting problem. They don't come up frequently on the Math Subject Tests, but when they do come up they're easy pickings if you're prepared for them. If you're not, they can be a mite confusing. Here's a sample group question:

34. At Bedlam Music School, 64 students are enrolled in the gospel choir, and 37 students are enrolled in the handbell choir. Fifteen students are enrolled in neither group. If there are 100 students at Bedlam, how many students are enrolled in both the gospel choir and the handbell choir?

   (A) 12
   (B) 16
   (C) 18
   (D) 21
   (E) 27

As you can see, part of the difficulty of such problems lies in reading them—they're confusing. The other trick lies in the actual counting. If there are students in both the gospel choir and the handbell choir, then when you count the members of both groups, you're counting some kids twice—the kids who are in both groups. To find out how many students are in both groups, just use this formula:

> **Group Problem Formula**
>
> Total = Group 1 + Group 2 + Neither − Both

For question 34, this formula gives you 64 + 37 + 15 − Both = 100. Solve this, and you get Both = 16. The correct answer is (B).

This formula will work for any group question with two groups. Just plug in the information you know, and solve the piece that's missing.

Drill

Use the group formula on the following practice questions. The answers to all drills are found in chapter 12.

17. Of the students who made the honor roll at Buford Prep School, 23 got straight A's. If 253 students did not make the honor roll, and the total number of students enrolled at Buford Prep is 370, how many students made the honor roll but did not get straight A's?

    (A) 38
    (B) 56
    (C) 74
    (D) 94
    (E) 140

28. On the Leapwell gymnastics team, 14 gymnasts compete on the balance beam, 12 compete on the uneven bars, and 9 compete on both the balance beam and the uneven bars. If 37 gymnasts compete on neither the balance beam nor the uneven bars, how many gymnasts are on the Leapwell team?

    (A) 45
    (B) 51
    (C) 54
    (D) 63
    (E) 72

42. In a European tour group, $\frac{1}{3}$ of the tourists speak Spanish, $\frac{2}{5}$ of the tourists speak French, and $\frac{1}{2}$ of the tourists speak neither language. What fraction of the tourists in the tour group speak both Spanish and French?

    (A) $\frac{2}{15}$
    (B) $\frac{7}{30}$
    (C) $\frac{1}{3}$
    (D) $\frac{1}{2}$
    (E) $\frac{14}{15}$

Union and Intersection

The terms *union* and *intersection* are ways of discussing the combination of different sets. The union of two or more sets is the combination of them; it is the set of all values contained in *any one* of the individual sets. The intersection of two or more sets is their overlap; it is the set of values contained in *all* of the individual sets.

This can be represented algebraically or graphically. For example, if set S = {2, 3, 4} and set T = {4, 5, 6}, then the union of S and T (which can be written S∪T) is the set {2, 3, 4, 5, 6}. The intersection of S and T (which can be written S∩T) is the set {4}.

Similarly, if set A contains all even numbers, and set B contains all prime numbers, then the set A∪B would contain an infinite number of values—all even numbers and all primes. By contrast, the set A∩B would contain only one value, {2}—the only number that is both even and prime.

Here is a graphical representation of the union and intersection of sets O and P:

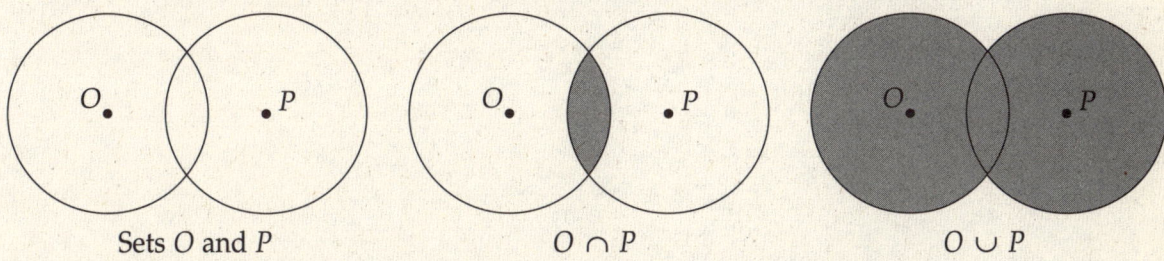

Sets O and P   O ∩ P   O ∪ P

Questions about union and intersection generally test simple comprehension of the definitions of the two terms, and sometimes the symbols, ∩ and ∪, that represent them.

Drill

Try the following practice questions. The answers to all drills are found in chapter 12.

$F = \{1, 2, 3, 4, 5, 6, 7, 8\}$
$G = \{2, 4, 6, 8, 10, 12\}$
$H = F \cap G$

16. What is the mean of set H?

    (A) 1.25
    (B) 5.00
    (C) 5.50
    (D) 6.25
    (E) 7.33

$F = \{1, 2, 3, 4, 5, 6, 7, 8\}$
$G = \{2, 4, 6, 8, 10, 12\}$
$H = F \cup G$

17. What is the mean of set H?

    (A) 1.25
    (B) 5.00
    (C) 5.57
    (D) 6.25
    (E) 7.33

$S = \{3, 4, 5, 6, 7, 8\}$
$T = \{-3, 1, 11, 15, x\}$
$S \cap T = \emptyset$

34. Which of the following could not be the mean of set T?

    (A) 4.00
    (B) 4.80
    (C) 5.20
    (D) 6.20
    (E) 12.60

FLASHCARDS

The rules and formulas listed below represent the most important points to study in this chapter. Memorize them by covering up the right column and testing yourself on the left column, or, better still, by making real flashcards as directed.

Front of Card / Back of Card

Front of Card	Back of Card
What is the mean of a set?	The average of the elements in the set.
What is the median of a set?	The middle value of a set when arranged in order. If the set contains an even number of elements, the average of the two middle values.
What is the mode of a set?	The value that occurs most often in a set.
What is the range of a set?	The difference between the highest and lowest values in a set.
What is the probability that an event x will happen?	$\dfrac{\text{number of possible outcomes that are } x}{\text{total number of possible outcomes}}$
What is the formula for overlapping-group questions?	Total = Group$_1$ + Group$_2$ + Neither − Both

STATISTICS AND SETS ◆ 215

Miscellaneous

MATHEMATICAL ODDS AND ENDS

The techniques and rules covered in this chapter are relatively rare on the Math Subject Tests. They occur only on the Math IIC Test or the difficult third of the Math IC Test; if you're not supposed to be tackling those questions, don't waste your time in this chapter. If you will take the Math IIC Test, or will take the Math IC Test very aggressively, then it's a good idea to learn the rules in this chapter—but remember, the material in the preceding chapters is still more important.

ARITHMETIC AND GEOMETRIC SERIES

The average Math Subject Test has one question dealing with arithmetic or geometric series. They're very easy once you know how they work, so read the next few paragraphs and fear not.

Arithmetic series

The big-forehead people at ETS define an arithmetic sequence as "one in which the difference between successive terms is constant." Real human beings just say that an arithmetic sequence is what you get when you pick a starting value and add the same number again and again.

Here are some sample arithmetic series:

$$\{a_n\} = 1, 7, 13, 19, 25, 31...$$
$$\{b_n\} = 3, 13, 23, 33, 43, 53...$$
$$\{c_n\} = 12, 7, 2, -3, -8, -13...$$

It's not hard to figure out what difference separates any two terms in a series. To continue a series, you would just continue adding that difference. The larger letter in each case is the name of the series (these are series a, b, and c). The subscript, n, represents the number of the term in the series. The expression a_4, for example, represents the fourth term in the a series, which is 19. The expression b_7 means the seventh term in the b series, which would be 63.

The typical arithmetic-series question asks you to figure out the difference between any two successive terms in the series, and then calculate the value of a term much farther along. There's just one trick to that: To calculate the value of a_{26}, for example, start by figuring out the difference between any two consecutive terms. You'll find that the terms in the a series increase at intervals of 6. Now here's the trick: To get to the twenty-sixth term in the series, you'll start with a_1, which is 1, and increase it by 6 *twenty-five* times. The term $a_{26} = 1 + (25 \times 6)$, or 151. It's like climbing stairs in a building; to get to the fifth floor, you climb 4 flights. To get to the 12th floor, you climb 11 flights, and so on. In the same way, it takes 11 steps to get to the 12th term in a series from the first term. To get to the nth term in a series, take $(n - 1)$ steps from the first term.

Here's another example: To figure out the value of c_{17}, start with 12 and add –5 *sixteen* times. The value of $c_{17} = 12 + (16 \times -5)$, or –68. That's all there is to calculating values in arithmetic series.

Here's the algebraic definition of the nth term of an arithmetic series, if the starting value is a_1 and the difference between any two successive terms is d:

The nth term of an Arithmetic Series:

$$a_n = a_1 + (n-1)d$$

Finding the sum of an arithmetic series

You might be asked to figure out the sum of the first 37 terms of an arithmetic series, or the first 48 terms, and so on. To figure out the sum of a chunk of an arithmetic series, take the average of the first and last terms in that chunk, and multiply by the number of terms you're adding up. For example:

$$\{a_n\} = 5, 11, 17, 23, 29, 35...$$

What is the sum of the first 40 terms of a_n?

The first term of a_n is 5. The fortieth term is 239. The sum of these terms will be the average of these two terms, 122, multiplied by the number of terms, 40. The product of 122 and 40 is 4,880; that's the sum of the first 40 terms of the series. Here's the algebraic definition of the sum of the first n terms of an arithmetic series, where the difference between any two successive terms is d:

> **Sum of the First n Terms of an Arithmetic Series**
>
> $$sum = n\left(\frac{a_1 + a_n}{2}\right)$$

Geometric series

A geometric series is formed by taking a starting value and multiplying it by the same factor again and again. While any two successive terms in an arithmetic series are separated by a constant *difference*, any two successive terms in a geometric series are separated by a constant *factor*. Here are some sample geometric series:

$$\{a_n\} = 2, 6, 18, 54, 162, 486\ldots$$
$$\{b_n\} = 8, 4, 2, 1, 0.5, 0.25\ldots$$
$$\{c_n\} = 3, 15, 75, 375, 1875\ldots$$

Just like arithmetic-series questions, geometric-series questions most often test your ability to calculate the value of a term farther along in the series. As with arithmetic series, the trick to geometric series is that it takes 19 steps to get to the 20th term, 36 steps to get to the 37th term, and so on.

To find the value of a_{10}, for example, start with the basic information about the series: Its starting value is 2, and each term increases by a factor of 3. To get to the tenth term, start with 2 and multiply it by 3 *nine* times—that is, multiply 2 by 3^9. You get 39,366, which is the value of a_{10}. As you can see, geometric series tend to grow much faster than arithmetic series do.

Here's the algebraic definition of the nth term in a geometric series, where the first term is a_1 and the factor separating any two successive terms is r:

> **The nth term of a Geometric Series**
>
> $$a_n = a_1 r^{n-1}$$

The sum of a geometric series

You may also be asked to find the sum of part of a geometric series. This is a bit tougher than calculating the sum of an arithmetic series. To add up the first n terms of a geometric series, use this formula. Once again, the first term in the series is a_1, and the factor separating any two successive terms is r:

> **The Sum of the First n Terms of a Geometric Series**
>
> $$sum = \frac{a_1(1 - r^n)}{1 - r}$$

This is not a formula that is called upon very often, but it's good to know it if you're taking the Math IIC Test.

The sum of an infinite geometric series

Every now and then, a question will ask you to figure out the sum of an infinite geometric series—that's right, add up an *infinite* number of terms. There's a trick to this as well. Whenever the factor between any two terms is greater than 1, then the series keeps growing and growing. The sum of such a series is infinitely large: it never stops increasing, and its sum cannot be calculated.

An infinite geometric series has a finite sum only when its constant factor is a positive or negative fraction. When the constant factor of a geometric series is less than 1, the terms in the series continually decrease, and there exists some value that the sum of the series will never exceed. For example:

$$\{a_n\} = 1, 0.5, 0.25, 0.125, 0.0625...$$

The series a_n above will never be greater than 2. The more of its terms you add together, the closer the sum gets to 2. If you add *all* of its terms, all the way out to infinity, you get exactly 2. Here's the formula you use to figure that out. Once again, a_1 is the first term in the series, and r the factor between each two terms. Remember that r must be between −1 and 1:

Sum of an Infinite Geometric Series

$$sum = \frac{a_1}{1-r} \quad \text{for } -1 < r < 1$$

The five rules reviewed so far are all you'll ever need to work with arithmetic and geometric series on the Math Subject Tests.

Drill

Try the following practice questions about arithmetic and geometric series. The answers to all drills can be found in chapter 12.

14. In an arithmetic series, the second term is 4 and the sixth term is 32. What is the fifth term in the series?

    (A) 8
    (B) 15
    (C) 16
    (D) 24
    (E) 25

19. In the arithmetic series a_n, $a_1 = 2$ and $a_7 = 16$. What is the value of a_{33}?

    (A) 72.00
    (B) 74.33
    (C) 74.67
    (D) 75.14
    (E) 76.67

26. If the second term of a geometric series is 4, and the fourth term of the series is 25, then what is the ninth term in the series?

    (A) 804.43
    (B) 976.56
    (C) 1864.35
    (D) 2441.41
    (E) 6103.52

34. $3 + 1 + \dfrac{1}{3} + \dfrac{1}{9} + \dfrac{1}{27} + \dfrac{1}{81}\ldots =$

    (A) 4.17
    (B) 4.33
    (C) 4.50
    (D) 5.00
    (E) ∞

LIMITS

A limit is the value a function approaches as its independent variable approaches a given constant. That may be confusing to read, but the idea is really fairly simple. A limit can be written in different ways, as these examples show:

$$\lim_{x \to 2} \frac{2x^2 + x - 10}{x - 2} =$$

What is the limit of $\dfrac{2x^2 + x - 10}{x - 2}$ as x approaches 2?

If $f(x) = \dfrac{2x^2 + x - 10}{x - 2}$, then what value does $f(x)$ approach as x approaches 2?

These three questions are exactly equivalent. The first of the three is in limit notation and is read exactly like the question, "What is the limit of $\dfrac{2x^2 + x - 10}{x - 2}$ as x approaches 2?"

Finding a limit is very simple. Just take the value that x approaches and plug it into the expression. The value you get is the limit. It's so simple that you just know there's got to be a hitch—and there is. The limits that appear on the Math Subject Tests share a common problem: Tricky denominators. The question introduced above is no exception. Let's take a look at it again:

$$\lim_{x \to 2} \frac{2x^2 + x - 10}{x - 2} =$$

You can find the value of this limit just by plugging 2 into the expression as x. But there's a hitch. When $x = 2$, the fraction's denominator is undefined, and it seems that the limit does not exist. The same solution always applies to such questions. You need to factor the top and bottom of the fraction and see whether there's anything that will make the denominator cancel out and stop being such a nuisance. Let's see how this expression factors out:

$$\lim_{x \to 2} \frac{2x^2 + x - 10}{x - 2} = \lim_{x \to 2} \frac{(x - 2)(2x + 5)}{x - 2} =$$

Now, you can cancel out that pesky $(x-2)$:

$$\lim_{x \to 2}(2x+5) =$$

Now the expression is no longer undefined when you plug 2 in. It simply comes out to $2(2) + 5$, or 9. The limit of $\dfrac{2x^2 + x - 10}{x-2}$ as x approaches 2 is 9.

That's all there is to limit questions; just factor the top and bottom of the expression as much as possible, and try to get the problematic terms to cancel out so that the limit is no longer undefined. When it's no longer undefined, just plug in the constant value to find the limit.

One more dirty trick: You might run into a limit problem in which it's impossible to cancel out the term that makes the expression undefined. Take a look at this example:

$$\lim_{x \to -3} \frac{3x^2 + 3x - 36}{x^2 - 9}$$

Because the constant that x approaches, -3, makes the limit undefined, you've got to factor the expression and try to cancel out the problematic part of the denominator:

$$\lim_{x \to -3} \frac{3x^2 + 3x - 36}{x^2 - 9}$$

$$\lim_{x \to -3} \frac{3(x-3)(x+4)}{(x-3)(x+3)}$$

$$\lim_{x \to -3} \frac{3(x+4)}{(x+3)}$$

The expression can be factored, and you can even cancel out a term in the denominator. When the dust clears, however, you find that the denominator of the fraction still approaches zero, and that the limit remains undefined. When this happens, it's said that the limit does not exist, and that would be the correct answer.

Drill

Try the following practice questions involving limits. The answers to all drills are found in chapter 12.

30. What value does the expression $\dfrac{4x^2 - x - 5}{16x^2 - 25}$ approach as x approaches 1.25?

   (A) 0
   (B) 0.225
   (C) 0.625
   (D) 1.275
   (E) 2.250

38. $\lim_{x \to 3} \dfrac{x^2 + x - 12}{2x^2 - 6x} =$

   (A) 1.17
   (B) 2.25
   (C) 3.33
   (D) 6.67
   (E) The limit does not exist.

40. $\lim\limits_{x \to -3} \dfrac{x^3 + 4x^2 - 21x}{x^2 + 10x + 21}$

(A) –3.00
(B) 2.46
(C) 7.50
(D) 10.33
(E) The limit does not exist.

Vectors

A vector is a visual representation of something that has both direction and magnitude. A vector can represent a force, a velocity, a distance traveled, or any of a variety of physical quantities. On the Math Subject Tests, vectors usually represent travel.

A vector-arrow's orientation indicates the direction of travel. Its length represents the distance traveled (this is the magnitude of the vector). Sometimes, test questions will deal with vectors without telling you what they represent.

Basically, there are only two things you have to do with vectors on the Math Subject Tests: Compute their lengths, and add and subtract them. Computing their lengths is generally done on the coordinate plane, where it's just a matter of using the Pythagorean theorem. Adding and subtracting vectors is also pretty simple. Here's how it's done.

Adding vectors

Suppose you wanted to add these two vectors together:

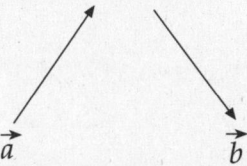

To add them, redraw the second vector so that its tail stands on the tip of the first vector. Then draw the resulting vector, closing the triangle (make sure that the resulting vector's direction is in agreement with the vectors you added). This is what the addition of vectors a and b looks like:

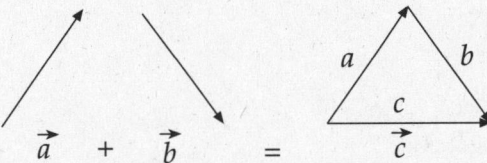

Vector c is the sum of vectors a and b.

Subtracting vectors

To subtract vectors, you'll use the same technique you used to add them, with one extra step. First, reverse the sign of the vector that's being subtracted. You do this by simply moving the arrowhead to the other end of the vector. Then add the two vectors as you usually would. Here's an example of subtraction using the two vectors we just added together. First, reverse the sign of the subtracted vector:

And then, add them up:

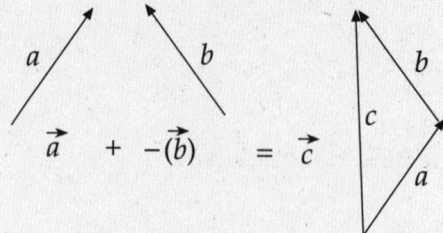

Vector *c* is the vector produced by subtracting vector *b* from vector *a*.

Drill

The answers to all drills are found in chapter 12.

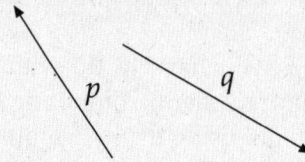

18. Which of the following could be the sum of vectors *p* and *q*?

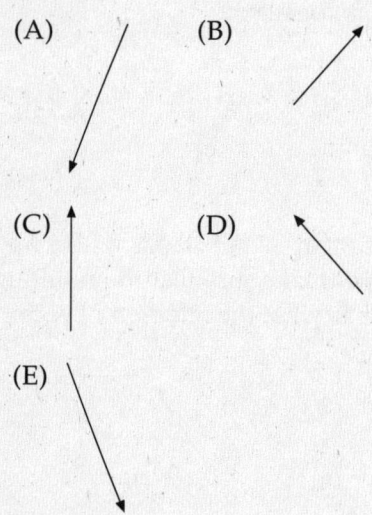

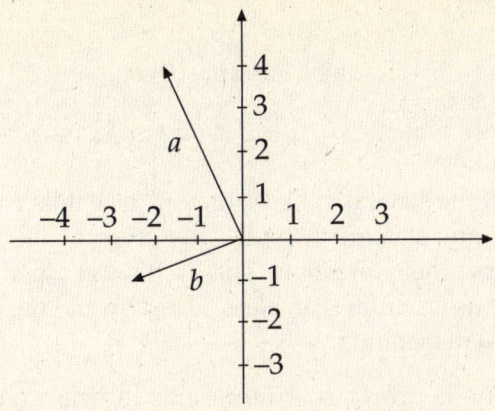

36. If $\vec{c} = \vec{a} + \vec{b}$, then what is the magnitude of $\vec{c}$?

   (A) 2
   (B) 3
   (C) 4
   (D) 5
   (E) 6

39. Let m, n, p, and q represent the magnitudes of vectors $\vec{m}, \vec{n}, \vec{p},$ and $\vec{q}$, respectively, where $mn \neq 0$. If $\vec{m} + \vec{n} = \vec{p}$, $\vec{m} - \vec{n} = \vec{q}$, and $m^2 + n^2 = p^2$, then which of the following statements must be true?

   (A) $m + n = p$
   (B) $m + n = p + q$
   (C) $m = n$
   (D) $p - m = q - n$
   (E) $p = q$

Logic

Every now and then, as you proceed innocently through a Math Subject Test, you will come upon a question asked in simple English that seems to have nothing at all to do with math. This is a logic question. Here's a typical example:

24. If it's true that every precious stone is harder than glass, which of the following statements must also be true?

   (A) Glass can be a precious stone.
   (B) Every stone harder than glass is a precious stone.
   (C) No stone is exactly as hard as glass.
   (D) Some stones softer than glass are precious stones.
   (E) Every stone softer than glass is not a precious stone.

This is madness. There's no math here at all. However, there is a rule here for you to work with. The rule states that given one statement, there's only one other statement that is logically necessary, the *contrapositive*. This is what the contrapositive states:

The Contrapositive

Given the statement $A \rightarrow B$, you also know $\sim B \rightarrow \sim A$.

In English, that means that the statement "If A, then B" also tells you that "If not B, then not A." To find the contrapositive of any statement, switch the order of the first and second parts of the original statement, and reverse their direction. This is how you'd find the contrapositive of the statement, "Every precious stone is harder than glass." Start by making sure that you clearly see what the two parts of the original statement are:

$$\text{stone is precious} \rightarrow \text{stone harder than glass}$$

Then switch the order of the statement's parts, and reverse their meanings, like so:

$$\text{stone not harder than glass} \rightarrow \text{stone is not precious}$$

This is the contrapositive. Once you've found it, just check the answer choices for a statement with an equivalent meaning. In this case, answer choice (E) is equivalent to the contrapositive. Almost all logic questions test your understanding of the contrapositive. There are just a couple of other points that might come up in logic questions:

- To disprove the claim, "X might be true," or "X is possible," you must show that X is never, ever true, in any case, anywhere.

- To disprove the claim, "X is true," you only need to show that there's one exception, somewhere, sometime.

In other words, a statement that something *may* be true is very hard to disprove; you've got to demonstrate conclusively that there's no way it could be true. On the other hand, a statement that something is *definitely* true is easy to disprove; all you have to do is find one exception. If you remember these two rules and the contrapositive, you'll be prepared for any logic question on the Math Subject Tests.

Drill

Exercise your powers of logic on these practice questions. The answers to all drills are found in chapter 12.

28. At Legion High School in a certain year, no sophomore received failing grades. Which of the following statements must be true?

    (A) There were failures in classes other than the sophomore class.
    (B) Sophomores had better study skills than other students that year.
    (C) No student at Legion High School received failing grades that year.
    (D) Any student who received failing grades was not a sophomore.
    (E) There were more passing grades in the sophomore class than in other classes.

33. "If one commits arson, a building burns." Which of the following is a contradiction of this statement?

(A) Many people would refuse to commit arson.
(B) A building did not burn, and yet arson was committed.
(C) Some buildings are more difficult to burn than others.
(D) A building burned, although no arson was committed.
(E) Arson is a serious crime.

35. In a necklace of diamonds and rubies, some stones are not genuine. If every stone that is not genuine is a ruby, which of the following statements must be true?

(A) There are more diamonds than rubies in the necklace.
(B) The necklace contains no genuine rubies.
(C) No diamonds in the necklace are not genuine.
(D) Diamonds are of greater value than rubies.
(E) The necklace contains no genuine diamonds.

Imaginary numbers

Almost all math on the Math Subject Tests is confined to real numbers. Only a few questions deal with the square roots of negative numbers: Imaginary numbers. For the sake of simplicity, imaginary numbers are expressed in terms of i. The quantity i is equal to the square root of -1. It's used to simplify the square roots of negative numbers. For example, here's how i can be used to simplify square roots of negative numbers:

$$\sqrt{-25} = \sqrt{25}\sqrt{-1} = 5\sqrt{-1} = 5i$$
$$\sqrt{-48} = \sqrt{48}\sqrt{-1} = \sqrt{16}\sqrt{3}\sqrt{-1} = 4i\sqrt{3}$$
$$\sqrt{-7} = \sqrt{7}\sqrt{-1} = i\sqrt{7}$$

There are three basic kinds of questions on the Math Subject Tests that require you to work with imaginary numbers.

Computing powers of i

You may run into a question that asks you to find the value of i^{34}, or something equally outrageous. This may seem difficult or impossible at first, but, as usual, there's a trick to it. The powers of i repeat in a cycle of 4 values, over and over:

$i^1 = i$ $i^5 = i$
$i^2 = -1$ $i^6 = -1$
$i^3 = -i$ $i^7 = -i$
$i^4 = 1$ $i^8 = 1$

And so on. These are the only four values that can be produced by raising i to a power. To find the value of i^{34}, either write out the cycle of four values up to the 34th power, which would take less than a minute, or, more simply, divide 34 by 4. You find that 34 contains eight cycles of 4, with a remainder of 2. The eight cycles of 4 just bring you back to where you started. It's the remainder that's important. The remainder of 2 means that the value of i^{34} is equal to the value of i^2, or –1. In order to raise i to any power, just divide the exponent by 4 and use the remainder as your exponent.

Doing algebra with i

Algebra that includes complex numbers is no different from ordinary algebra. You just need to remember that i raised to an exponent changes in value, which can have some odd effects in algebra.

Here's an example:

$$(x - 3i)(2x + 6i) =$$
$$2x^2 - 6ix + 6ix - 18i^2 =$$
$$2x^2 - 18i^2 =$$
$$2x^2 - 18(-1) =$$
$$2x^2 + 18$$

As you can see, i sometimes has a way of dropping out of algebraic expressions. ETS likes this trick, so keep an eye out for it.

The complex plane (IIC ONLY)

A complex number is a specific kind of imaginary number—specifically, the sum of a real number and an imaginary number, such as 5 + 3i. A complex number is one that takes the form $a + bi$, where a and b are real numbers and i is the imaginary unit, the square root of –1. On the Math Subject Tests, the principal importance of complex numbers is that they can be represented on the complex plane. This is what the complex plane looks like:

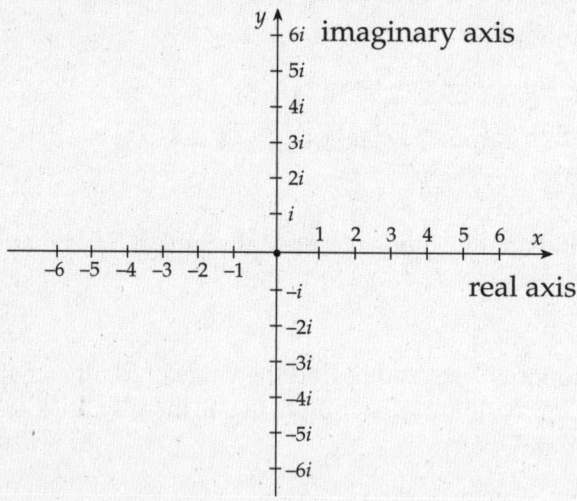

The complex plane looks just like the ordinary coordinate plane, but the axes have different meanings. On the complex plane, the x-axis is referred to as the *real axis*. The y-axis is referred to as the *imaginary axis*. Each unit on the x-axis equals 1—a real unit. Each unit on the imaginary axis equals i—the imaginary unit. Any complex number in the form $a + bi$, such as 5 + 3i, can be plotted on the complex plane almost like a coordinate pair. Just plot a, the real component of the complex number, on the x-axis; and bi, the imaginary component, on the y-axis.

Here are several complex numbers plotted on the complex plane:

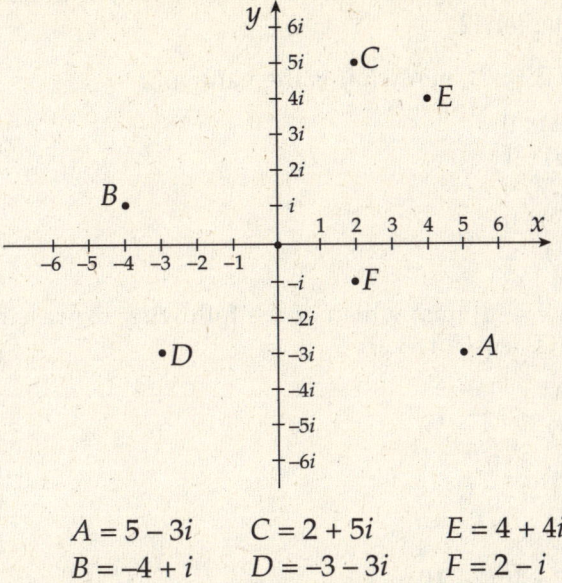

$A = 5 - 3i \quad C = 2 + 5i \quad E = 4 + 4i$
$B = -4 + i \quad D = -3 - 3i \quad F = 2 - i$

Once you've plotted a complex number on the complex plane, you can use all of the usual coordinate-geometry techniques on it, including the Pythagorean Theorem and even right-triangle trigonometry. The most common complex-plane question asks you to find the distance between a complex number and the origin, using the Pythagorean Theorem. This distance is most often referred to as the magnitude or absolute value of a complex number. If you're asked to compute $|4 + 3i|$, just plot the number on the complex plane and use the Pythagorean Theorem to find its distance from the origin. This distance is the absolute value of the complex number:

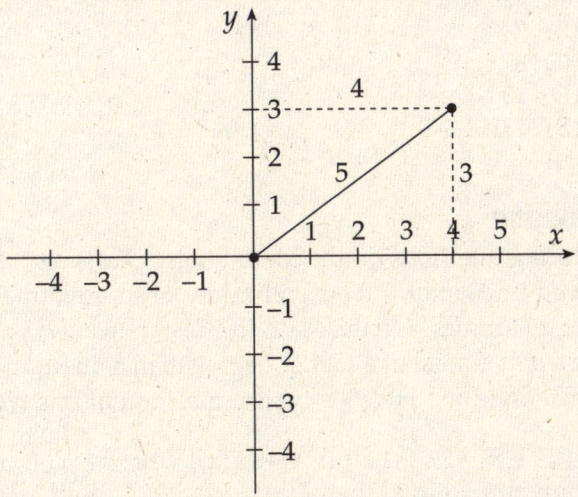

The Pythagorean Theorem will quickly show you that $|4 + 3i| = 5$.

MISCELLANEOUS ♦ 229

Drill

Test your understanding of imaginary numbers with the following practice questions. The answers to all drills are found in chapter 12.

25. If $i^2 = -1$, then what is the value of i^{51}?

    (A) 0
    (B) –1
    (C) –i
    (D) i
    (E) 1

36. If $i^2 = -1$, then which of the following expressions is NOT equal to zero?

    (A) $i^0 - i^{12}$
    (B) $i + i^3$
    (C) $i^4 + i^{10}$
    (D) $i^{11} - i^9$
    (E) $i^8 - i^{12}$

40. $\dfrac{(2+4i)(2-4i)}{5} =$

    (A) 2.2
    (B) 4.0
    (C) 4.6
    (D) 5.0
    (E) 8.4

43. $|5 - 12i| =$

    (A) $7i$
    (B) 7
    (C) 8
    (D) 13
    (E) $13i$

Polynomial division

IIC ONLY

Most of the factoring questions on the Math Subject Tests are very traditional, using only the tools reviewed in chapter 4. You will rarely need anything more advanced than the reverse FOIL technique for quadratics. On the Math IIC Test, however, you may run into a question that requires you to factor a polynomial of a higher degree than a quadratic. Then you'll need to use the technique of polynomial division. Here's a typical question of this type:

21. If $x^3 + x^2 - 7x + 20 = (x + 4) \cdot F(x)$, where $F(x)$ is a polynomial in x, then $F(x) =$

    (A) $x + 20$
    (B) $x^2 + 5$
    (C) $x^2 - 2x$
    (D) $x^2 - 3x + 5$
    (E) $x^2 - 7x + 20$

To figure out what F(x) is, you must divide $x^3 + x^2 - 7x + 20$ by $x + 4$. That's polynomial division. Polynomial division is actually just like ordinary division. You set it up like this:

$$x+4 \overline{\smash{)}x^3 + x^2 - 7x + 20}$$

Then just move through the process of long division as you would with numbers. At each step, your goal is to make the first terms match:

$$\begin{array}{r}
x^2 - 3x + 5 \\
x+4 \overline{\smash{)}x^3 + x^2 - 7x + 20} \\
\underline{x^3 + 4x^2 } \\
-3x^2 - 7x \\
\underline{-3x^2 - 12x } \\
5x + 20 \\
\underline{5x + 20} \\
0
\end{array}$$

To make the first term of $x + 4$ match the first term of $x^3 + x^2 - 7x + 20$, you have to multiply by x^2. That gives you $x^3 + 4x^2$, which you subtract as in ordinary long division. Subtraction gives you $-3x^2$; you then complete the binomial by bringing down the $-7x$, producing $-3x^2 - 7x$. To match the first terms again, you must multiply $x + 4$ by $-3x$, which gives you $-3x^2 - 12x$. Subtraction then gives you $5x$, and bringing down the 20 completes the binomial. To make the first terms match this time, you must multiply $x + 4$ by 5. That gives you $5x + 20$, which is exactly what remains. There is no remainder.

30. What is the remainder when $x^4 - 5x^2 + 12x + 18$ is divided by $(x + 1)$?

(A) $x^2 - 1$
(B) $x - 6$
(C) 6
(D) 3
(E) 2

This piece of polynomial division is a little weirder. For one thing, there will be a remainder; in addition, there's a missing term in $x^4 - 5x^2 + 12x + 18$. To do polynomial division, you must write the polynomials with their terms in descending order of degree. If there's a term missing (here, there's no x^3 term between the x^4 and the x^2 terms) you must fill it in with a coefficient of zero to hold its place:

$$\begin{array}{r}
x^3 - x^2 - 4x + 16 \\
x+4 \overline{\smash{)}x^4 + 0x^3 - 5x + 12x + 18} \\
\underline{x^4 + x^3 } \\
-x^3 - 5x^2 \\
\underline{-x^2 - x^2 } \\
-4x^2 + 12x \\
\underline{-4x^2 - 4x } \\
16x + 18 \\
\underline{16x + 16} \\
2
\end{array}$$

In this case, the remainder is 2, so (E) is the correct answer.
That's all there is to polynomial division.

Drill

Try your talents on these practice questions. The answers to all drills are found in chapter 12.

21. If $x^4 - 5x^3 - 2x^2 + 24x = G(x) \cdot (x + 2)$, then which of the following is $G(x)$?

    (A) $x + 12$
    (B) $x^2 + 3x - 18$
    (C) $x^3 - 7x^2 + 12x$
    (D) $x^3 + 10x^2 + 6x$
    (E) $x^4 - 3x^3 + 2x^2 - 6$

27. What is the remainder when $x^3 + 2x^2 - 27x + 40$ is divided by $(x - 3)$?

    (A) 4
    (B) 16
    (C) $2x + 2$
    (D) $x^2 - 5$
    (E) $x^2 + 5x - 12$

FLASHCARDS

The rules and formulas listed below represent the most important points to study in this chapter. Memorize them by covering up the right column and testing yourself on the left column, or, better still, by making real flashcards as directed.

Front of Card — Back of Card

What is the contrapositive of the statement, "If A, then B"?

"If not B, then not A."

What are the first 4 powers of i?

$i^1 = i$
$i^2 = -1$
$i^3 = -i$
$i^4 = 1$

What is the formula for the nth term of an arithmetic series?

$a_n = a_1 + (n-1)d$

What is the formula for the nth term of a geometric series?

$a_n = a_1 r^{n-1}$

What is the formula for the sum of an infinite geometric series?

$$\text{sum} = \frac{a_1}{1-r}$$
(decreasing series only)

12

Answers and Explanations

3 ARITHMETIC

Factorizations, p. 17–18

1. $64 = 2 \cdot 2 \cdot 2 \cdot 2 \cdot 2 \cdot 2$
2. $70 = 2 \cdot 5 \cdot 7$
3. $18 = 2 \cdot 3 \cdot 3$
4. $98 = 2 \cdot 7 \cdot 7$
5. $68 = 2 \cdot 2 \cdot 17$
6. $51 = 3 \cdot 17$

3. **(B)** 21 can also be written as $3 \cdot 7$, and 18 as $2 \cdot 3 \cdot 3$. The smallest number that contains both is $2 \cdot 3 \cdot 3 \cdot 7$, or 126.

7. **(E)** 53 is a prime number; its largest prime factor is 53. None of the other answers contains so large a prime factor.

9. **(C)** The smallest integer divisible by 10 and 32 is 160, and the smallest integer divisible by 6 and 20 is 60; the difference between them is 100.

Even and odd, positive and negative, pp. 19–20

15. **(A)** Plug in numbers and use the results to eliminate answers. If you plug in 1 and 3, for example, only I produces an odd result.

18. **(D)** The statement $cd < 0$ means that one of the numbers is positive and one negative; that's the only way to get a negative product. Once again, plug in numbers to eliminate answer choices. If you plug in 1 and –3, only (A) and (D) are true. Switch them and plug in –3 and 1, and only (D) remains true.

20. **(C)** Plug in numbers. Plugging in 2 and –3 shows you that only (C) must be odd and positive.

PEMDAS and your calculator, p. 22

1. 5
2. 35
3. 12
4. 35
5. 0

Fractions, decimals, and percentages, p. 23

Fractions	Decimals	Percentages
$\frac{1}{2}$	0.5	50%
$\frac{1}{4}$	0.25	25%
$\frac{1}{8}$	0.125	12.5%
$\frac{3}{8}$	0.375	37.5%
$\frac{1}{3}$	0.33...	$33\frac{1}{3}\%$
$\frac{2}{3}$	0.66...	$66\frac{2}{3}\%$
$\frac{1}{6}$	0.166...	$16\frac{2}{3}\%$
$\frac{1}{10}$	0.1	10%
$\frac{1}{100}$	0.01	1%
$\frac{3}{4}$	0.75	75%
$\frac{1}{5}$	0.2	20%

Word-problem translation, p. 24

1. $6.5 = \frac{x}{100} \times 260$
2. $20 = n \times 180$
3. $\frac{30}{100} \times \frac{40}{100} \times 25 = x$
4. $x = \sqrt{\frac{1}{3} \times 48}$
5. $\sqrt{y} = \frac{1}{8} \times y$

Percent change, p. 25

2. **(B)** 25 is 16.67% of 150.

5. **(B)** The decrease from 5 to 4 is a 20% decrease, and the increase from 4 to 5 is a 25% increase.

12. **(C)** 150 is 12% of 1,250, so 1,250 must have been the original amount. $1{,}250 + 150 = 1{,}400$.

Averages, pp. 27–28

1. There were 9 people at dinner.
2. All told, 4,500 apples were picked.
3. The average height of a chess club member is 5.5 feet.

Averages, p. 29

33. **(D)** Nineteen donations averaging $485 total $9,215. Twenty donations averaging $500 total $10,000. The difference is $785.

35. **(A)** The Tribune received 80 letters in the first 20 days and 70 for the last 10 days. That's 150 in 30 days, or 5 letters per day on average.

36. **(A)** One day in five out of a year is 73 days. At 12 a day, that's 876 umbrellas sold on rainy days. The rest of the year (292 days) is clear. At 3 umbrellas a day, that's another 876 umbrellas. A total of 1,752 umbrellas in 365 days makes a daily average of 4.8 umbrellas.

Exponents, p. 30

1. $b = 3$ (1 root)
2. $x = 11, -11$ (2 roots)
3. $n = 2$ (1 root)
4. $c = \sqrt{10}, -\sqrt{10}$ (2 roots)
5. $x = 3, -3$ (2 roots)
6. $x = -2$ (1 root)
7. $d = 3, -3$ (2 roots)
8. n = any real number; everything to the 0 power is 1.

Special exponents, pp. 34–35

1. **(C)**
2. **(D)**
3. **(A)**
4. **(E)**
5. **(A)**
6. **(D)**

Repeated percent change, p. 37

35. **(D)** To compute 12 annual increases of 5%, multiply the starting amount by 1.05 12 times, or 1.05^{12}. $1,000 \times 1.05^{12} =$ 1,795.856.

40. **(E)** For this question, you need to compute 100 annual increases of 8%, so you must multiply the starting amount, 120,000, by 1.08^{100}, which gives you 263,971,350.8—which can also be written as 2.6×10^8.

43. **(C)** This question is a little trickier. For each annual decrease of 4%, you must multiply by 0.96; the easiest way to solve the question is to start with 2,000 and keep multiplying until the result is less than 1,000—just count the number of decreases it takes. The seventeenth annual decrease makes it less than 1,000, so the sixteenth is the last one that is not less than 1,000—and 1995 + 16 = 2011.

Scientific notation, p. 38

13. **(D)** $3^{70} = 2.5 \times 10^{33}$. That means 2.5 with the decimal point moved to the right 33 places, for a total of 34 digits.

14. **(E)** $6^{44} = 1.7 \times 10^{34}$, which has 35 digits, and $6^{55} = 6.3 \times 10^{42}$, which has 43 digits. That's an increase of 8 digits.

19. **(C)** $10^{18} = 1.0 \times 10^{18}$, and $9^{19} = 1.35085 \times 10^{18}$. The difference between them is 0.35085×10^{18}, or 3.5×10^{17}.

Logarithms, p. 39

1. 5
2. 81
3. 3
4. 4
5. y

6. 0
7. x^2
8. 12
9. 1.5682—use your calculator
10. 0.6990—use your calculator

Logarithmic rules, p. 40

1. $\log 20$
2. $\log_5 2$
3. $\log 3$
4. $\log_4 16 = 2$
5. $\log 75$

Logarithms in exponential equations, p. 42

1. $x = 2.52$
2. $\log_5 18 = 1.7959$
3. $n = 2.1367$
4. $\log_{12} 6 = 0.7210$
5. $4^{x+2} = 80$
6. $\log_2 50 = 5.6439$
7. $3^{x+1} = 21$
8. $x = 22.9$

Natural logarithms, p. 44

18. **(B)** This is a very simple example. The equation $e^z = 8$ converts simply into a natural logarithm: $\ln 8 = z$. To find the value of z, just fire $\ln 8$ into your calculator and see what happens. You'll get 2.07944.

23. **(C)** This is just a matter of knowing the decimal values of e and π. All you need to know is that $\pi \approx 3.14$ and $e \approx 2.718$. Then it's easy to put the quantities in order; just remember that you're supposed to put them in descending order.

38. **(A)** To solve this equation, start by isolating the e term:

$$6e^{\frac{n}{3}} = 5$$

$$e^{\frac{n}{3}} = \frac{5}{6}$$

Then, use the definition of a logarithm to change the form of the equation:

$$\ln \frac{5}{6} = \frac{n}{3}$$

$$n = 3 \ln \frac{5}{6}$$

Finally, use your calculator to evaluate the logarithm:

$n = 3(-0.18232)$

$n = -0.54696$

The correct answer is (A).

4 ALGEBRA

Solving equations, p. 49

1. $x = \{5, -5\}$
2. $n = \{0, 5\}$
3. $a = 0.75$
4. $s = 12$
5. $x = 0.875$

Factoring and distributing, p. 51

3. **(D)** This equation can be rearranged to look like this: $50x(11 + 29) = 4000$. This is done simply by factoring out 50 and x. Once you've done that, you can add 11 and 29 to produce a very simple piece of math: $50x(40) = 2000x = 4000$. Therefore $x = 2$.

17. **(E)** Here, distributing makes your math easier. Distributing $-3b$ into the expression $(a + 2)$ on top of the fraction gives you $\frac{-3ab - 6b + 6b}{-ab}$, which simplifies to $\frac{-3ab}{-ab}$, which simply equals 3.

36. **(B)** The trap in this question is to try and cancel similar terms on the top and bottom—but that's not possible, because these terms are being added together, and you can cancel only in multiplication. Instead, factor out x^2 on top of the fraction. That gives you $\frac{x^2(x^3 + x^2 + x + 1)}{x^3 + x^2 + x + 1}$. The whole mess in parentheses cancels out (now that it's being multiplied), and the answer is x^2.

Algebraic functions, p. 52

27. **(B)** Just follow instructions on this one, and you get –64 – (–27), or –64 + 27, which is –37.

30. **(A)** You've just got to plug through this one. The original expression ¥5 + ¥6 becomes $5(3)^2 + 5(4)^2$, which equals 125. Work through the answer choices from the top to find the one that gives you 125.

34. **(B)** The function §x leaves even numbers alone and flips the signs of odd numbers. That means that the series §1 + §2 + §3...§100 + §101 will become (–1) + 2 + (–3) + 4 + (–5)...+ 100 + (–101). Rather than adding up all those numbers, find the pattern: –1 and 2 add up to 1; –3 and 4 add up to 1; and so on, all the way up to –99 and 100. That means fifty pairs that add up to 1, plus the –101 left over. 50 + –101 = –51.

Plugging in, p. 55

15. **(E)** Plug in numbers that make the math easy. For example, $p = 20$, $t = 10$, and $n = 3$. That's 3 items for $20.00 each with 10% tax. Each item would then cost $22.00, and three could cost $66.00. Only answer choice (E) equals $66.00.

23. **(C)** Plug in easy numbers: $x = 5$, $a = 2$, $b = 3$. That means Vehicle A travels at 5 mph for 5 hours, or 25 miles. Vehicle B travels at 7 mph for 8 hours, or 56 miles. That's a difference of 31 miles. Only (C) equals 31.

33. **(C)** Plug in anything, though it's preferable to choose a number between zero and 5. If $n = 3$, then 5 – 3 = 2 and 3 – 5 = –2. These numbers have the same absolute value, so the difference between them is zero.

38. **(B)** Plug in easy numbers, such as these: $a = 10$, $b = 1$, $m = 4$. That means that Company A builds 10 skateboards a week, and 40 skateboards in 4 weeks. Company B builds 7 skateboards in a week (1 per day), or 28 in 4 weeks. That's a difference of 12 between the two companies. Only (B) equals 12 when you plug in $a = 10$, $b = 1$, $m = 4$.

Backsolving, pp. 57–58

11. **(D)** The answer choices represent Michael's hats. Start with answer (C); if Michael has 12 hats, then Matt has 6 hats and Aaron has 2. That adds up to 20, not 24—you need more hats, so move on the next bigger answer, (D). Michael now has 14 hats, meaning Matt has 7 and Aaron has 3. That adds up to 24, so you're done.

17. **(D)** Work through the answer choices, starting with (C). A ratio of 2 : 5 has 7 parts. That would mean each part was $\frac{3200}{7}$, or 457.14—it doesn't work out with whole numbers, so it can't be right. Then move on to (D); a ratio of 3 : 5 has 8 parts, each of which would be 400. That means the shipment is divided into shares of 1,200 and 2,000. Their difference is 800, and their average is 1,600. The difference is therefore half the mean, and this is the correct answer.

27. **(D)** Start with (C). If 5 is the biggest of the three integers, then the others would have to be smaller—so they couldn't ever

add up to 15, which is 3 times 5. Try a bigger number, as in (D). If 9 is the biggest of the three numbers, then there are two factors of 45 which would bring the total to 15: 1 and 5. Since 1 + 5 + 9 = 15, and 1 × 5 × 9 = 45, this is the correct answer.

Inequalities, pp. 58–59

1. $n \geq 3$
2. $r < 7$
3. $x \geq -\frac{1}{2}$
4. $x < \frac{1}{8}$
5. $t \leq 3$
6. $n \leq 4$
7. $p > \frac{1}{5}$
8. $s \geq 1$
9. $x \geq -7$
10. $s \geq \frac{2}{5}$

Working with ranges, p. 59

1. $5 > -x > -8$
2. $-20 < 4x < 32$
3. $1 < (x + 6) < 14$
4. $7 > 2 - x > -6$
5. $-2.5 < \frac{x}{2} < 4$

Working with ranges, p. 61

1. $-4 \leq b - a \leq 11$
2. $-2 \leq x + y \leq 17$
3. $0 \leq n^2 \leq 64$
4. $3 < x - y < 14$
5. $-13 \leq r + s \leq 13$
6. $-126 < cd < 0$

Direct and indirect variation, p. 63

15. **(C)** Plug in! Quantities in indirect variation always have the same product. That means that $ab = 3 \times 5$, or 15, always. Plug in an easy number for x, like 1. If $ab = 15$, then when $b = x$, $ax = 15$. If $x = 1$, then $a = 15$. Find the answer choice that gives you 15. Only (C) does it.

18. **(D)** Quantities in direct variation always have the same proportion. In this case, that means that $\frac{n}{m} = \frac{5}{4}$, or 1.25. When $m = 5$, just solve the simple equation $\frac{n}{5} = 1.25$. Multiply both side by 5 and you'll find that $n = 6.25$.

24. **(A)** Direct variation means the proportion is constant, so that $\frac{p}{q} = \frac{3}{10}$, or 0.3. To find the value of p when $q = 1$, just solve the equation $\frac{p}{1} = 0.3$; p must be 0.3.

26. **(D)** Inversely proportional quantities have a constant product, so $rs = 21$, always. To find the value of r when $s = 4$, just solve the equation $4r = 21$. Divide each side by 4, and you find that $r = 5.25$.

Work and travel questions, p. 64

11. **(C)** The important thing to remember here is that when two things or people work together, their work-rates are added up. Pump #1 can fill 12 tanks in 12 hours, and Pump #2 can fill 11 tanks in 12 hours. That means that together, they could fill 23 tanks in 12 hours. To find the work they would do in 1 hour, just divide 23 by 12. You get 1.91666....

12. **(A)** To translate feet per second to miles per hour, take it one step at a time. First find the feet per hour by multiplying 227 feet per second by the number of seconds in an hour (3,600). You find that the

projectile travels at a speed of 817,200 feet per hour. Then divide by 5,280 to find out how many miles that is. You get 154.772.

18. **(B)** The train travels a total of 400 miles (round trip) in 5.5 hours. Now that you know distance and time, plug them into the formula and solve to find the rate: $400 = r \times 5.5$, so $r = 72.73$.

25. **(D)** Plug in! Say Jules can make 3 muffins in 5 minutes ($m = 3$, $s = 5$). Say Alice can make 4 muffins in 6 minutes ($n = 4$, $t = 6$). That means that Jules can make 18 muffins in 30 minutes, and Alice can make 20 muffins in 30 minutes. Together, they make 38 muffins in 30 minutes. That's your target number. Take the numbers you plugged in to the answers and find the one that gives you 38. Answer choice (D) does the trick.

Average speed, pp. 65–66

19. **(D)** Find the total distance and total time. The round trip distance is 12 miles. It takes $\frac{1}{2}$ hour to jog 6 miles at 12 mph, and $\frac{2}{3}$ hour to jog back at 9 mph, for a total of $1\frac{1}{6}$ hours. Do the division, and you get 10.2857 mph.

24. **(D)** This one is deceptively easy. Fifty miles in 50 minutes is a mile a minute, or 60 mph. Forty miles in 40 minute is also 60 mph. The whole trip is made at one speed, 60 mph.

33. **(B)** Plug in an easy number for the unknown distance, like 50 miles. It takes 2 hours to travel 50 miles at 25 mph, and 1 hour to return across 50 miles at 50 mph. That's a total distance of 100 miles in 3 hours, for an average speed of $33\frac{1}{3}$ mph.

Simultaneous equations, pp. 68–69

26. **(C)** Here, you want to make all of the b terms cancel out. Add the two equations, and you get $5a = 20$, so $a = 4$.

31. **(D)** Here, you need to get x and y terms with the same coefficient. If you subtract the second equation from the first, you get $10x - 10y = 10$, so $x - y = 1$.

34. **(D)** The question is solvable as the example on the previous page was, by multiplication. Multiplying all three equations together gives you $a^2b^2c^2 = 2.25$. Take the positive square root of both sides, and you get $abc = 1.5$.

37. **(E)** Here, you need to get rid of the z term and cancel out a y. The way to do it is to divide the first equation by the second one: $\frac{xyz}{y^2z} = \frac{4}{5}$. The z and a y cancel out, and you're left with $\frac{x}{y} = \frac{4}{5}$, or 0.8.

FOIL, p. 70

1. $x^2 + 9x - 22$
2. $b^2 + 12b + 35$
3. $x^2 - 12x + 27$
4. $2x^2 - 3x - 5$
5. $n^3 - 3n^2 + 5n - 15$
6. $6a^2 - 11a - 35$
7. $x^2 - 9x + 18$
8. $c^2 + 7c - 18$
9. $d^2 + 4d - 5$

Factoring quadratics, p. 71

1. $a = \{1, 2\}$
2. $d = \{-7, -1\}$

3. $x = \{-7, 3\}$
4. $x = \{-5, 2\}$
5. $x = \{-11, -9\}$
6. $p = \{-13, 3\}$
7. $c = \{-4, -5\}$
8. $s = \{-6, 2\}$
9. $x = \{-1, 4\}$
10. $n = \pm\sqrt{5}$

Special quadratic identities, pp. 72–73

17. **(A)** Use $(n - m)(n + m) = n^2 - m^2$.
19. **(B)** Use $(x + y)^2 = x^2 + 2xy + y^2$.
24. **(D)** Use $(x - y)^2 = x^2 + 2xy + y^2$.

The quadratic formula, p. 74

1. 2 real roots; $x = \{0.81, 6.19\}$
2. no real roots
3. 2 real roots; $s = \{0.76, 5.24\}$
4. 2 real roots; $x = \{-1.41, 1.41\}$
5. 1 real root; $n = -2.5$

5 PLANE GEOMETRY

Basic rules of lines and angles, p. 80–81

1. $x = 50°$; $y = 130°$; $z = 130°$;
   $a = 50°$; $b = 130°$; $c = 50°$
2. $x = 105°$; $y = 75°$
3. $a = 120°$; $b = 60°$; $c = 120°$;
   $d = 60°$; $e = 120°$; $f = 60°$
8. **(B)** When parallel lines are crossed by a third line, any small angle plus any big angle equals 180 degrees.

13. **(E)** Plug in! Remember that where the line intersects the parallel sides of the rectangle, any small angle plus any big angle is 180 degrees. Plug in, for example, 80° for the smaller angles s and v, and 100° for the larger angles, t and u. The angles r and w must measure 90° each. Once you've plugged in these value's you'll quickly see that only **(E)** must be equivalent to angle t.

16. **(E)** The parallel-lines theorem tells you that the sum of $\angle DBC$ and $\angle BDE$ is 180°. You don't, however, know anything about the *difference* between them. Plugging in various numbers should soon convince you that the difference cannot be determined.

Third-side rule, p. 83

12. **(E)** The length of the unknown side, ST, must be between the sum and difference of the other two lengths—that is, between 3 and 19. That makes the *perimeter* anywhere between 22 and 38.

17. **(A)** Since the triangle is isosceles, the unknown side must be either 5 or 11. But a 5-5-11 triangle violates the Third-Side Rule—it's not possible. That leaves only 11 as a possible value of the missing side.

18. **(E)** This is just another application of the Third-Side Rule. The third distance must be between 2 and 10. Only **(E)** violates the rule.

The Pythagorean theorem, p. 84

1. $x = \sqrt{89}$, or 9.434
2. $n = \sqrt{85}$, or 9.220
3. $a = \sqrt{15}$, or 3.873
4. $d = \sqrt{50} = 5\sqrt{2}$, or 7.071
5. $x = 7$
6. $r = \sqrt{145}$, or 12.042

Pythagorean triplets, p. 85

1. $x = 9$
2. $d = 26$
3. $n = 0.5$

The 45-45-90 triangle, p. 85

1. $x = 6\sqrt{2}$, or 8.485
2. $n = 1.5\sqrt{2}$, or 2.121
3. $s = \dfrac{7}{\sqrt{2}}$, or 4.950

The 30-60-90 triangle, p. 86

1. $x = 3\sqrt{3}$, or 5.196
2. $n = \dfrac{5}{\sqrt{3}}$, or 2.887
3. $d = 4\sqrt{3}$, or 6.928

Right triangle drills, pp. 86–87

7. **(A)** The Pythagorean Theorem tells you that the only possible length of AB is 11.4. There's only one possible perimeter.

13. **(E)** You can't use the Pythagorean Theorem unless you know which side of a right triangle is the hypotenuse. Since you don't know whether 8 or the unknown side is the hypotenuse, it's impossible to know the length of the missing side.

16. **(C)** Draw it! You'll find that the floor, wall, and ladder form a 30-60-90 triangle. The ladder itself is the hypotenuse; the distance from the ladder's foot to the wall is the short leg, $\frac{1}{2}$ the length of the hypotenuse; multiply by $\sqrt{3}$ to get the other leg of the triangle, the height of the ladder.

19. **(B)** An isosceles right triangle must be a 45-45-90 triangle, with sides of x, x, and $x\sqrt{2}$. Since the perimeter equals 23.9, that means $2x + x\sqrt{2} = 23.9$. When solved, this tells you that $x = 7$, and the area of the triangle $= \frac{1}{2}x^2$, or 24.5.

Similar triangles, pp. 88–90

1. $x = 2$; $y = 1.73$; $a = 4$; $b = 8$
2. $a = 5$; $s = 15$
3. $c = 11$; $m = 12.25$

37. **(A)** These triangles are similar because they have identical angles. If one length of triangle ABC is half of a corresponding length of triangle FGH, that means that *all* of the lengths in ABC will be half the corresponding lengths in FGH. The easiest way to solve the problem is by plugging in a base and height for FGH which would yield an area of 0.5; a base of 1 and a height of 1 would be the easiest. Then just plug in the corresponding dimensions of ABC, making sure that they're half of the dimensions in triangle FGH. The area comes out to $\frac{1}{8}$, or 0.125, which rounds to 0.13.

40. **(C)** This question is much like question #37, except that you have more freedom to plug in numbers. Once again, the triangles are similar; Fred's Theorem tells you that they have identical angles. Then just plug in easy numbers. Suppose AD = 4, and DE = 2. This shows you that the lengths of the smaller triangle are $\frac{2}{3}$ the lengths of the larger. Suppose AE = 6, and AC = 9. The area of triangle ADE would then be 12, while the area of triangle ABC would be 27. Twelve is $\frac{4}{9}$ of 27.

45. **(A)** Use the Pythagorean Theorem to complete the dimensions of triangle LMN: It has sides of length 4, 8, and $\sqrt{48}$. If you simplify $\sqrt{48}$, you'll find it's equal to $4\sqrt{3}$. That makes the lengths 4, 8, and $4\sqrt{3}$, which should look familiar: It's a 30-60-90 triangle. Since a right triangle divided by an altitude from its right angle forms three similar triangles, the little triangle LPN must also be a 30-60-90 triangle. If its hypotenuse has a length of 4, then its legs have lengths of 2 and $2\sqrt{3}$. Those legs are the base and height of triangle LPN. Use them to calculate the triangle's area, and you get 3.464.

Area of triangles, pp. 91–92

9. **(E)** Just plug the values given into the formula for the area of a triangle, and solve for a.

15. **(D)** Both triangles have the same height, and if their areas are equal, they must have the same base as well. Since triangle OAB has a base of 8, triangle ADC must also have a base of 8.

37. **(B)** Before you go crazy with your right-triangle rules, notice that you can compute the area of this triangle two ways; with AC as the base or with BC as the base. Either way, the base and height must multiply to the same number, since a triangle can have only one area. That means that 12 times BE must equal 9 times 10.

38. **(C)** If the perimeter of an equilateral triangle is 24, then each side has a length of 8. Plug that in to the formula for the area of an equilateral triangle, and you get 27.71.

44. **(D)** Just set the formula for the area of an equilateral triangle equal to 12, and solve for s. You get 5.26. The perimeter is $3s$, or 15.79.

Quadrilaterals, pp. 95–96

22. **(A)** Use the cool $\frac{d^2}{2}$ formula for the area of a square.

34. **(D)** You can use the proportions of the 30-60-90 triangle to find the height of the parallelogram; it's $3\sqrt{3}$, or 5.196. Then just multiply by the base, and you get 51.96.

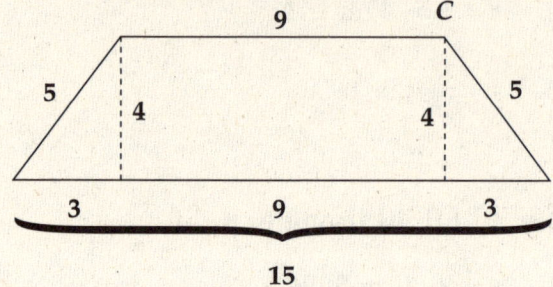

40. **(B)** Draw a vertical line down from C, and you've formed a 3-4-5 triangle. The bases of this trapezoid have lengths of 9 and 15, and the height is 4. Plug those numbers into the formula for a trapezoid's area, and you get 48.0.

45. **(B)** Each diagonal cuts the rectangle into two 30-60-90 triangles, with the diagonal as hypotenuse. That makes the base and height of the rectangle the x and $x\sqrt{3}$ legs of the triangle. If $x \cdot x\sqrt{3} = 62.35$, then $x = 6.0$, and the hypotenuse of the triangle ($2x$) has a length of 12.0. The sum of the two diagonals is 24.0.

Circles, pp. 98 and 100–101

1. $C = 8\pi; A = 16\pi$
2. $C = 15.85; r = 2.52$
3. $A = 5.09; r = 1.27$
12. **(D)** Remember that a sector of a circle takes equal portions of angle, circumference, and area. 4.71 is one fourth of the circle's circumference of 6π, or 18.85. That means it also takes one-fourth of 360°, or 90°.

29. **(A)** The shaded region is a quarter-circle plus a 45-45-90 triangle. The triangle has an area of $\frac{25}{2}$, or 12.5. The quarter-circle has an area of 19.64. They add up to 31.24.

31. **(B)** As in most shaded-area problems, the trick is to subtract one shape from another to get the shaded region itself. Here, you need to subtract the hexagon's area from that of the circle. To compute the area of the hexagon, divide it into 6 equilateral triangles, each with a side of 4—use the equilateral triangle formula to find their areas. Then subtract the hexagon's area (41.57) from the circle's area (50.27) to get the answer, 8.7.

43. **(C)** The triangle OAC must be equilateral, so all of its angles measure 60°. The radius OA must be perpendicular to line l, making a 90° angle; that makes triangle OAC a 30–60–90 triangle. The length of AB is $2\sqrt{3}$, or 3.46.

45. **(A)** One statement at a time: I must be true, since BC is the hypotenuse of a right triangle, and must be longer than the triangle's legs. Choice II *could* be true, but doesn't *have* to be; and III is impossible, since l and m must be parallel.

6 SOLID GEOMETRY

Triangles in rectangular solids, pp. 111–112

32. **(D)** Use the formula for the long diagonal of a cube. Given that the cube's edge is 3, the cube's long diagonal must be 5.2.

36. **(E)** Be careful here; you won't be using the "long diagonal" formula. The sides of this triangle are the diagonals of three of the solid's faces. Just use the Pythagorean Theorem three times and add up the results.

39. **(A)** This is a long-diagonal question, with a twist. Each edge of the cube is 1, but you're actually finding the long diagonal of a quarter of the cube. Think of it as finding the long diagonal of a rectangular solid of dimensions $1 \times \frac{1}{2} \times \frac{1}{2}$. Plug those three numbers into the Super Pythagorean Theorem.

Volume, pp. 112–113

17. **(D)** Backsolve! Quickly move through the answer choices (starting in this case with the smallest, easiest numbers), calculating the volumes and surface areas of each. The only answer choice that makes these quantities equal is (D).

24. **(E)** This one's a pain. The only way to do it is try out the various possibilities. The edges of the solid must be three factors which multiply to 30, such as 2, 3, and 5. That solid would have a surface area of 62. The solid could also have dimensions of 1, 5, and 6, which would give it a surface area of 82. Keep experimenting, and you find that the solid with greatest surface area has the dimensions 1, 1, and 30, giving it a surface area of 122.

43. **(B)** The sphere has a volume of 4.19. When submerged, it will push up a layer of water having equal volume. The volume of this layer of water is the product of the area of the circular surface (50.27) and the height to which it's lifted—it's like calculating the volume of a very flat cylinder. You get this equation: 50.27h = 4.19. Solve, and you find that h = 0.083.

Volume and plugging in, pp. 113–114

17. **(B)** If the cube's surface area is 6x, then x is the area of one face. Pick an easy number, and plug in! Suppose x is 4. That means that the length of any edge of the cube is 2, and that the cube's volume is 8.

ANSWERS AND EXPLANATIONS ◆ 245

Just plug $x = 4$ into all of the answer choices, and find the one that gives you 8. (B) does the trick.

36. **(C)** Plug in! Suppose the sphere's original radius $r = 2$, which would give it a surface area of 50.3. Suppose the radius is then increased by $b = 1$, making the new radius 3. The sphere would then have a surface area of 113.1. That's an increase of 62.8. Plug $r = 2$ and $b = 1$ into the answer choices; the one that gives you 62.8 is correct. That's (C).

40. **(B)** Just plug in any values for b and h that obey the proportion $b = 2h$. Then just plug those values into the formula for the volume of a pyramid.

Inscribed solids, pp. 115–116

32. **(B)** The long diagonal of the rectangular solid is the diameter of the sphere. Just find the length of the long diagonal and divide it in half.

35. **(C)** Calculate the volume of each shape separately. The cube's volume is 8; the cylinder, with a radius of 1 and a height of 2, has a volume of 6.28. The difference between them is 1.72.

38. **(B)** The cube must have the dimensions 1 by 1 by 1. That means that the cone's base has a radius of 0.5, and that the cone's height is 1. Plug these numbers into the formula for the volume of a cone, and you should get 0.26.

Rotation solids, pp. 117–118

34. **(C)** This rotation will generate a cylinder with a radius of 2 and a height of 5. Its volume is 62.83.

39. **(D)** This rotation will generate a cone lying on its side, with a height of 5 and a radius of 3. Its volume is 47.12.

46. **(D)** Here's an odd one. The best way to think about this one is as two triangles, base to base, being rotated. The rotation will generate two cones places base-to-base, one right-side-up and one upside-down. Each cone has a radius of 3 and a height of 3. The volume of each cone is 28.27. The volume of the two together is 56.55.

Changing dimensions, p. 119

13. **(E)** If the factor separating the lengths of these solids is $\frac{1}{3}$, then the factor separating their volumes will be the cube of that, or $\frac{1}{27}$.

18. **(D)** If the factor separating the lengths of these solids is $\frac{1}{2}$, then the factor separating their volumes will be the cube of that, or $\frac{1}{8}$. One-eighth of 24 is 3.

21. **(B)** If the surface area increases by a factor of 2.25, then the lengths of the edges of the cube increase by the square root of that, or 1.5. The volume increases by the cube of this factor, or 3.375.

7 COORDINATE GEOMETRY

The coordinate plane, p. 125

1. Point E, quadrant II
2. Point A, quadrant I
3. Point C, quadrant IV
4. Point D, quadrant III
5. Point B, quadrant I

The equation of a line, pp. 126–128

7. **(A)** You can plug the line's slope and the given point into the point-slope equation of a line: $y - 1 = \frac{3}{5}(x - 3)$. You can then simplify that into the shorter

slope-intercept form: $y = \frac{3}{5}x - \frac{4}{5}$. To find the point that is also on this line, go to each answer choice and plug the x-coordinate into the formula. You'll have the right answer when the formula produces a y-coordinate that matches the given one.

10. **(E)** Once again, get the line into slope-intercept form: $y = 5x - 4$. Then plug in zero. You get a y-value of −4.

11. **(B)** Put the line into the slope-intercept formula by isolating y. The coefficient of x is then the slope of the line.

19. **(D)** You can figure out the line formula ($y = mx + b$) from the graph. The line has a y-intercept (b) of −2, and it rises 6 as it runs 2, giving it a slope (m) of 3. Use those values of m and b to test the statements in the answer choices.

23. **(A)** Once again, the slope-intercept formula is your most powerful tool. Isolate y, and you get $y = -3x + 5$. The line must then have a slope of −3 and a y-intercept of 5. Only (A) and (D) show lines with negative slope, and the line in (D) has a slope which is between −1 and 0, because it forms an angle with the x-axis that is less than 45°.

More about slope, p. 130

4. **(C)** Use the slope formula on the point (0,0) and (−3,2).

17. **(D)** Draw it. Remember that perpendicular lines have slopes that are negative reciprocals of each other. A line containing the origin and the point (2,−1) has a slope of $-\frac{1}{2}$. The perpendicular line must then have a slope of 2. Quickly move through the answer choices, determining the slope of a line passing through the given point and the origin. The one that gives a slope of 2 is correct.

47. **(D)** Remember that perpendicular lines have slopes that are negative reciprocals of each other. The slope x and y are therefore negative reciprocals—you can think of them as x and $-\frac{1}{x}$. The difference between them will therefore be the sum of a number and its reciprocal: $x - \left(-\frac{1}{x}\right) = x + \frac{1}{x}$. If $x = 1$, then the sum of x and its reciprocal is 2; if $x = 5$, then the sum of x and its reciprocal is 5.2; but no sum of a number and its reciprocal can be less than 2; a sum of 0.8 is impossible.

Line segments, p. 132

12. **(D)** Use the distance formula on the points (−5, 9) and (0, 0).

19. **(E)** Plug the points you know into the midpoint formula. The average of −4 and the x-coordinate of B is 1; the average of 3 and the y-coordinate of B is −1. That makes B the point (6, −5).

27. **(D)** You'll essentially be using the distance formula on (2, 2) and the points in the answer choices. (−5, −3) is the point at the greatest distance from (2, 2).

General equations (parabolas), p. 135

34. **(C)** This is a quadratic function, which always produces a parabola. If a parabola has a maximum or minimum, then that extreme value is the parabola's vertex. Just find the vertex: The x-coordinate is $-\frac{b}{2a}$, which is 3 in this case. The y-coordinate will be $f(3)$, or −1.

37. **(B)** Use that vertex formula again. The x-coordinate is $-\frac{b}{2a}$, which is −1 in this

case. That's enough to get you the right answer. (If you needed the y-coordinate as well, you'd just find $f(-1)$.)

38. **(D)** At every point on the x-axis, $y = 0$. Just set y equal to zero in the equation and solve for x. You'll find that x has two possible values, 1 and 5, only one of which is present in the answer choices.

General equations (circles), pp. 136–137

30. **(B)** Just plug each point in to the equation. The one that does *not* make the equation true is not on the circle.

33. **(A)** Get the circle into its general form, and then the center will be obvious. The general form of the equation is $(x-h)^2 + (y-k)^2 = r^2$. In its general form, this equation looks like this: $(x+3)^2 + (y-2)^2 = 25$. The circle's center, (h, k), is $(-3, 2)$.

34. **(E)** If S and T are the endpoints of a diameter, then the distance between them is 8. If they are very close to each other on the circle, then the distance between them approaches zero; the distance between S and T cannot be determined.

General equations (ellipses), p. 138

15. **(E)** When the equation of an ellipse takes this form, you know that the center of the ellipse is at the origin, and the major and minor axes of the ellipse lie along the x- and y- axes. This makes everything much easier. By setting x equal to zero and solving for y, and then setting y equal to zero and solving for x, you can easily find the end points of both axes and sketch the ellipse. You'll find that when $x = 0$, $y = -5$ or 5, and when $y = 0$, $x = -4$ or 4. Here's a sketch of the ellipse.

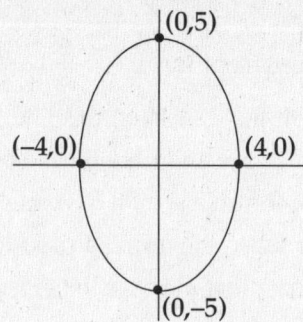

As you can see, the vertical axis is the longer one. It stretches (0,5) to (0,–5), and must therefore have a length of 10.

40. **(C)** For an ellipse in its general form, the center is (h, k), which in this case is $(-5, 3)$.

43. **(B)** This time, you've got to get the equation into its general form. Factor the 9 out of the x-terms, and the 4 out of the y-terms. Complete the squares inside the parentheses, being careful to add corresponding numbers to the other side of the equation: $9(x^2 + 2x + 1) + 4(y^2 + 4y + 4) = 36$. Express the trinomials as the squares of binomials: $9(x - 1)^2 + 4(y + 2)^2 = 36$. Then divide everything by 36 (4×9) to get rid of the coefficients. Then complete the squares in parentheses. In its general form, the equation looks like this: $\dfrac{(x-1)^2}{4} + \dfrac{(y+2)^2}{9} = 1$. The center is $(1, -2)$.

General equations (hyperbolas), pp. 139–140

38. **(B)** Like circles and ellipses, hyperbolas in general form have their centers at (h, k). This one is centered at $(-4, -5)$.

43. **(A)** First, get the equation into its general form: $\dfrac{(x-3)^2}{1} - \dfrac{(y+3)^2}{4} = 1$. The center is then (h, k)—which is $(3, -3)$.

Triaxial coordinates, p. 141

29. **(C)** This is once again a job for the Super Pythagorean Theorem; it's just like finding the long diagonal of a box which is 5 by 6 by 7. Set up this equation: $d^2 = 5^2 + 6^2 + 7^2$, and solve.

34. **(B)** A point will be outside the sphere if the distance between it and the origin is greater than 6. Use the Super Pythagorean Theorem to measure the distance of each point from the origin.

8 TRIGONOMETRY

Trig functions in right triangles, pp. 147–148

1. $\sin\theta = \dfrac{3}{5} = 0.6$; $\cos\theta = \dfrac{4}{5} = 0.8$; $\tan\theta = \dfrac{3}{4} = 0.75$

2. $\sin\theta = \dfrac{5}{13} = 0.385$; $\cos\theta = \dfrac{12}{13} = 0.923$; $\tan\theta = \dfrac{5}{12} = 0.417$

3. $\sin\theta = \dfrac{24}{25} = 0.96$; $\cos\theta = \dfrac{7}{25} = 0.28$; $\tan\theta = \dfrac{24}{7} = 3.429$

4. $\sin\theta = \dfrac{6}{10} = 0.6$; $\cos\theta = \dfrac{8}{10} = 0.8$; $\tan\theta = \dfrac{6}{8} = 0.75$

Completing triangles, pp. 149–150

1. AB = 3.381; CA = 7.25; $\angle B = 65°$
2. EF = 2.517; FD = 3.916; $\angle D = 40°$
3. HJ = 41.41; JK = 10.72; $\angle J = 75°$
4. LM = 5.736; MN = 8.192; $\angle N = 35°$
5. TR = 4.0; $\angle S = 53.13°$; $\angle T = 36.87°$
6. YW = 13; $\angle W = 22.62°$; $\angle Y = 67.38°$

Trigonometric identities, pp. 152–153

1. **(B)** Express $\tan x$ as $\dfrac{\sin x}{\cos x}$. The cosine then cancels out, and you're left with $(\sin x)(\sin x)$, or $\sin^2 x$.

2. **(D)** Use FOIL on these binomials, and you get $1 - \sin^2 x$. Because $\sin^2 x + \cos^2 x = 1$, you know that $1 - \sin^2 x = \cos^2 x$.

3. **(C)** Express $\tan x$ as $\dfrac{\sin x}{\cos x}$. The cosine then cancels out on the top of the fraction, and you're left with $\dfrac{\sin x}{\sin x}$, or 1.

4. **(A)** The term $(\sin x)(\tan x)$ can be expressed as $(\sin x)\left(\dfrac{\sin x}{\cos x}\right)$, or $\dfrac{\sin^2 x}{\cos x}$. The first and second terms can then be combined, like this:
$$\dfrac{1}{\cos x} - \dfrac{\sin^2 x}{\cos x} = \dfrac{1 - \sin^2 x}{\cos x}.$$
Because $1 - \sin^2 x = \cos^2 x$, this expression simplifies to $\cos x$.

5. **(E)** Break the fraction into two terms, like this: $\dfrac{\tan x}{\tan x} - \dfrac{\sin x \cos x}{\tan x}$. The first term simplifies to 1, and the second term becomes easier to work with when you express the tangent in terms of the sine and cosine:
$$\dfrac{\sin x \cos x}{\tan x} = \dfrac{\sin x \cos x}{\dfrac{\sin x}{\cos x}} = \cos^2 x.$$

The whole expression then equals $1 - \cos^2 x$, or $\sin^2 x$.

Other trig functions, p. 154

19. **(E)** Express the function as a fraction: $\dfrac{1}{\cos^2 x} - 1$. You can then combine the terms by changing the form of the second term: $\dfrac{1}{\cos^2 x} - \dfrac{\cos^2 x}{\cos^2 x}$. This allows you to combine the terms, like this: $\dfrac{1 - \cos^2 x}{\cos^2 x} = \dfrac{\sin^2 x}{\cos^2 x} = \tan^2 x$.

23. **(D)** Express the cotangent as a fraction, like this: $\dfrac{1}{\sin x \cot x} = \dfrac{1}{\sin x \dfrac{\cos x}{\sin x}}$. The $\sin x$ then cancels out, leaving you with $\dfrac{1}{\cos x}$, or $\sec x$.

24. **(A)** Express the cotangent as a fraction, and the second term can be simplified, like this:

$$(\cos x)(\cot x) = \cos x \dfrac{\cos x}{\sin x} = \dfrac{\cos^2 x}{\sin x}.$$

Express both terms as fractions, and the terms can be combined:

$$\dfrac{\sin^2 x}{\sin x} + \dfrac{\cos^2 x}{\sin x} = \dfrac{\sin^2 x + \cos^2 x}{\sin x} = \dfrac{1}{\sin x},$$

or $\csc x$.

Angle equivalencies, p. 159

18. **(A)** Draw the unit circle. $-225°$ and $135°$ are equivalent angle measures, because they are separated by $360°$.

21. **(D)** Draw the unit circle. $300°$ and $60°$ are not equivalent angles, but they have the same cosine: It's a simple matter to check with your calculator.

26. **(B)** Use your calculator!

30. **(C)** Use your calculator!

36. **(D)** Draw the unit circle. If the product of the sine and cosine is negative, then the sine and cosine must have different signs—one positive and one negative. The signs of the sine and cosine differ in quadrants II and IV, which are described by **(D)**.

Degrees and Radians, p. 161

Degrees	Radians
30°	$\dfrac{\pi}{6}$
45°	$\dfrac{\pi}{4}$
60°	$\dfrac{\pi}{3}$
90°	$\dfrac{\pi}{2}$
120°	$\dfrac{2\pi}{3}$
135°	$\dfrac{3\pi}{4}$
150°	$\dfrac{5\pi}{6}$
180°	π
225°	$\dfrac{5\pi}{4}$
240°	$\dfrac{4\pi}{3}$
270°	$\dfrac{3\pi}{2}$
300°	$\dfrac{5\pi}{3}$
315°	$\dfrac{7\pi}{4}$
330°	$\dfrac{11\pi}{6}$
360°	2π

Non-right triangles, pp. 165–166

1. $a = 8.25$, $B = 103.4°$, $C = 34.6°$
2. $A = 21.79°$, $B = 120.0°$, $C = 38.21°$
3. $c = 9.44$, $B = 57.98°$, $C = 90.02°$

Polar coordinates, p. 168

39. **(C)** The x-coordinate of the point is $6 \cos \frac{\pi}{3}$, or 3. The y-coordinate is $6 \sin \frac{\pi}{3}$, which is 5.196.

42. **(B)** The y-value of a point is its distance from the x-axis. The y-coordinate of this point is $7 \sin \frac{3\pi}{4}$, which equals 4.949.

45. **(B)** In rectangular coordinates, A, B, and C have x-coordinates of 3. This means that they are placed in a straight vertical line. They therefore define a straight line, but not a plane or space.

9 FUNCTIONS

Basic functions, pp. 174–175

14. **(E)** $f(-1) = (-1)^2 - (-1)^3 = 1 - (-1) = 1 + 1 = 2$.
17. **(B)** $f(7) = 10.247$. $f(8) = 11.314$. That's a difference of 1.067.
26. **(B)** $g(3) = 3^3 + 3^2 - 9(3) - 9 = 0$.
29. **(D)** $f(3, -6) = \frac{3(-6)}{3+(-6)} = \frac{-18}{-3} = 6$
30. **(A)** Make the function equal to 10, and solve for n: $h(n) = n^2 + n - 2 = 10$. This can be rearranged to $n^2 + n - 12 = 0$, which can in turn be factored to $(n + 4)(n - 3) = 0$. There are two possible values of n: 3 and -4, only one of which is among the answer choices.
33. **(E)** The greatest factor of 75 not equal to 75 is 25. Therefore, $f(75) = 75 \cdot 25 = 1875$.

34. **(E)** If $y = 3$, then $g(-y) = g(-3)$. Because $-3 < 0$, $g(-3) = 2 |-3| = 2(3) = 6$.

Compound functions, p. 177

17. **(D)** The compound function $f(g(x))$ equals $3(x + 4)$, or $3x + 12$. The compound function $g(f(x))$ equals $3x + 4$. There's a difference of 8 between them.

24. **(E)** To evaluate $f(g(-2))$, first find the value of $g(-2)$, which equals $(-2)^3 - 5$, or -13. Then put that result into $f(x)$: $f(-13) = |-13| - 5 = 13 - 5 = 8$.

32. **(D)** To evaluate the compound function $g(f(x))$ algebraically, take the definition of $f(x)$ and put it into $g(x)$: $g(f(x)) = \sqrt{x^2 + 10x + 25} + 4$

$g(f(x)) = \sqrt{(x+5)^2} + 4$

$g(f(x)) = x + 5 + 4 = g(f(x)) = x + 9$.

36. **(C)** $f(g(3)) = 5$. $g(f(3)) = 3.196$. The difference between them is 1.804.

Inverse functions, p. 179

22. **(B)** Plug a number into $f(x)$. For example, $f(2) = 1.5$. The correct answer is the function that turns 1.5 back into 2. (B) does the trick.

33. **(B)** $f(x) = 4x^2 - 12x + 9$, which equals $(2x - 3)^2$. The inverse function will reverse the steps of $f(x)$: where $f(x)$ multiplies x by 2, subtracts 3, and squares the result, the inverse function must take the square root of x, add 3, and divide by 2.

$f^{-1}(x) = \frac{\sqrt{x}+3}{2}$. Therefore

$f^{-1}(9) = \frac{\sqrt{9}+3}{2} = \frac{6}{2} = 3$

35. **(E)** The fact that $f(3) = 9$ doesn't tell you what $f(x)$ is. It's possible that $f(x) = x^2$, or that $f(x) = 3x$, or that $f(x) = 2x + 3$, and so on. Each of these functions would have a different inverse function. The definition of the inverse function cannot be determined.

Domain and range, pp. 184–185

24. **(A)** This function factors to $f(x) = \dfrac{1}{x(x-3)(x+2)}$. Three values of x will make this fraction undefined: –2, 0, and 3. The function's domain must exclude these values.

27. **(E)** This function factors to $g(x) = \sqrt{(x+2)(x-6)}$. The product of these binomials must be nonnegative (that means positive or zero), since a square root of a negative number is not a real number. The product will be nonnegative when both binomials are negative ($x \leq -2$) or when both are nonnegative ($x \geq 6$). The function's domain is $\{x : x \leq -2 \text{ or } x \geq 6\}$.

30. **(D)** Take this one step at a time. Because a number raised to an even power can't be negative, the range of a^2 is the set of nonnegative numbers—that is, $\{y : y \geq 0\}$. The range of a^2+5 is found by simply adding 5 to the range of a^2: $\{y : y \geq 5\}$. Finally, to find the range of $\dfrac{a^2+5}{3}$, divide the range of a^2+5 by 3: $\left\{y : y \geq \dfrac{5}{3}\right\}$, or $\{y : y \geq 1.67\}$. The correct answer is (D).

34. **(D)** Because this is a linear function (without exponents) you can find its range over the given interval by plugging in the bounds of the domain. $f(-1) = -1$, and $f(4) = 19$. Therefore the range of f is $\{y: y \ -1 \leq y \leq 19\}$

Identifying graphs of functions, pp. 188–189

9. **(D)** It's possible to intersect the graph shown in (D) twice with a vertical line, where the point duplicates an x-value on the curve.

15. **(B)** It's possible to intersect the graph shown in (B) more than once with a vertical line, at each point where the graph becomes vertical.

Range and domain in graphs, pp. 191–192

17. **(A)** The graph has a vertical asymptote at $x = 0$, so 0 must be excluded from the domain of f.

24. **(D)** Only two x-values are absent from the graph, $x = 2$ and $x = -2$. The domain must exclude these values. This can be written as $\{x: x \neq -2, 2\}$ or $\{x: |x| \neq 2\}$.

28. **(C)** The graph extends upward forever, but never goes lower than –3. Its range is therefore $\{y: y \geq -3\}$.

Roots of functions, pp. 193–194

16. **(D)** Use the quadratic formula to find the two possible values of x. Only one is represented in the answer choices.

19. **(C)** The function $g(x)$ can be factored this way: $g(x) = x(x + 3)(x - 2)$. Set this function equal to zero and solve for x: you'll find the function has three distinct roots: –3, 0, and 2.

25. **(D)** The roots of a function are the x-values at which $f(x) = 0$. In short, the roots are the x-intercepts—in this case, –4, –1, and 2.

Symmetry in functions, pp. 196–197

6. **(D)** "Symmetrical with respect to the x-axis" means reflected as though the x-axis were a mirror.

17. **(E)** An even function is one for which $f(x) = f(-x)$. This is true by definition of an absolute value. Confirm by plugging in numbers.

22. **(D)** Radial symmetry is the same thing as origin symmetry. A function has origin symmetry when $-f(x) = f(-x)$.

Degrees of functions, p. 199

31. **(E)** The graph shown has four relative extreme values: two local maxima (the "hills" on the graph) and two local minima (the "valleys" on the graph). A function with four relative extreme values must be at least a fifth-degree function. The degree of a function is determined by its greatest exponent. Only the function in answer choice (E) is at least a fifth-degree function.

35. **(D)** Since the degree of a function is determined by its greatest exponent, all you need to do in order to find the fourth-degree function is figure out the greatest exponent in each answer choice when it's multiplied out. Remember, you don't need to do all of the algebra; just see what the greatest exponent will be. Answer choice (A) is a second-degree function, because its highest-order term is x^2. Answer choices (B) and (C) are third-degree functions, because the highest-order term in each function is x^3. Answer choice (E) is a fifth-degree function, since $x \cdot x \cdot x^3 = x^5$. Only answer choice (D) is a fourth-degree function.

10 STATISTICS AND SETS

Statistics, p. 203

21. **(B)** Not counting 1, 21 and 48 have only one factor in common—3. The only duplicated element in set C would be 3; therefore 3 would be the mode.

25. **(B)** If the sum of a set's elements is zero, then the mean must also be zero. It's impossible to know what the median is; the set could be {0, 0, 0, 0, 0, 0, 0, 0, 0, 0} or {–9, 1, 1, 1, 1, 1, 1, 1, 1, 1}. Both sets add up to zero, but have different medians. The mode is not necessarily zero for the same reasons.

Probability, pp. 206–207

13. **(A)** There are only two things that can happen, rain or not rain. If 5 out of 12 possible outcomes mean rain, then the other 7 of the 12 possible outcomes must mean no rain.

16. **(A)** The probability that two events will occur together is the product of the chances that each will happen individually. The probability that these two events will happen together is $\frac{2}{5} \times \frac{1}{3}$, or $\frac{2}{15}$.

20. **(B)** Out of a total of 741 cookies, 114 are burned. The probability of getting a burned cookie is therefore $\frac{114}{741}$. That reduces to $\frac{1}{6.5}$, which is equivalent to $\frac{2}{13}$.

24. **(B)** For the product of the numbers to be odd, both numbers must be odd themselves. There's a $\frac{3}{6}$ chance of getting an odd number on each die. The odds of getting odd numbers on both dice are $\frac{3}{6} \times \frac{3}{6} = \frac{9}{36} = \frac{1}{4}$.

44. **(C)** This one's pretty tricky. It's difficult to compute the odds of getting "at least one basket" in three tries, since there are so many different ways to do it (basket-basket-basket, miss-miss-basket, basket-basket-miss, and so on). It's a simpler solution to calculate the odds of Heather's

missing all three times. If the probability of her making a basket on any given try is $\frac{4}{5}$, then the probability of her missing is $\frac{1}{5}$. The probability of her missing three times in a row is $\frac{1}{5} \times \frac{1}{5} \times \frac{1}{5}$, or $\frac{1}{125}$. That means that Heather makes no baskets in 1 out of 125 possible outcomes. The other 124 possible outcomes must involve her making at least 1 basket.

Permutations and combinations, pp. 210–211

27. **(B)** All committee questions are combination questions, because different arrangements of the same people don't count as different committees. The number of permutations of 12 items in 4 spaces is 12 · 11 · 10 · 9, or 11,880. To find the number of combinations, divide this number by 4 · 3 · 2 · 1, or 24. You get 495.

32. **(D)** The number of permutations of 6 items in 6 spaces is 6 · 5 · 4 · 3 · 2 · 1, or 720.

45. **(A)** Compute the number of combinations of the females and males separately. The number of combinations of 17 females in 4 spaces is $\frac{17 \cdot 16 \cdot 15 \cdot 14}{4 \cdot 3 \cdot 2 \cdot 1}$, or 2,380. The number of combinations of 12 males in 3 spaces is $\frac{12 \cdot 11 \cdot 10}{3 \cdot 2 \cdot 1}$, or 220. The total number of combinations is the product of these two numbers, 220 · 2,380, or 523,600.

Group questions, p. 212

17. **(D)** If 253 students out of 370 did not make the honor roll, then 117 students did make the honor roll. If 23 honor-roll students got straight A's, then 94 did not.

28. **(C)** Remember the group-problem formula: Total = Group 1 + Group 2 + Neither − Both. Then plug in the number from the question: T = 14 + 12 + 37 − 9. The total T therefore equals 54.

42. **(B)** Plug in a number of tourists that you can easily take $\frac{1}{3}$, $\frac{2}{5}$, and $\frac{1}{2}$ of—like 30. If the total number of tourists is 30, then 10 speak Spanish, 12 speak French, and 15 speak neither language. Once again, plug these numbers into the group formula: 30 = 10 + 12 + 15 − N. This simplifies to 30 = 37 − N, so N = 7. Seven tourists speak neither Spanish nor French. That's $\frac{7}{30}$ of the whole group.

Union and intersection, pp. 213–214

16. **(B)** The intersection of F and G is the set {2, 4, 6, 8}. That's set H. The mean of the set is 5.

17. **(C)** The union of sets F and G is {1, 2, 2, 3, 4, 4, 5, 6, 6, 7, 8, 8, 10, 12}. That's now set H. The mean of this set is 5.57.

34. **(D)** The statement that $S \cap T = \emptyset$ means that none of the members of set S are members of set T. The mean cannot be 6.2, since that would make x equal to 7, and 7 is not a member of set T, since there is no intersection between set S and set T.

11 MISCELLANEOUS

Arithmetic and geometric series, pp. 220–221

14. **(E)** An arithmetic series is formed by adding a value again and again to an original term. From the second term to a sixth term is 4 steps; going from 4 to 32 is a change of 28 in four steps, making each

step an increase of 7. The fifth term must then be 7 less than the sixth term, or 25.

19. **(E)** Going from the first to the seventh term of a series is 6 steps, and going from 2 to 16 is a difference of 14. That makes each step an increase of $\frac{7}{3}$, or 2.33. To find the thirty-third term, plug these numbers into the formula for the nth term of an arithmetic series: $a_{33} = 2 + (33-1)\frac{7}{3}$ $= 2 + 74\frac{2}{3} = 76.67$.

26. **(D)** From the second term of a series to the fourth is two steps—that is, 4 is multiplied by the factor r twice to get to 25. That means that $4 \cdot r^2 = 25$. Solve for r and you'll find that $r = 2.5$. Just punch 4 into your calculator as the second term, and keep multiplying by 2.5 until you've counted up to the ninth term. OR figure out that the first term in the series is $\frac{4}{2.5}$, or 1.6. Then just plug those values into the formula for the nth term of a geometric series: $a_9 = 1.6 \cdot 2.5^8 = 2441.41$.

34. **(C)** Because no end term is given, this is an infinite geometric series. It's decreasing, not increasing, so its sum is finite. Its first term is 3, and each successive term is multiplied by a factor of $\frac{1}{3}$. Plug those values into the formula for the sum of an infinite geometric series:
$$sum = \frac{3}{1-\frac{1}{3}} = \frac{3}{\frac{2}{3}} = 4.5.$$

Limits, pp. 222–223

30. **(B)** This expression factors into $\frac{(4x-5)(x+1)}{(4x-5)(4x+5)}$. The binomial $4x-5$ cancels out, leaving you with $\frac{(x+1)}{(4x+5)}$. This expression is no longer undefined when x equals 1.25, so just plug 1.25 into the expression—the result is the limit.

38. **(A)** This expression factors into $\frac{(x+4)(x-3)}{2x(x-3)}$. The binomial $(x-3)$ cancels out, leaving you with $\frac{x+4}{2x}$. This expression is no longer undefined when $x = 3$, so just plug $x = 3$ into the expression to obtain the limit.

40. **(E)** This expression factors into $\frac{x(x+7)(x-3)}{(x+7)(x+3)}$. The binomial $x+7$ factors out, leaving you with $\frac{x(x-3)}{(x+3)}$. Notice, however, that the expression is *still* undefined when $x = -3$. The limit remains undefined and does not exist.

Vectors, pp. 224–225

18. **(B)** The vectors in the answer choices are all very different, so all you need to do is get a basic idea of what the sum of vectors p and q would look like. Do that by sketching the addition:

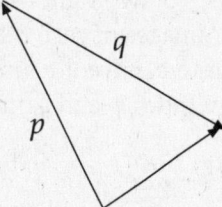

Only answer choice (B) looks anything like the result. That's the correct answer.

36. **(D)** Add the two vectors by sketching:

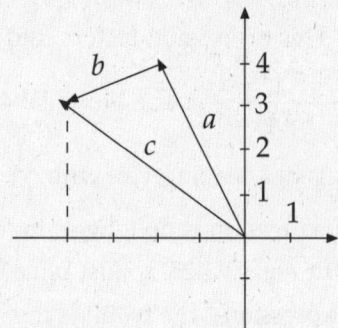

The resulting vector travels four units to the right from the origin, and three units up. That makes the vector the hypotenuse of a 3-4-5 triangle. The vector's length is 5, and that makes its magnitude 5 as well.

39. **(E)** The most important piece of information given to you in this question is the equation $m^2 + n^2 = p^2$. Two vectors and their sum can be thought of as the sides of a triangle. The equation $m^2 + n^2 = p^2$ tells you that the lengths of the vectors can be related by the Pythagorean theorem, which means that vector m and vector n are at right angles to one another. You can then sketch the situation:

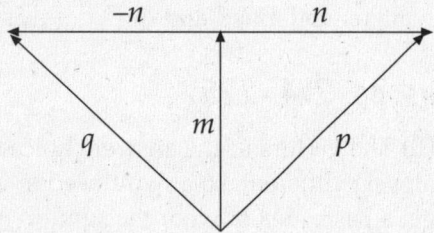

As you can see, vectors p and q are the hypotenuses of two right triangles which are mirror images of each other. They must, therefore, have the same length, and their magnitudes, p and q, must be equal.

Logic, pp. 226–227

28. **(D)** The basic statement here is: sophomore → not failing. The contrapositive of this statement would be: failing → not sophomore. Answer choice (D) is the contrapositive.

33. **(B)** The basic statement here is: arson → building burns. The contrapositive would be: no building burns → no arson. Answer choice (B) directly contradicts the contrapositive.

35. **(C)** The basic statement here is: not genuine → ruby. The contrapositive would be: not ruby → genuine. Answer choice (C) paraphrases the contrapositive.

Imaginary numbers, pp. 230–231

25. **(C)** Remember that the powers of i repeat in a cycle of four. Divide 51 by 4, and you'll find that the remainder is 3. i^{51} will be equal to i^3, which is $-i$.

36. **(D)** Powers of i separated by 4 powers will be equal. Powers of i separated by 2 powers will be negative versions of each other. Only (D) contains two values that do not cancel each other out.

40. **(B)** Use FOIL on the top of the fraction, and you get $\frac{4-16i^2}{5}$, or $\frac{4+16}{5} = \frac{20}{5} = 4$.

43. **(D)** Plot the point on the complex plane; it will have a real coordinate of 5 and an imaginary coordinate of -12. The Pythagorean Theorem will give you the point's distance from the origin: 13.

Polynomial division, p. 232

21. **(C)** Set up your polynomial division like ordinary long division, making sure your terms are in descending order of degree, with no degrees skipped. Then divide carefully. You can check your result by multiplying the terms together again. $(x + 2)(x^3 - 7x^2 + 12x) = x^4 - 5x^3 - 2x^2 + 24x$, so (C) is correct.

27. **(A)** For this one, you've got to do the polynomial division. The polynomial $x^3 + 2x^2 - 27x + 40$ divided by $(x - 3)$ is $x^2 + 5x - 12$, with a remainder of 4.

13
The Princeton Review Mathematics Subject Tests

MATHEMATICS LEVEL IC TEST FORM A

Directions: For each of the following problems, decide which is the BEST of the choices given. If the exact numerical value is not one of the choices, select the choice that best approximates this value. Then fill in the corresponding oval on the answer sheet.

Notes: (1) A calculator will be necessary for answering some (but not all) of the questions in this test. For each question you will have to decide whether or not you should use a calculator. The calculator you use must be at least a scientific calculator; programmable calculators and calculators that can display graphs are permitted.

(2) The only angle measure used on this test is degree measure. Make sure that your calculator is in the degree mode.

(3) Figures that accompany problems in this test are intended to provide information useful in solving the problems. They are drawn as accurately as possible EXCEPT when it is stated in a specific problem that its figure is not drawn to scale. All figures lie in a plane unless otherwise indicated.

(4) Unless otherwise specified, the domain of any function f is assumed to be the set of all real numbers x for which $f(x)$ is a real number.

(5) Reference information that may be useful in answering the questions in this test can be found below.

Reference Information: The following information is for your reference in answering some of the questions in this test.

Volume of a right circular cone with radius r and height h: $V = \frac{1}{3}\pi r^2 h$

Lateral Area of a right circular cone with circumference of the base c and slant height ℓ: $S = \frac{1}{2}c\ell$

Volume of a sphere with radius r: $V = \frac{4}{3}\pi r^3$

Surface Area of a sphere with radius r: $S = 4\pi r^2$

Volume of a pyramid with base area B and height h: $V = \frac{1}{3}Bh$

USE THIS SPACE FOR SCRATCHWORK.

1. If $3 - x = 5$, then $x =$

    (A) -8
    (B) -2
    (C) 2
    (D) 5
    (E) 8

2. If $\frac{a}{b} = 0.625$, then $\frac{b}{a}$ is equal to which of the following?

    (A) 1.60
    (B) 2.67
    (C) 2.70
    (D) 3.33
    (E) 4.25

GO ON TO THE NEXT PAGE

MATHEMATICS LEVEL IC TEST FORM A—*Continued*

USE THIS SPACE FOR SCRATCHWORK.

3. If $x - 3 = 3(1 - x)$, then what is the value of x?

    (A) 0.33
    (B) 0.67
    (C) 1.50
    (D) 1.67
    (E) 2.25

4. Points A, B, C, and D are arranged on a line in that order. If $AC = 13$, $BD = 14$, and $AD = 21$, then $BC =$

    (A) 12
    (B) 9
    (C) 8
    (D) 6
    (E) 3

5. A group of z people buys x widgets at a price of y dollars each. If the people divide the cost of this purchase evenly, then how much must each person pay, in dollars?

    (A) $\dfrac{xy}{z}$
    (B) $\dfrac{yz}{x}$
    (C) $\dfrac{xz}{y}$
    (D) $xy + z$
    (E) xyz

GO ON TO THE NEXT PAGE

MATHEMATICS LEVEL IC TEST FORM A—*Continued*

USE THIS SPACE FOR SCRATCHWORK.

6. At what coordinates does the line $3y + 5 = x - 1$ intersect the *y*-axis?

   (A) $(0, -2)$
   (B) $(0, -1)$
   (C) $\left(0, \frac{1}{3}\right)$
   (D) $(-2, 0)$
   (E) $(-6, 0)$

7. In Figure 1, if $m \parallel n$ and $b = 125$, then $d + f =$

   (A) 50
   (B) 55
   (C) 110
   (D) 130
   (E) 180

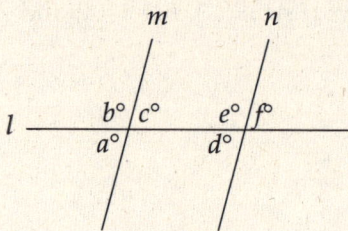

Figure 1

8. Twenty-five percent of 28 is what percent of 4?

   (A) 7%
   (B) 57%
   (C) 70%
   (D) 128%
   (E) 175%

9. If $7a + 2b = 11$ and $a - 2b = 5$, then what is the value of *a*?

   (A) −2.0
   (B) −0.5
   (C) 1.4
   (D) 2.0
   (E) It cannot be determined from the information given.

GO ON TO THE NEXT PAGE

MATHEMATICS LEVEL IC TEST FORM A—Continued

USE THIS SPACE FOR SCRATCHWORK.

10. Which of the following is equal to $a + c$?

    (A) $2b$
    (B) $b + 90$
    (C) $d + e$
    (D) $b + d - e$
    (E) $180 - (d + e)$

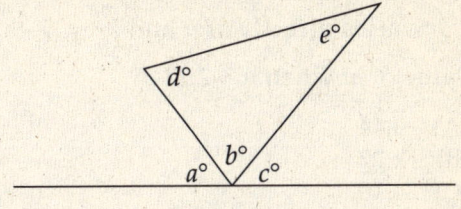

Figure 2

11. Which of the following could be the graph of $|y| \geq 3$?

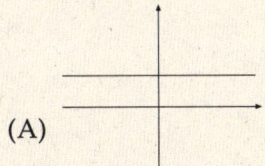

(A)

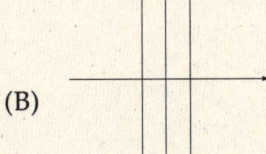

(B)

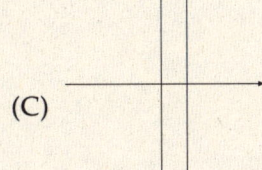

(C)

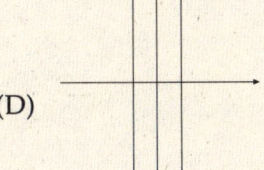

(D)

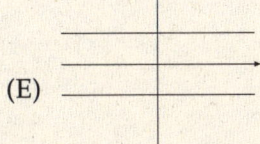

(E)

GO ON TO THE NEXT PAGE

MATHEMATICS LEVEL IC TEST FORM A—Continued

USE THIS SPACE FOR SCRATCHWORK.

12. If m varies directly as n and $\frac{m}{n} = 5$, then what is the value of m when $n = 2.2$?

    (A) 0.44
    (B) 2.27
    (C) 4.10
    (D) 8.20
    (E) 11.00

13. What is the slope of the line $3y - 5 = 7 - 2x$?

    (A) -2
    (B) $-\frac{2}{3}$
    (C) $\frac{3}{2}$
    (D) 2
    (E) 6

14. $\dfrac{(n^3)^6 \times (n^4)^5}{n^2} =$

    (A) n^9
    (B) n^{16}
    (C) n^{19}
    (D) n^{36}
    (E) n^{40}

GO ON TO THE NEXT PAGE

MATHEMATICS LEVEL IC TEST FORM A—Continued

15. If $\|x\| = x^2 - 3x$, then $\|-3\| =$

    (A) 0
    (B) 3.3
    (C) 6.0
    (D) 9.0
    (E) 18.0

16. Students in a certain research program are either engineers or doctoral candidates; some students graduate each year. In a certain year, no doctoral candidates graduate. Which of the following statements must be true?

    (A) The program then contains more engineers than doctoral candidates.
    (B) Doctoral candidates are poorer students than engineers.
    (C) More doctoral candidates will graduate in following years.
    (D) Every student graduating in that year is an engineer.
    (E) All engineers in the program graduate in that year.

17. In Figure 3, if segments PA and PB are tangent to circle O at A and B, respectively, then which of the following must be true?

    I. $PB > PO$
    II. $\angle APO = \angle BPO$
    III. $\angle APB + \angle AOB = \angle PAO + \angle PBO$

    (A) I only
    (B) II only
    (C) I and II only
    (D) II and III only
    (E) I, II, and III

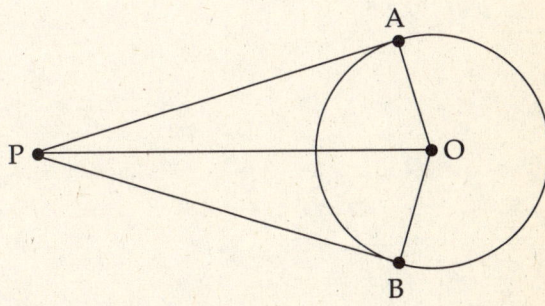

Figure 3

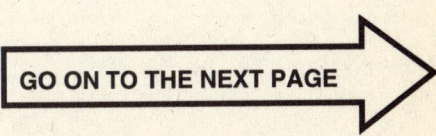

MATHEMATICS LEVEL IC TEST FORM A—Continued

USE THIS SPACE FOR SCRATCHWORK.

18. If $2x^2 - 5x - 9 = 0$, then x could equal which of the following?

    (A) −1.12
    (B) 0.76
    (C) 1.54
    (D) 2.63
    (E) 3.71

19. If the ratio of QR to RS is 2:3 and the ratio of PQ to QS is 1:2, then what is the ratio of PQ to RS?

    (A) 5:6
    (B) 7:6
    (C) 5:4
    (D) 11:6
    (E) 11:5

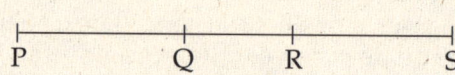

Figure 4

20. If $3^x = 54$, then $3^{x-2} =$

    (A) 3.8
    (B) 4.5
    (C) 6.0
    (D) 7.3
    (E) 9.0

GO ON TO THE NEXT PAGE

MATHEMATICS LEVEL IC TEST FORM A—Continued

USE THIS SPACE FOR SCRATCHWORK.

Questions 21-22 refer to the chart at right.

21. According to the chart in Figure 5, exactly how many club members spotted at least 3 hawks?

    (A) 6
    (B) 13
    (C) 20
    (D) 23
    (E) 26

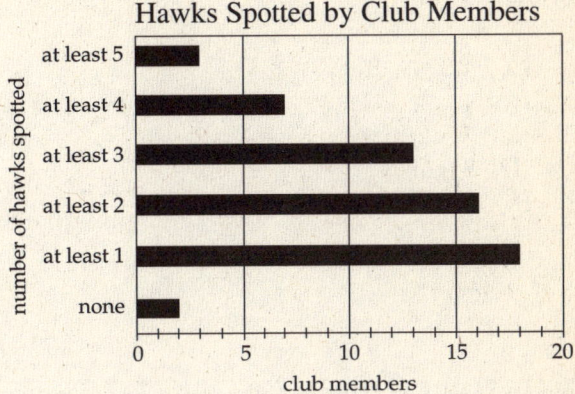

Figure 5

22. If the results for all club members are shown in the chart, then the club must have how many members?

    (A) 18
    (B) 20
    (C) 38
    (D) 57
    (E) 59

23. If the surface area of a cube may be expressed as $6x^2 - 36x + 54$, then which of the following correctly expresses the volume of the cube?

    (A) $x^3 - 3$
    (B) $x^3 - 27$
    (C) $x^3 - 6x^2 + 9x$
    (D) $6x^3 - 36x^2 + 54x$
    (E) $x^3 - 9x^2 + 27x - 27$

24. If A, B, and C are points on a circle, AC bisects the circle, and $AB = BC$, then the area of triangle ABC is most nearly what percent of the circle's area?

    (A) 16%
    (B) 24%
    (C) 32%
    (D) 48%
    (E) 66%

GO ON TO THE NEXT PAGE

MATHEMATICS LEVEL IC TEST FORM A—Continued

USE THIS SPACE FOR SCRATCHWORK.

25. $\dfrac{(3-i)^2}{2} =$

    (A) $3 - 2i$
    (B) $4 - 3i$
    (C) $7 + 2i$
    (D) $8 - 6i$
    (E) $9 + 6i$

26. In Figure 6, what is the value of x?

    (A) 0.62
    (B) 0.79
    (C) 2.46
    (D) 3.13
    (E) 3.15

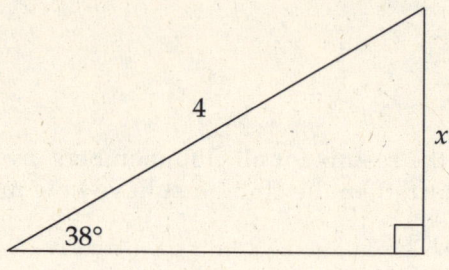

Figure 6

27. If Figure 7 shows part of the graph of $f(x)$, then which of the following could be the range of $f(x)$?

    (A) $\{y:\ y \le 2\}$
    (B) $\{y:\ y = -2, 3\}$
    (C) $\{y:\ y = 1, 2\}$
    (D) $\{y:\ -2 \le y \le 3\}$
    (E) $\{y:\ 1 \le y \le 2\}$

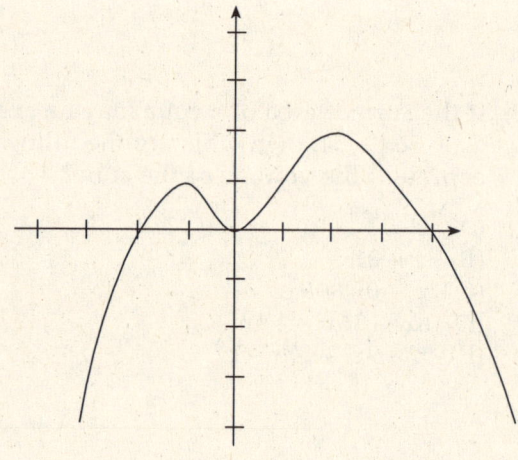

Figure 7

GO ON TO THE NEXT PAGE

28. Which of the following flat shapes could be folded into a pyramid, if folds are made only along the dotted lines?

(A)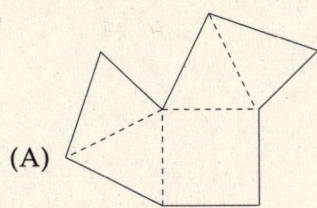

(B)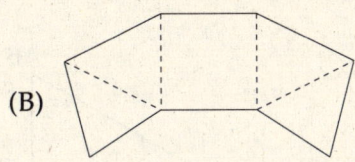

(C)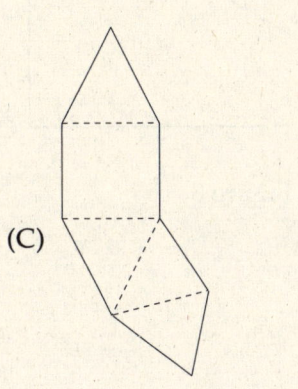

(D)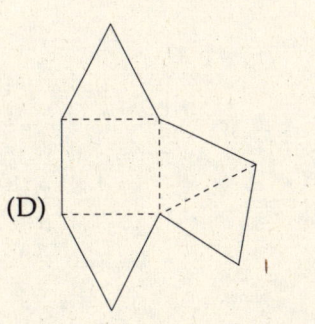

(E)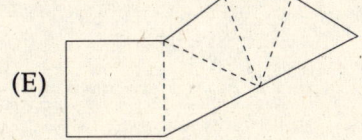

MATHEMATICS LEVEL IC TEST FORM A—Continued

USE THIS SPACE FOR SCRATCHWORK.

29. If a cube and a sphere intersect at exactly eight points, then which of the following must be true?

   (A) The sphere is inscribed in the cube.
   (B) The cube is inscribed in the sphere.
   (C) The sphere's diameter is equal in length to an edge of the cube.
   (D) The sphere and the cube have equal volumes.
   (E) The sphere and the cube have equal surface areas.

30. If $\sin \theta = 0.6$ and $AB = 12$, then what is the area of $\triangle ABC$?

   (A) 28.80
   (B) 31.18
   (C) 34.56
   (D) 42.33
   (E) 69.12

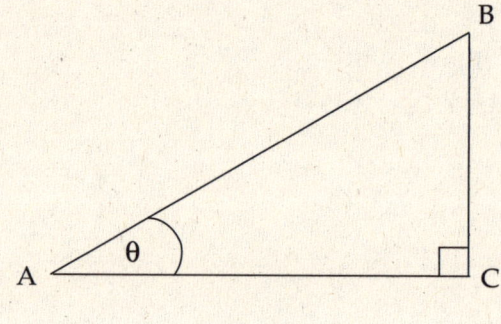

Figure 8

31. In the table of values above, what is the value of s?

   (A) 6
   (B) 8
   (C) 9
   (D) 15
   (E) It cannot be determined from the information given.

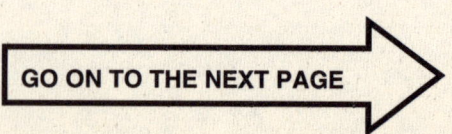

MATHEMATICS LEVEL IC TEST FORM A—Continued

32. If $\dfrac{9(\sqrt{x}-2)^2}{4} = 6.25$, then x could equal which of the following?

    (A) 0.11
    (B) 2.42
    (C) 9.00
    (D) 10.24
    (E) 13.76

33. $\left(\dfrac{1}{\cos\theta} - \dfrac{\sin\theta}{\tan\theta}\right)(\cos\theta) =$

    (A) $\cos\theta$
    (B) $\sin\theta$
    (C) $\tan\theta$
    (D) $\sin^2\theta$
    (E) $\tan^2\theta$

34. If $f(x) = 2x^5$, then which of the following must be true?

    I. $f(x) = f(-x)$
    II. $f(-x) = -f(x)$
    III. $\dfrac{1}{2}f(x) = f\left(\dfrac{1}{2}x\right)$

    (A) I only
    (B) II only
    (C) I and III only
    (D) II and III only
    (E) I, II, and III

MATHEMATICS LEVEL IC TEST FORM A—Continued

USE THIS SPACE FOR SCRATCHWORK.

35. If $f(x) = x^3 - 6$, then $-f(-x) =$

    (A) $-x^3 - 6$
    (B) $-x^3 + 6$
    (C) 0
    (D) $x^3 - 6$
    (E) $x^3 + 6$

36. How many distinct 3-digit numbers contain only nonzero digits?

    (A) 909
    (B) 899
    (C) 789
    (D) 729
    (E) 504

37. At what points does the circle $(y - 3)^2 + (x - 2)^2 = 16$ intersect the y-axis?

    (A) $(0, -5.66)$ and $(0, 5.66)$
    (B) $(0, -0.46)$ and $(0, 6.46)$
    (C) $(0, -1.00)$ and $(0, 7.00)$
    (D) $(-0.65, 0)$ and $(4.65, 0)$
    (E) $(-2.00, 0)$ and $(6.00, 0)$

38. A rectangular solid of dimensions $4 \times 5 \times 7$ is inscribed in a sphere. What is the radius of this sphere?

    (A) 4.74
    (B) 5.66
    (C) 7.29
    (D) 9.49
    (E) 11.83

GO ON TO THE NEXT PAGE

MATHEMATICS LEVEL IC TEST FORM A—Continued

USE THIS SPACE FOR SCRATCHWORK.

39. In circle O, which of the following is equal to c?

   (A) $\dfrac{b}{2}$
   (B) d
   (C) $2a$
   (D) $\dfrac{a+d}{2}$
   (E) $b - 90$

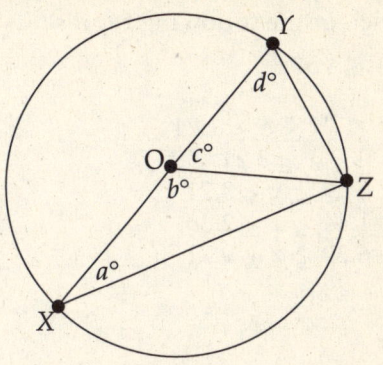

Figure 9

40. A researcher finds that an ant colony's population increases by almost exactly 8% each month. If the colony has an initial population of 1,250 insects, which of the following is the nearest approximation of the population of the colony 2 years later?

   (A) 7,926
   (B) 5,832
   (C) 3,650
   (D) 2,400
   (E) 1,458

41. If circle O has a radius of 4 and $OD = 3DB$, then $\sin \angle A =$

   (A) 0.60
   (B) 0.71
   (C) 0.80
   (D) 0.87
   (E) 1.00

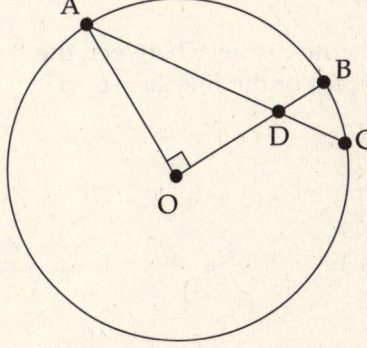

Figure 10

GO ON TO THE NEXT PAGE

MATHEMATICS LEVEL IC TEST FORM A—Continued

USE THIS SPACE FOR SCRATCHWORK.

42. Which of the following represents the solution set of $|x^3 - 8| \leq 5$?

    (A) $-1.71 \leq x \leq 1.71$
    (B) $0 \leq x \leq 3.21$
    (C) $0.29 \leq x \leq 3.71$
    (D) $1.44 \leq x \leq 2.35$
    (E) $6.29 \leq x \leq 9.71$

43. Six congruent circle's areas are arranged so that each circle touches at least two others without overlapping. The centers of these six circles are then connected to form a polygon. If each circle has a radius of 2, then what is the perimeter of this polygon?

    (A) 6
    (B) 12
    (C) 24
    (D) 36
    (E) It cannot be determined from the information given.

44. What is the distance between the x-intercept and the y-intercept of the line $2y = 6 - x$?

    (A) 3.67
    (B) 6.32
    (C) 6.71
    (D) 7.29
    (E) 8.04

GO ON TO THE NEXT PAGE

MATHEMATICS LEVEL IC TEST FORM A—Continued

45. The point (5, –10) is at a distance of 26 from Q, and the point (2, –10) is at a distance of 25 from Q. Which of the following could be the coordinates of Q?

    (A) (–5, 14)
    (B) (–3, 18)
    (C) (–1, 19)
    (D) (0, 21)
    (E) (2, 16)

46. In Figure 10, what is the area of isosceles trapezoid *RSTU*?

    (A) 19.31
    (B) 19.80
    (C) 24.24
    (D) 27.80
    (E) 56.00

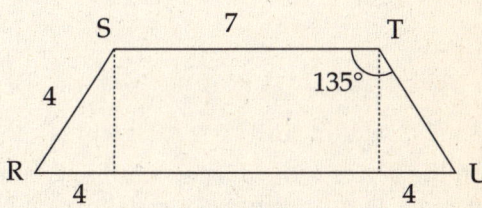

Figure 10

47. Which of the following expressions has the greatest value?

    (A) 1.73^{999}
    (B) 2^{799}
    (C) 3^{500}
    (D) 4^{400}
    (E) 250^{100}

MATHEMATICS LEVEL IC TEST FORM A—Continued

USE THIS SPACE FOR SCRATCHWORK.

48. If a cube has a volume of v, then which of the following represents the length of a diagonal drawn from corner to corner through the center of the cube?

   (A) $\dfrac{\sqrt{3}}{3}v^3$

   (B) $\dfrac{1}{3}v^{\frac{1}{3}}$

   (C) $v^{\frac{2}{3}}\sqrt{3}$

   (D) $v^{\frac{2}{3}}$

   (E) $v^{\frac{1}{3}}\sqrt{3}$

49. In the series $a_1, a_2, a_3, \ldots$, $a_1 = 1$, $a_2 = 2$, and thereafter $a_n = a_{n-1} + a_{n-2}$. What is the value of a_8?

   (A) 21
   (B) 34
   (C) 56
   (D) 90
   (E) 146

50. If $z = \log_x(y^x)$, then $x^z =$

   (A) x^x
   (B) y^x
   (C) xy^x
   (D) y^{2x}
   (E) It cannot be determined from the information given.

STOP

IF YOU FINISH BEFORE TIME IS CALLED, YOU MAY CHECK YOUR WORK ON THIS TEST ONLY.
DO NOT WORK ON ANY OTHER TEST IN THIS BOOK.

HOW TO SCORE THE PRINCETON REVIEW MATH SUBJECT TEST

When you take the real exam, the proctors will collect your text booklet and bubble sheet and send your answer sheet to New Jersey where a computer (yes, a big old-fashioned one that has been around since the 1960s) looks at the pattern of filled-in ovals on your answer sheet and gives you a score. We couldn't include even a small computer with this book, so we are providing this more primitive way of scoring your exam.

DETERMINING YOUR SCORE

STEP 1 Using the answers on the next page, determine how many questions you got right and how many you got wrong on the test. Remember, questions that you do not answer don't count as either right answers or wrong answers.

STEP 2 List the number of right answers here. (A) _____

STEP 3 List the number of wrong answers here. Now divide that number by 4. (Use a calculator if you're feeling particularly lazy.) (B) _____ ÷ 4 = (C) _____

STEP 4 Subtract the number of wrong answers divided by 4 from the number of correct answers. Round this score to the nearest whole number. This is your raw score. (A) _____ − (C) _____ = _____

STEP 5 To determine your real score, take the number from Step 4 above and look it up in the left column of the Score Conversion Table on page 277; the corresponding score on the right is your score on the exam.

MATHEMATICS LEVEL IC FORM A

1. B
2. A
3. C
4. D
5. A
6. A
7. C
8. E
9. D
10. C
11. E
12. E
13. B
14. D
15. E
16. D
17. D
18. E
19. A
20. C
21. B
22. B
23. E
24. C
25. B
26. C
27. A
28. E
29. B
30. C
31. E
32. A
33. D
34. B
35. E
36. D
37. B
38. A
39. C
40. A
41. A
42. D
43. C
44. C
45. A
46. D
47. D
48. E
49. B
50. B

MATHEMATICS LEVEL IC SUBJECT TEST SCORE CONVERSION TABLE

Raw Score	College Board Scaled Score	Raw Score	College Board Scaled Score
50	800	15	450
49	780	14	440
48	770	13	430
47	760	12	430
46	740	11	420
45	730	10	410
44	720	9	400
43	710	8	390
42	700	7	380
41	690	6	370
40	680	5	370
39	670	4	360
38	660	3	350
37	650	2	340
36	640	1	340
35	630	0	330
34	610	−1	320
33	600	−2	310
32	590	−3	300
31	580	−4	300
30	570	−5	290
29	560	−6	280
28	550	−7	270
27	550	−8	260
26	540	−9	260
25	530	−10	250
24	520	−11	240
23	510	−12	230
22	510		
21	500		
20	490		
19	480		
18	480		
17	470		
16	460		

MATHEMATICS LEVEL IC TEST FORM B

Directions: For each of the following problems, decide which is the BEST of the choices given. If the exact numerical value is not one of the choices, select the choice that best approximates this value. Then fill in the corresponding oval on the answer sheet.

Notes: (1) A calculator will be necessary for answering some (but not all) of the questions in this test. For each question you will have to decide whether or not you should use a calculator. The calculator you use must be at least a scientific calculator; programmable calculators and calculators that can display graphs are permitted.

(2) The only angle measure used on this test is degree measure. Make sure that your calculator is in the degree mode.

(3) Figures that accompany problems in this test are intended to provide information useful in solving the problems. They are drawn as accurately as possible EXCEPT when it is stated in a specific problem that its figure is not drawn to scale. All figures lie in a plane unless otherwise indicated.

(4) Unless otherwise specified, the domain of any function f is assumed to be the set of all real numbers x for which $f(x)$ is a real number.

(5) Reference information that may be useful in answering the questions in this test can be found below.

Reference Information: The following information is for your reference in answering some of the questions in this test.

Volume of a right circular cone with radius r and height h: $V = \frac{1}{3}\pi r^2 h$

Lateral Area of a right circular cone with circumference of the base c and slant height l: $S = \frac{1}{2}cl$

Volume of a sphere with radius r: $V = \frac{4}{3}\pi r^3$

Surface Area of a sphere with radius r: $S = 4\pi r^2$

Volume of a pyramid with base area B and height h: $V = \frac{1}{3}Bh$

USE THIS SPACE FOR SCRATCHWORK.

1. Rob and Sherry together weigh 300 pounds. Sherry and Heather together weigh 240 pounds. If all three people together weigh 410 pounds, then what is Sherry's weight?

    (A) 110
    (B) 115
    (C) 120
    (D) 130
    (E) 145

GO ON TO THE NEXT PAGE

MATHEMATICS LEVEL IC TEST FORM B—Continued

USE THIS SPACE FOR SCRATCHWORK.

2. If the point $(x, 4)$ is the intersection of the graphs of $y = 4$ and $y = \dfrac{x^3}{2}$, then $x =$

   (A) -2
   (B) -1
   (C) 2
   (D) 4
   (E) 8

3. If $r = \dfrac{2}{3}$ and $s = 6$, then $\dfrac{s}{r} + \dfrac{4}{r^2} =$

   (A) 4
   (B) 6
   (C) 9
   (D) 12
   (E) 18

4. In Figure 1, what is the value of p in terms of m and n?

   (A) $m + n - 180$
   (B) $m + n + 180$
   (C) $m - n + 360$
   (D) $360 - (m - n)$
   (E) $360 - (m + n)$

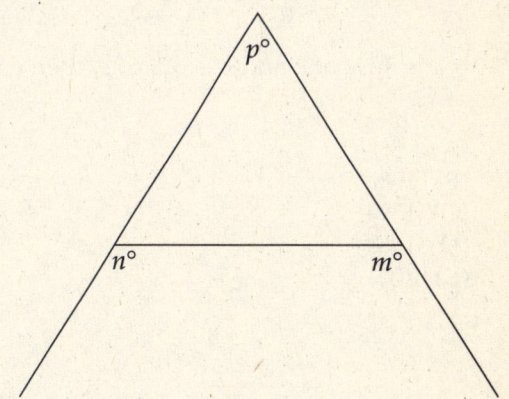

Figure 1

GO ON TO THE NEXT PAGE

MATHEMATICS LEVEL IC TEST FORM B—Continued

USE THIS SPACE FOR SCRATCHWORK.

5. After 8:00 P.M., a ride in a taxi costs $2.50 plus $0.30 for every fifth of a mile traveled. If a passenger travels b miles, then what is the cost of the trip, in dollars, in terms of b?

   (A) $2.5 + 0.3b$
   (B) $2.5 + 1.5b$
   (C) $2.8b$
   (D) $30 + 250b$
   (E) $250 + 30b$

6. If x is an even integer, then which of the following must be equal to an odd integer?

   (A) $4x - 2$
   (B) $5x$
   (C) $4(x - 1)$
   (D) $x^2 - 3$
   (E) $3x^2 - 2$

7. If a is 30% of c and b is 6% of c, then b is what percent of a?

   (A) 5%
   (B) 10%
   (C) 15%
   (D) 18%
   (E) 20%

GO ON TO THE NEXT PAGE

MATHEMATICS LEVEL IC TEST FORM B—Continued

$$\begin{cases} x(x-1) = 6 \\ x^2 + 1 = 5 \end{cases}$$

8. Which of the following is the solution set of this system of equations?

   (A) $\{x: x = -3, -2, 2\}$
   (B) $\{x: x = -2, 2, 3\}$
   (C) $\{x: x = -2\}$
   (D) $\{x: x = -2, 2\}$
   (E) $\{x: x = 2, 3\}$

9. In Figure 2, if every angle in the polygon is a right angle, then what is the polygon's perimeter?

   (A) 34
   (B) 42
   (C) 47
   (D) 52
   (E) 60

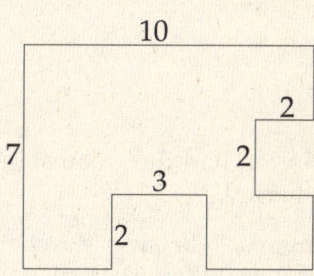

Figure 2

10. For which of the points shown in Figure 3 is $|x + y| > 5$?

    (A) A
    (B) B
    (C) C
    (D) D
    (E) E

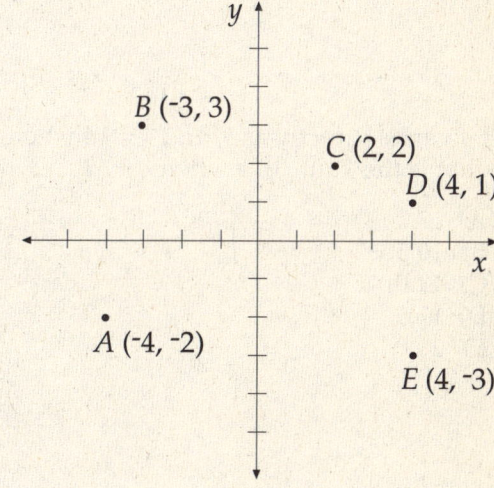

Figure 3

MATHEMATICS LEVEL IC TEST FORM B—Continued

USE THIS SPACE FOR SCRATCHWORK.

Questions 11-12 refer to the chart at right, which shows the monthly sales made by a salesperson in 1996.

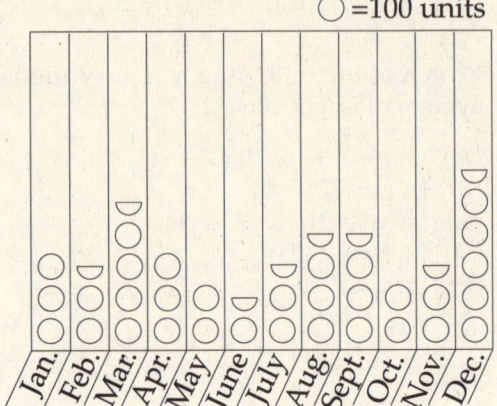

11. As a saleswoman, Keri receives a $10.00 commission for each unit she sells. In any month in which she sells more than 300 units, she receives an additional bonus of $1,000.00. What was the total amount Keri received in bonuses in 1996?

    (A) $3,000.00
    (B) $4,000.00
    (C) $5,000.00
    (D) $6,000.00
    (E) $8,000.00

12. In 1996, Keri had the greatest income in what three-month period?

    (A) January, February, March
    (B) February, March, April
    (C) March, April, May
    (D) July, August, September
    (E) October, November, December

13. If a varies directly as b, and $a = 14$ when $b = 8$, then what is the value of a when $b = 6$?

    (A) 7.0
    (B) 10.5
    (C) 12.0
    (D) 17.5
    (E) 21.0

GO ON TO THE NEXT PAGE

MATHEMATICS LEVEL IC TEST FORM B—Continued

USE THIS SPACE FOR SCRATCHWORK.

14. If $\dfrac{1}{x} = \dfrac{4}{5}$, then $\dfrac{x}{3} =$

   (A) 0.27
   (B) 0.33
   (C) 0.42
   (D) 0.66
   (E) 1.25

15. In Figure 4, sin ∠RSU must be equal to which of the following?

   (A) cos ∠RTU
   (B) cos ∠TSU
   (C) sin ∠SRT
   (D) sin ∠STR
   (E) sin ∠TRU

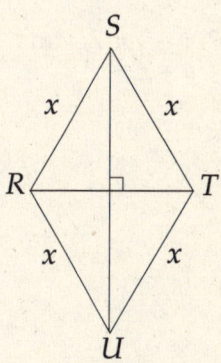

Figure 4

16. If y is a real number and $y = \sqrt{x} + \dfrac{1}{x-3}$, then which of the following statements must be true?

   I. $x > 1$
   II. $x \neq 3$
   III. $x \neq -3$

   (A) I only
   (B) II only
   (C) I and III only
   (D) II and III only
   (E) I, II, and III

GO ON TO THE NEXT PAGE

MATHEMATICS LEVEL IC TEST FORM B—Continued

USE THIS SPACE FOR SCRATCHWORK.

17. Sphere O is inscribed within cube A, and cube B is inscribed within sphere O. Which of the following quantities must be equal?

    (A) An edge of A and the radius of O.
    (B) The diameter of O and the longest diagonal in A.
    (C) An edge of B and the diameter of O.
    (D) An edge of B and the radius of O.
    (E) An edge of A and the longest diagonal in B.

18. If $a - x = 12$, $b - y = 7$, and $c - z = 15$, and $a + b + c = 50$, then $x + y + z =$

    (A) 16
    (B) 18
    (C) 34
    (D) 66
    (E) 84

19. A jeep has four seats, including one driver's seat and three passenger seats. If Amber, Bunny, Cassie, and Donna are going for a drive in the jeep, and only Cassie can drive, then how many different seating arrangements are possible?

    (A) 3
    (B) 6
    (C) 12
    (D) 16
    (E) 24

20. If $\frac{1}{2}x - 3 = 2\left(\frac{x-1}{5}\right)$, then $x =$

    (A) 9
    (B) 11
    (C) 13
    (D) 22
    (E) 26

GO ON TO THE NEXT PAGE

MATHEMATICS LEVEL IC TEST FORM B—Continued

USE THIS SPACE FOR SCRATCHWORK.

21. Line l passes through the origin and point (a, b). If $ab \neq 0$ and line l has a slope greater than 1, then which of the following must be true?

    (A) $a = b$
    (B) $a < b$
    (C) $a^2 < b^2$
    (D) $b - a < 0$
    (E) $a + b > 0$

22. In Figure 5, points A, B, and C are three vertices of a parallelogram, and point D (not shown) is the fourth vertex. How many points could be D?

    (A) 1
    (B) 2
    (C) 3
    (D) 4
    (E) 5

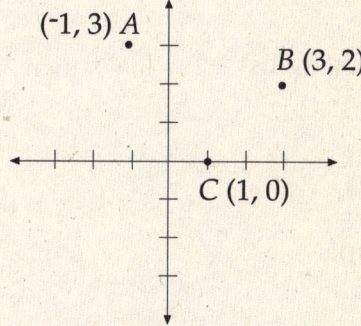

Figure 5

23. In Figure 6, if $y = \frac{2}{3}x$ and $w = 2z$, then $x =$

    (A) 30
    (B) 40
    (C) 48
    (D) 60
    (E) 72

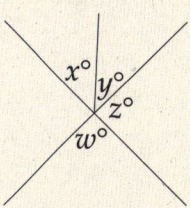

Figure 6

Note: Figure Not Drawn To Scale

24. Circle O has a radius of r. If this radius is increased by t, then which of the following correctly expresses the new area of circle O?

    (A) πt^2
    (B) $2\pi(r + t)$
    (C) $\pi(t^2 + r^2)$
    (D) $\pi(r^2 + 2rt + t^2)$
    (E) $4\pi(r^2 + 2rt + t^2)$

GO ON TO THE NEXT PAGE

MATHEMATICS LEVEL IC TEST FORM B—Continued

USE THIS SPACE FOR SCRATCHWORK.

25. In Figure 7, AC and BD are perpendicular diameters of circle O. If the circle has an area of 9π, what is the length of AB?

    (A) 2.12
    (B) 3.36
    (C) 4.24
    (D) 6.36
    (E) 8.48

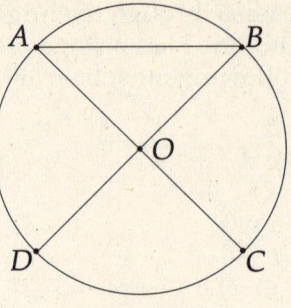

Figure 7

26. If $x < |x|$ and $x^2 + 2x - 3 = 0$, then $2x + 4 =$

    (A) −2
    (B) 2
    (C) 6
    (D) 8
    (E) 10

27. In Figure 8, triangles ABC and CBD are similar. What is the area of triangle CBD?

    (A) 3.07
    (B) 3.84
    (C) 5.24
    (D) 7.68
    (E) 9.60

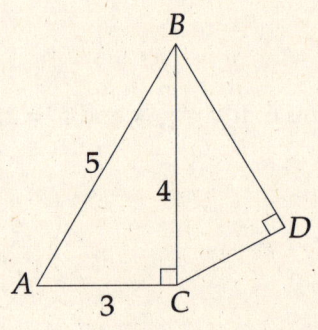

Figure 8

28. $(5 - 3i)(4 + 2i) =$

    (A) $14 - 2i$
    (B) 16
    (C) 24
    (D) $26 - 2i$
    (E) 28

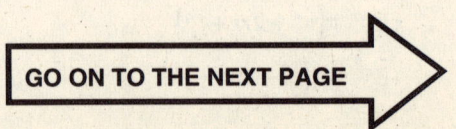

MATHEMATICS LEVEL IC TEST FORM B—Continued

29. If $f(x) = x^2 - 5x$ and $f(n) = -4$, then which of the following could be n?

    (A) -5
    (B) -4
    (C) -1
    (D) 1
    (E) 5

30. Two identical rectangular solids, each of dimensions $3 \times 4 \times 5$, are joined face-to-face to form a single rectangular solid with a length of 8. What is the length of the longest line that can be drawn within this new solid?

    (A) 8.60
    (B) 9.90
    (C) 10.95
    (D) 11.40
    (E) 12.25

31. Which of the following most closely approximates $(5.5 \times 10^4)^2$?

    (A) 3.0×10^5
    (B) 3.0×10^6
    (C) 3.0×10^7
    (D) 3.0×10^8
    (E) 3.0×10^9

MATHEMATICS LEVEL IC TEST FORM B—Continued

32. Line l passes through $(-2, 2)$ and $(2, -3)$. What is the slope of this line?

    (A) $-\dfrac{5}{4}$

    (B) -1

    (C) $-\dfrac{4}{5}$

    (D) $\dfrac{2}{3}$

    (E) $\dfrac{4}{5}$

33. A sample of metal is heated to 698° C and then allowed to cool. The temperature of the metal over time is given by the formula $n = 698 - 2t - 0.5t^2$, where t is the time in seconds after the start of the cooling process, and n is the sample's temperature in degrees. After how many seconds will the sample's temperature be 500° C?

    (A) 16
    (B) 18
    (C) 20
    (D) 22
    (E) 24

34. Perpendicular lines l and m intersect at $(4, 5)$. If line m has a slope of $-\dfrac{1}{2}$, what is the equation of line l?

    (A) $y = \dfrac{1}{2}x - 1$

    (B) $y = \dfrac{1}{2}x + 3$

    (C) $y = \dfrac{1}{2}x + 5$

    (D) $y = 2x - 1$

    (E) $y = 2x - 3$

MATHEMATICS LEVEL IC TEST FORM B—Continued

USE THIS SPACE FOR SCRATCHWORK.

35. In Figure 9, points A, B, C, and D are all on circle O. If $\angle BDA$ measures 25°, and $\angle CAD$ measures 32°, what is the measure of $\angle BOC$ in degrees?

    (A) 33
    (B) 66
    (C) 123
    (D) 147
    (E) 303

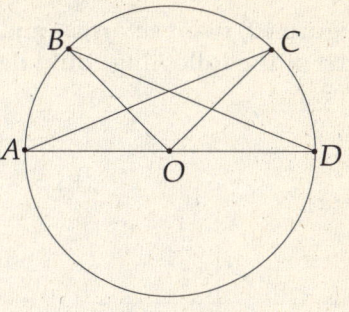

Figure 9

36. If $4^{x+2} = 48$, then $4^x =$

    (A) 3.0
    (B) 6.4
    (C) 6.9
    (D) 12.0
    (E) 24.0

37. If $r(x) = 6x + 5$ and $s(r(x)) = 2x - 1$, then $s(x) =$

    (A) $-4x - 6$

    (B) $\dfrac{x-2}{3}$

    (C) $\dfrac{x-8}{3}$

    (D) $3x - 6$

    (E) $4x + 4$

GO ON TO THE NEXT PAGE

MATHEMATICS LEVEL IC TEST FORM B—Continued

USE THIS SPACE FOR SCRATCHWORK.

38. In Figure 10, two vectors, $\vec{a}$ and $\vec{b}$, are shown. Which of the following could be the sum of these two vectors?

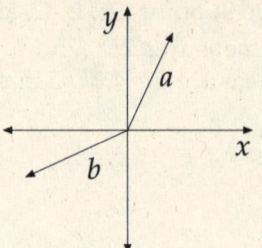

Figure 10

(A)   (B)

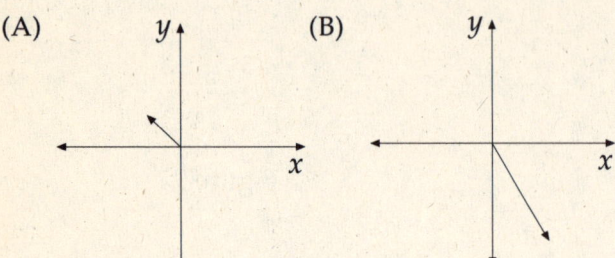

(C)   (D)

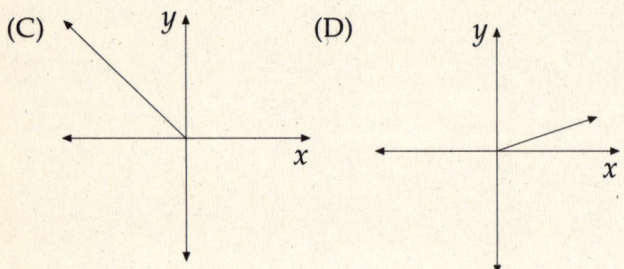

(E)

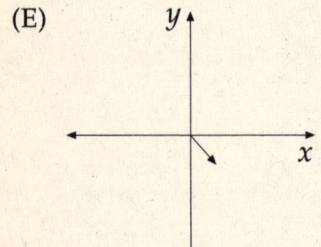

GO ON TO THE NEXT PAGE →

MATHEMATICS LEVEL IC TEST FORM B—Continued

USE THIS SPACE FOR SCRATCHWORK.

39. If $\log_3 6 = n$, then $n =$

    (A) 0.37
    (B) 0.61
    (C) 1.63
    (D) 1.82
    (E) 2.33

40. A cylindrical cup has a height of 3 inches and a radius of 2 inches. How many such cups may be completely filled from a full rectangular tank whose dimensions are $6 \times 7 \times 8$ inches?

    (A) 8
    (B) 9
    (C) 12
    (D) 17
    (E) 28

41. Line segments AC and BD intersect at point O, such that each segment is the perpendicular bisector of the other. If $AC = 7$ and $BD = 6$, then $\sin \angle ADO =$

    (A) 0.16
    (B) 0.24
    (C) 0.39
    (D) 0.76
    (E) 0.85

GO ON TO THE NEXT PAGE

MATHEMATICS LEVEL IC TEST FORM B—Continued

USE THIS SPACE FOR SCRATCHWORK.

42. The first and second terms of a geometric sequence are 24 and 6, respectively. What is the sixth term of this sequence?

    (A) $\dfrac{3}{128}$
    (B) $\dfrac{3}{64}$
    (C) $\dfrac{3}{32}$
    (D) $\dfrac{3}{16}$
    (E) $\dfrac{3}{8}$

43. If $f(x) = kx$, where k is a nonzero constant, and $g(x) = x + k$, then which of the following statements must be true?

    I. $f(2x) = 2f(x)$
    II. $f(x + 2) = f(x) + 2$
    III. $f(g(x)) = g(f(x))$

    (A) I only
    (B) II only
    (C) I and II only
    (D) I and III only
    (E) I, II, and III

44. A rectangular room has walls facing due north, south, east, and west. On the southern wall, a tack is located 85 inches from the floor and 38 inches from the western wall, and a nail is located 48 inches from the floor and 54 inches from the western wall. What is the distance in inches between the tack and the nail?

    (A) 21.0
    (B) 26.4
    (C) 32.6
    (D) 37.0
    (E) 40.3

GO ON TO THE NEXT PAGE

MATHEMATICS LEVEL IC TEST FORM B—Continued

45. If $s(x) = \sqrt{12 - x^2}$, then which of the following is the domain of s?

    (A) $\{x: x \neq \sqrt{12}\}$
    (B) $\{x: x > 0\}$
    (C) $\{x: -\sqrt{12} < x < \sqrt{12}\}$
    (D) $\{x: 0 < x < \sqrt{12}\}$
    (E) $\{x: 0 < x < 144\}$

"If a tree falls in the forest, a sound is heard."

46. If the statement above is true, then which of the following situations is logically impossible?

    (A) No tree falls in the forest, but a sound is heard.
    (B) A sound is heard as a tree falls in the forest.
    (C) No sound is heard as a tree falls in the forest.
    (D) No tree falls in the forest, and no sound is made.
    (E) A sound is heard in the forest as no tree falls.

47. $\dfrac{8!}{5! \times 3!} =$

    (A) 1
    (B) 6
    (C) 26
    (D) 56
    (E) 336

MATHEMATICS LEVEL IC TEST FORM B—Continued

USE THIS SPACE FOR SCRATCHWORK.

48. In Figure 11, $AB = 4$, $BC = 7$, and $CD = 1$. If AC is a diameter of the circle, then what is the length of AD?

    (A) 3.0
    (B) 6.0
    (C) 8.0
    (D) $\sqrt{65}$
    (E) It cannot be determined from the information given.

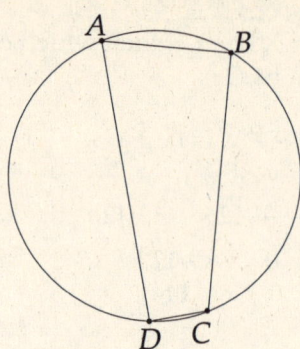

Figure 11

49. The points (0, 0) and (-2, 2) are two vertices of an equilateral triangle. Which of the following could be the coordinates of the third vertex?

    (A) (−2.0, 0)
    (B) (−0.73, 2.73)
    (C) (-0.73, 0.73)
    (D) (0, 2.0)
    (E) (0.73, 2.73)

GO ON TO THE NEXT PAGE

MATHEMATICS LEVEL IC TEST FORM B—Continued

USE THIS SPACE FOR SCRATCHWORK.

50. If $f(x) = 1 - x$ and $g(x) = \begin{cases} x-1, x \geq 2 \\ x+1, x < 2 \end{cases}$, then which of the following could be the graph of $g(f(x))$?

(A)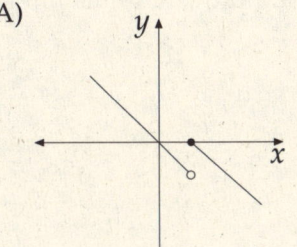

(D)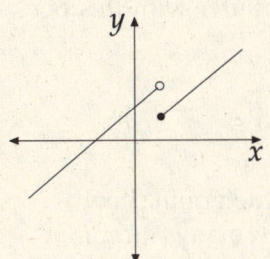

(B)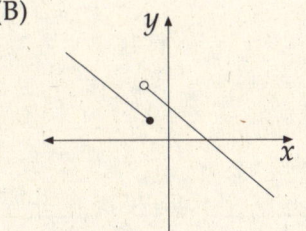

(E)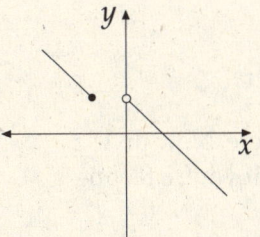

(C)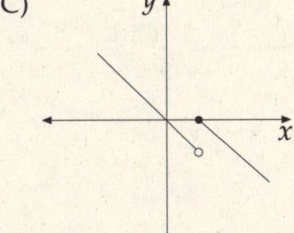

STOP

IF YOU FINISH BEFORE TIME IS CALLED, YOU MAY CHECK YOUR WORK ON THIS TEST ONLY.
DO NOT WORK ON ANY OTHER TEST IN THIS BOOK.

HOW TO SCORE THE PRINCETON REVIEW MATH SUBJECT TEST

When you take the real exam, the proctors will collect your text booklet and bubble sheet and send your answer sheet to New Jersey where a computer (yes, a big old-fashioned one that has been around since the '60s) looks at the pattern of filled-in ovals on your answer sheet and gives you a score. We couldn't include even a small computer with this book, so we are providing this more primitive way of scoring your exam.

DETERMINING YOUR SCORE

STEP 1   Using the answers on the next page, determine how many questions you got right and how many you got wrong on the test. Remember, questions that you do not answer don't count as either right answers or wrong answers.

STEP 2   List the number of right answers here.   (A) _____

STEP 3   List the number of wrong answers here. Now divide that number by 4. (Use a calculator if you're feeling particularly lazy.)   (B) _____ ÷ 4 = (C) _____

STEP 4   Subtract the number of wrong answers divided by 4 from the number of correct answers. Round this score to the nearest whole number. This is your raw score.   (A) _____ − (C) _____ = ____

STEP 5   To determine your real score, take the number from Step 4 above and look it up in the left column of the Score Conversion Table on page 337; the corresponding score on the right is your score on the exam.

MATHEMATICS LEVEL IC TEST FORM B

1. D
2. C
3. E
4. A
5. B
6. D
7. E
8. C
9. B
10. A
11. B
12. D
13. B
14. C
15. A
16. D
17. E
18. A
19. B
20. E
21. C
22. C
23. E
24. D
25. C
26. A
27. B
28. D
29. D
30. B
31. E
32. A
33. B
34. E
35. B
36. A
37. C
38. A
39. C
40. A
41. D
42. A
43. A
44. E
45. C
46. C
47. D
48. C
49. E
50. B

MATHEMATICS LEVEL IC SUBJECT TEST SCORE CONVERSION TABLE

Raw Score	College Board Scaled Score	Raw Score	College Board Scaled Score
50	800	15	450
49	780	14	440
48	770	13	430
47	760	12	430
46	740	11	420
45	730	10	410
44	720	9	400
43	710	8	390
42	700	7	380
41	690	6	370
40	680	5	370
39	670	4	360
38	660	3	350
37	650	2	340
36	640	1	340
35	630	0	330
34	610	−1	320
33	600	−2	310
32	590	−3	300
31	580	−4	300
30	570	−5	290
29	560	−6	280
28	550	−7	270
27	550	−8	260
26	540	−9	260
25	530	−10	250
24	520	−11	240
23	510	−12	230
22	510		
21	500		
20	490		
19	480		
18	480		
17	470		
16	460		

MATHEMATICS LEVEL IIC TEST FORM A

Directions: For each of the following problems, decide which is the BEST of the choices given. Then fill in the corresponding oval on the answer sheet.

Notes: (1) Figures that accompany problems in this test are intended to provide information useful in solving the problems. They are drawn as accurately as possible EXCEPT when it is stated in a specific problem that its figure is not drawn to scale. All figures lie in a plane unless otherwise indicated.

(2) Unless otherwise specified, the domain of a function f is assumed to be the set of all real numbers x for which $f(x)$ is a real number.

(3) Reference information that may be useful in answering the questions in this test can be found to the right.

Reference Information: The following information is for your reference in answering some of the questions in this test.

Volume of a right circular cone with radius r and height h: $V = \frac{1}{3}\pi r^2 h$

Lateral Area of a right circular cone with circumference of the base c and slant height ℓ: $S = \frac{1}{2}c\ell$

Volume of a sphere with radius r: $V = \frac{4}{3}\pi r^3$

Surface Area of a sphere with radius r: $S = 4\pi r^2$

Volume of a pyramid with base area B and height h: $V = \frac{1}{3}Bh$

USE THIS SPACE FOR SCRATCHWORK.

1. If $r - s > r + s$, then

   (A) $r > s$
   (B) $s < 0$
   (C) $r < 0$
   (D) $r < s$
   (E) $s > 0$

2. If $f(x) = |x| + 10$, for which of the following values of x does $f(x) = f(-x)$?

   (A) -10 only
   (B) -10 and 10 only
   (C) All real x
   (D) All real x except 10
   (E) All real x except -10 and 10

GO ON TO THE NEXT PAGE

3. $\dfrac{15!}{13!\,2!} =$

(A) 0 (B) 0.58 (C) 1 (D) 105 (E) 210

4. In Figure 1, $\sin \angle ABC =$

(A) $\dfrac{5}{13}$

(B) $\dfrac{5}{12}$

(C) $\dfrac{12}{13}$

(D) $\dfrac{12}{5}$

(E) $\dfrac{13}{5}$

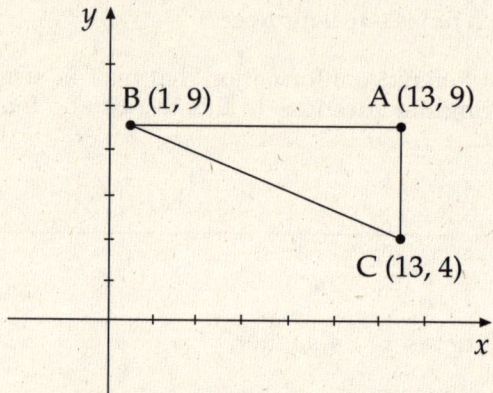

Figure 1

5. Which of the following is the complete solution set of the system:

$A = \{(x, y) : x^2 + y^2 = 25\}$ and
$B = \{(x, y) : y = x + 1\}$?

(A) $\{(5, 5)\}$
(B) $\{(16, 9)\}$
(C) $\{(-4, -3)\}$
(D) $\{(-4, -3), (3, 4)\}$
(E) $\{(-3, -4), (4, 3)\}$

MATHEMATICS LEVEL IIC TEST FORM A—Continued

6. If $jk \neq 0$, then $\dfrac{jk - \dfrac{j}{k}}{\dfrac{j}{k}} =$

    (A) $k^2 - \dfrac{j}{k}$

    (B) $j^2 - \dfrac{j^2}{k^2}$

    (C) $jk - 1$

    (D) $j^2 - 1$

    (E) $k^2 - 1$

7. All of the following can be formed by the intersection of a cube and a plane EXCEPT

    (A) a triangle
    (B) a point
    (C) a rectangle
    (D) a line
    (E) a circle

8. If $f(x) = \sqrt[3]{x}$ and $g(x) = \dfrac{1}{2}\sqrt{x} + 1$, then $f(g(2.3)) =$

    (A) 0.08 (B) 1.2 (C) 1.3 (D) 1.8 (E) 2.3

MATHEMATICS LEVEL IIC TEST FORM A—Continued

USE THIS SPACE FOR SCRATCHWORK.

9. If x mod y is the remainder when x is divided by y, then (61 mod 7) − (5 mod 5) =

   (A) 2
   (B) 3
   (C) 4
   (D) 5
   (E) 6

10. Which of the following must be true?

    I. $\sin(-\theta) = -\sin\theta$
    II. $\cos(-\theta) = -\cos\theta$
    III. $\tan(-\theta) = -\tan\theta$

    (A) I only
    (B) II only
    (C) III only
    (D) I and III only
    (E) I, II, and III

11. If for all real numbers x, a function $f(x)$ is defined by
    $f(x) = \begin{Bmatrix} 2, & x \neq 13 \\ 4, & x = 13 \end{Bmatrix}$, then $f(15) - f(14) =$

    (A) −2 (B) 0 (C) 1 (D) 2 (E) 4

12. If $\dfrac{x^5}{25} = 25$, then $x =$

    (A) 1.00
    (B) 1.90
    (C) 2.19
    (D) 3.62
    (E) 5.00

GO ON TO THE NEXT PAGE

MATHEMATICS LEVEL IIC TEST FORM A—Continued

USE THIS SPACE FOR SCRATCHWORK.

13. If the ratio of sec x to csc x is $1:4$, then the ratio of tan x to cot x is

    (A) $1:16$
    (B) $1:4$
    (C) $1:1$
    (D) $4:1$
    (E) $16:1$

14. In Figure 2, rectangle J consists of all points (x, y). What is the area of a rectangle that consists of all points $(2x, y - 1)$?

    (A) 12
    (B) 18
    (C) 24
    (D) 36
    (E) 48

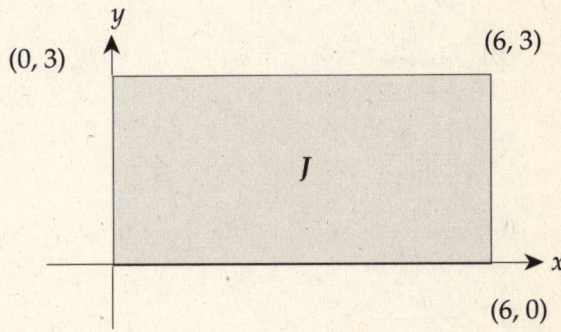

Figure 2

15. In right triangle ABC, $\angle B = 90°$, $\angle C = 27°$, and $AB = 9$. What is the length of the hypotenuse of $\triangle ABC$?

    (A) 4.1
    (B) 10.1
    (C) 17.7
    (D) 19.8
    (E) 21.2

GO ON TO THE NEXT PAGE

MATHEMATICS LEVEL IIC TEST FORM A—Continued

USE THIS SPACE FOR SCRATCHWORK.

16. Which of the following is a zero of $f(x) = x^2 + 6x - 12$?

    (A) −15.16   (B) −7.58   (C) 0.67   (D) 3.16   (E) 7.58

17. If $\sin x = m$ and $0 < x < 90°$, then $\tan x =$

    (A) $\dfrac{1}{m^2}$

    (B) $\dfrac{m}{\sqrt{1-m^2}}$

    (C) $\dfrac{1-m^2}{m}$

    (D) $\dfrac{m}{1-m^2}$

    (E) $\dfrac{m^2}{\sqrt{1-m^2}}$

18. If $\log_y 2 = 8$, then $y =$

    (A) 0.25
    (B) 1.04
    (C) 1.09
    (D) 2.83
    (E) 3.00

GO ON TO THE NEXT PAGE

MATHEMATICS LEVEL IIC TEST FORM A—Continued

USE THIS SPACE FOR SCRATCHWORK.

19. If $\sin \theta = \frac{1}{3}$ and $-\frac{\pi}{4} \leq \theta \leq \frac{\pi}{4}$, then $\cos(2\theta) =$

    (A) $-\frac{7}{9}$
    (B) $-\frac{2}{3}$
    (C) $\frac{2}{3}$
    (D) $\frac{7}{9}$
    (E) 1

20. Which of the following is the inverse of the function $f(x) = \sqrt{x} - 1$, for all $x > 0$?

    (A) $(x + 1)^2$
    (B) $x^2 + 2$
    (C) $x^2 + 1$
    (D) $(x - 1)^2$
    (E) $(x + 2)^2$

GO ON TO THE NEXT PAGE

MATHEMATICS LEVEL IIC TEST FORM A—Continued

USE THIS SPACE FOR SCRATCHWORK.

21. When $4x^2 + 6x + L$ is divided by $x + 1$, the remainder is 2. Which of the following is the value of L?

    (A) 4
    (B) 6
    (C) 10
    (D) 12
    (E) 15

22. What is the length of the major axis of the ellipse $\frac{x^2}{10} + \frac{y^2}{20} = 1$?

    (A) 3.2
    (B) 4.5
    (C) 8.9
    (D) 10.0
    (E) 20.0

GO ON TO THE NEXT PAGE

23. If $f(x) = [x]$, where $[x]$ is the greatest integer less than or equal to x, which of the following is a graph of $f\left(\dfrac{x}{2}\right) - 1$?

(A)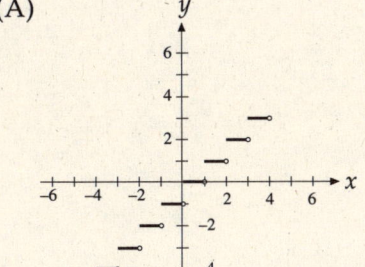

(B)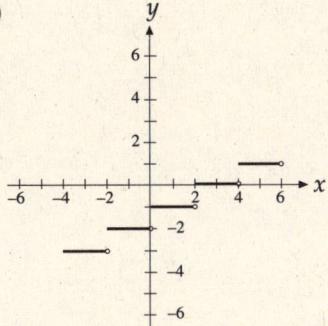

(C)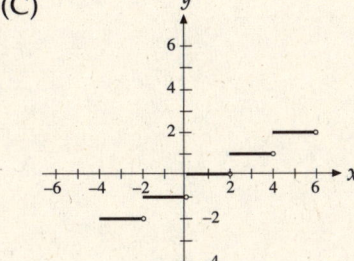

(D)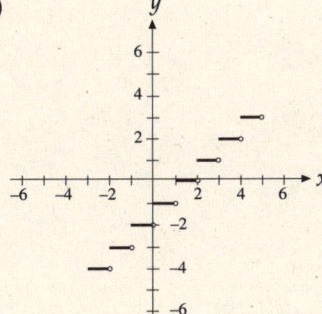

(E)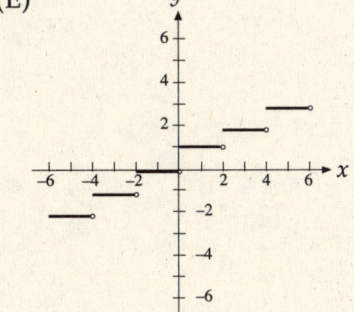

MATHEMATICS LEVEL IIC TEST FORM A—Continued

24. Which of the following is equal to the positive value of sec(arccos[0.3527])?

    (A) 0.01
    (B) 0.94
    (C) 1.69
    (D) 2.84
    (E) 69.35

25. For what value of x does $f(x) = x^2 + 5x + 6$ reach its minimum value?

    (A) -3
    (B) $-\frac{5}{2}$
    (C) -2
    (D) 0
    (E) $\frac{5}{2}$

26. If the 20th term of an arithmetic sequence is 20 and the 50th term is 100, what is the first term of the sequence?

    (A) −33.33
    (B) −30.67
    (C) 1.00
    (D) 2.00
    (E) 2.67

MATHEMATICS LEVEL IIC TEST FORM A—Continued

27. The polar equation $r \sin \theta = 1$ defines

    (A) a line
    (B) a circle
    (C) an ellipse
    (D) a parabola
    (E) a hyperbola

28. For which of the following functions is the inverse of the function also a function?

    I. $f(x) = x^2$
    II. $f(x) = x^3$
    III. $f(x) = |x|$

    (A) I only
    (B) II only
    (C) I and III only
    (D) II and III only
    (E) I, II, and III

29. What is $\lim\limits_{x \to -1} \dfrac{x^3 - x}{x + 1}$?

    (A) -2
    (B) -1
    (C) 1
    (D) 2
    (E) The limit does not exist.

MATHEMATICS LEVEL IIC TEST FORM A—Continued

30. If $f(x) = \dfrac{e^{7x} + \sqrt{3}}{2}$, and $g(f(x)) = x$, then $g(x) =$

    (A) $\dfrac{\ln(2x - \sqrt{3})}{7}$

    (B) $\dfrac{2x - \sqrt{3}}{e^7}$

    (C) $\dfrac{2x - \sqrt{3}}{7}$

    (D) $7 \ln(2x - \sqrt{3})$

    (E) $\dfrac{(2x - \sqrt{3}) \ln e}{7}$

31. A cube is inscribed in a sphere of radius 6. What is the volume of the cube?

    (A) $36\sqrt{3}$ (B) 36π (C) 216 (D) $192\sqrt{3}$ (E) $216\sqrt{3}$

32. A right circular cone has height h and radius r. If the cone is cut in two pieces by a plane parallel to the base and midway to the vertex, the volume of the larger of the two resulting figures is

    (A) $\dfrac{\pi r^2 h}{6}$

    (B) $\dfrac{\pi r^2 h}{3}$

    (C) $\dfrac{\pi r^2 h}{2}$

    (D) $\dfrac{2\pi r^2 h}{3}$

    (E) $\dfrac{7\pi r^2 h}{24}$

MATHEMATICS LEVEL IIC TEST FORM A—Continued

USE THIS SPACE FOR SCRATCHWORK.

33. If $e^x \neq 1$ and $e^{x^2} = \dfrac{1}{\sqrt{3}^x}$, then $x =$

    (A) −1.73
    (B) −0.55
    (C) 1.00
    (D) 1.10
    (E) 1.73

34. If the graph of the equation $y = 2x^2 − 6x + c$ is tangent to the x-axis, then the value of c is

    (A) 3 (B) $3\dfrac{1}{2}$ (C) 4 (D) $4\dfrac{1}{2}$ (E) 5

35. If $x = i − 1$, then $x^2 + 2x + 2 =$

    (A) $2i + 4$
    (B) $4 + 2i$
    (C) 0
    (D) i
    (E) −2

36. The curve shown in Figure 3 could represent a portion of the graph of which of the following functions?

    (A) $y = e^x$
    (B) $y = e^{-x}$
    (C) $y = 100 − x$
    (D) $y = x^2 − 3x + 2$
    (E) $xy = 3$

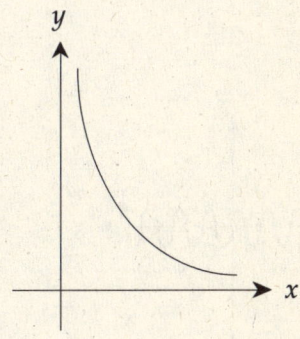

Figure 3

GO ON TO THE NEXT PAGE

MATHEMATICS LEVEL IIC TEST FORM A—Continued

USE THIS SPACE FOR SCRATCHWORK.

37. If two coins are removed at random from a purse containing three nickels and eight dimes, what is the probability that both coins will be dimes?

    (A) $\frac{14}{55}$ (B) $\frac{49}{110}$ (C) $\frac{28}{55}$ (D) $\frac{64}{121}$ (E) $\frac{32}{55}$

38. A function $g(x)$ is odd if $g(-x) = -g(x)$ for all x and even if $g(x) = g(-x)$ for all x. Which of the following functions is both odd and even?

(A) 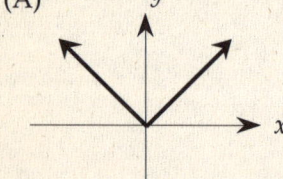   (B)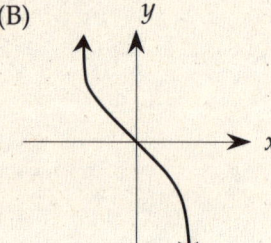

(C) 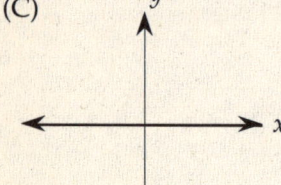   (D)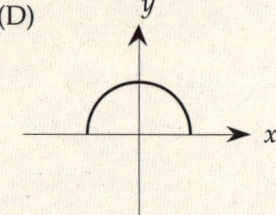

(E)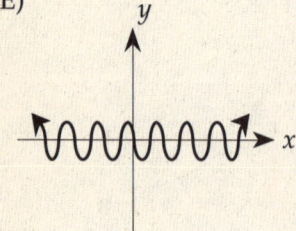

GO ON TO THE NEXT PAGE

MATHEMATICS LEVEL IIC TEST FORM A—Continued

USE THIS SPACE FOR SCRATCHWORK.

39. Points A and B lie on the edge of a circle with center O. If the circle has a radius of 5, and if the measure of $\angle AOB$ is 70°, what is the length of chord AB?

  (A) 2.9
  (B) 4.7
  (C) 5.0
  (D) 5.7
  (E) 9.4

40. Which of the equations below best represents the graph shown in Figure 4?

  (A) $f(x) = \tan\left(x - \dfrac{\pi}{4}\right)$

  (B) $f(x) = \cot\left(x - \dfrac{\pi}{4}\right)$

  (C) $f(x) = \tan\left(x + \dfrac{\pi}{2}\right)$

  (D) $f(x) = \cot\left(x + \dfrac{\pi}{4}\right)$

  (E) $f(x) = \tan\left(x + \dfrac{\pi}{4}\right)$

41. In Figure 5, vectors v and w both begin at the origin and end at (–3, 4) and (12, 5), respectively. If vector z is such that $z = -(v + w)$, and begins at the origin, at what point does it end?

  (A) (–9, –9)

  (B) (5, 13)

  (C) (–5, 13)

  (D) (9, 9)

  (E) $\left(-\dfrac{9}{2}, -\dfrac{9}{2}\right)$

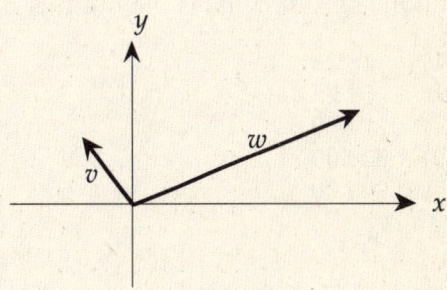

Figure 5

GO ON TO THE NEXT PAGE

42. If $f(x) = \dfrac{1}{\sqrt{2\pi}} e^{-\frac{x^2}{2}}$, what is the positive value of x for which $f(x) = 0.33$?

(A) 0.62
(B) 0.71
(C) 1.36
(D) 3.93
(E) 4.95

43. In Figure 6, which of the following must be true?

  I. $\angle SVT = \dfrac{1}{2} \angle SOT$

  II. arc ST = arc VT

  III. $\angle SVT = 120°$

(A) I only
(B) II only
(C) I and II only
(D) II and III only
(E) I, II, and III

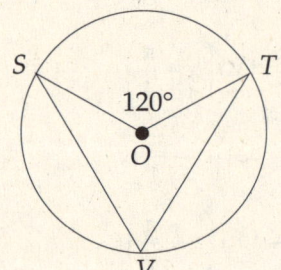

O is the center of the circle.

Figure 6

44. If $f(x, y) = \dfrac{xy}{3}$ for all x, y, and $f(a, b) = 15$, $f(b, c) = 20$, and $f(a, c) = 10$, what is the product of a, b, and c?

(A)   18.26
(B)   54.77
(C)   284.60
(D) 1,800.00
(E) 3,000.00

MATHEMATICS LEVEL IIC TEST FORM A—Continued

USE THIS SPACE FOR SCRATCHWORK.

45. If $x > 0$ and $y > 1$, then $\log_{x^2} y =$

   I. $\log_x y^2$
   II. $\log_x \sqrt{y}$
   III. $\log_x \left(\dfrac{y}{2}\right)$

   (A) I only
   (B) II only
   (C) III only
   (D) I and II only
   (E) II and III only

46. Figure 7 shows regular heptagon $ABCDEFG$ where the length of BD is 9. What is the length of diagonal GD?

   (A) 5.6
   (B) 7.0
   (C) 9.0
   (D) 11.2
   (E) 14.1

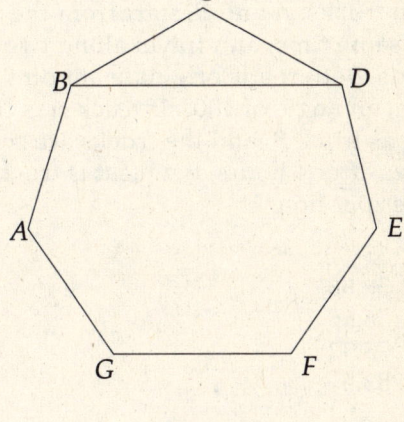

Figure 7

47. What is the value of $|6 - 3i|$?

   (A) -3 (B) $3\sqrt{2}$ (C) $3\sqrt{5}$ (D) 9   (E) 15

GO ON TO THE NEXT PAGE

MATHEMATICS LEVEL IIC TEST FORM A—Continued

USE THIS SPACE FOR SCRATCHWORK.

48. The menu of a certain restaurant lists 10 items in column A and 20 items in column B. A family plans to share 5 items from column A and 5 items from column B. How many different combinations of items could the family choose?

   (A)   25
   (B)   200
   (C)   3,425
   (D) 3,907,008
   (E) 5.63×10^{10}

49. Two trucks, A and B, start from the same point X at the same time, and travel along two distinct straight roads. Both roads originate at point X such that they form an angle of 100°. If truck A is traveling twice as fast as truck B, and the trucks are separated by 334 miles after 4 hours, how fast is truck B traveling in miles per hour?

   (A) 35
   (B) 36.66
   (C) 56.66
   (D) 73.32
   (E) 83.5

50. Seven blue marbles and six red marbles are held in a single container. Marbles are randomly selected one at a time and not returned to the container. If the first two marbles selected are both blue, what is the probability that *at least* two red marbles will be chosen in the next three selections?

   (A) $\dfrac{5}{33}$ (B) $\dfrac{5}{11}$ (C) $\dfrac{6}{11}$ (D) $\dfrac{19}{33}$ (E) $\dfrac{2}{3}$

STOP

IF YOU FINISH BEFORE TIME IS CALLED, YOU MAY CHECK YOUR WORK ON THIS TEST ONLY.
DO NOT WORK ON ANY OTHER TEST IN THIS BOOK.

HOW TO SCORE THE PRINCETON REVIEW MATH SUBJECT TEST

When you take the real exam, the proctors will collect your text booklet and bubble sheet and send your answer sheet to New Jersey where a computer (yes, a big old-fashioned one that has been around since the 1960s) looks at the pattern of filled-in ovals on your answer sheet and gives you a score. We couldn't include even a small computer with this book, so we are providing this more primitive way of scoring your exam.

DETERMINING YOUR SCORE

STEP 1 Using the answers on the next page, determine how many questions you got right and how many you got wrong on the test. Remember, questions that you do not answer don't count as either right answers or wrong answers.

STEP 2 List the number of right answers here.   (A) _____

STEP 3 List the number of wrong answers here. Now divide that number by 4. (Use a calculator if you're feeling particularly lazy.)   (B) _____ ÷ 4 = (C) _____

STEP 4 Subtract the number of wrong answers divided by 4 from the number of correct answers. Round this score to the nearest whole number. This is your raw score.   (A) _____ − (C) _____ = _____

STEP 5 To determine your real score, take the number from Step 4 above and look it up in the left column of the Score Conversion Table on page 359; the corresponding score on the right is your score on the exam.

MATHEMATICS LEVEL IIC TEST FORM A

1. B	11. B	21. A	31. D	41. A
2. C	12. D	22. C	32. E	42. A
3. D	13. A	23. B	33. B	43. A
4. A	14. D	24. D	34. D	44. C
5. D	15. D	25. B	35. C	45. B
6. E	16. B	26. B	36. E	46. D
7. E	17. B	27. A	37. C	47. C
8. B	18. C	28. B	38. C	48. D
9. D	19. D	29. D	39. D	49. A
10. D	20. A	30. A	40. D	50. D

MATHEMATICS LEVEL IIC SUBJECT TEST SCORE CONVERSION TABLE

Raw Score	College Board Scaled Score	Raw Score	College Board Scaled Score
50	800	15	560
49	800	14	540
48	800	13	530
47	800	12	520
46	800	11	500
45	800	10	490
44	800	9	470
43	800	8	460
42	790	7	440
41	780	6	430
40	770	5	400
39	760	4	400
38	750	3	380
37	740	2	370
36	730	1	360
35	720	0	350
34	710	−1	330
33	700	−2	320
32	690	−3	300
31	690	−4	280
30	680	−5	270
29	670	−6	250
28	660	−7	230
27	650	−8	220
26	640	−9	200
25	630	−10	200
24	630	−11	200
23	620	−12	200
22	610		
21	600		
20	590		
19	590		
18	580		
17	570		
16	560		

MATHEMATICS LEVEL IIC TEST FORM B

Directions: For each of the following problems, decide which is the BEST of the choices given. Then fill in the corresponding oval on the answer sheet.

Notes: (1) Figures that accompany problems in this test are intended to provide information useful in solving the problems. They are drawn as accurately as possible EXCEPT when it is stated in a specific problem that its figure is not drawn to scale. All figures lie in a plane unless otherwise indicated.

(2) Unless otherwise specified, the domain of a function f is assumed to be the set of all real numbers x for which $f(x)$ is a real number.

(3) Reference information that may be useful in answering the questions in this test can be found to the right.

Reference Information: The following information is for your reference in answering some of the questions in this test.

Volume of a right circular cone with radius r and height h: $V = \frac{1}{3}\pi r^2 h$

Lateral Area of a right circular cone with circumference of the base c and slant height ℓ: $S = \frac{1}{2}c\ell$

Volume of a sphere with radius r: $V = \frac{4}{3}\pi r^3$

Surface Area of a sphere with radius r: $S = 4\pi r^2$

Volume of a pyramid with base area B and height h: $V = \frac{1}{3}Bh$

USE THIS SPACE FOR SCRATCHWORK.

1. If $xy \neq 0$ and $3x = 0.3y$, then $\frac{y}{x} =$

   (A) 0.1
   (B) 1.0
   (C) 3.0
   (D) 9.0
   (E) 10.0

2. If $f(x) = \left(3\sqrt{x} - 4\right)^2$, then how much does $f(x)$ increase as x goes from 2 to 3?

   (A) 1.43
   (B) 1.37
   (C) 1.00
   (D) 0.74
   (E) 0.06

GO ON TO THE NEXT PAGE

MATHEMATICS LEVEL IIC TEST FORM B—Continued

USE THIS SPACE FOR SCRATCHWORK.

3. What is the equation of a line with a y-intercept of 3 and an x-intercept of -5?

   (A) $y = 0.6x + 3$
   (B) $y = 1.7x - 3$
   (C) $y = 3x + 5$
   (D) $y = 3x - 5$
   (E) $y = -5x + 3$

4. For what positive value of a does $a - \sqrt{5a + 18} = -4$?

   (A) 0.56
   (B) 1.00
   (C) 1.12
   (D) 2.06
   (E) 4.12

5. If the second term in an arithmetic sequence is 4, and the tenth term is 15, what is the first term in the sequence?

   (A) 1.18
   (B) 1.27
   (C) 1.38
   (D) 2.63
   (E) 2.75

6. If $g(x) = |5x^2 - x^3|$, then $g(6) =$

   (A) -54
   (B) -36
   (C) 36
   (D) 216
   (E) 396

GO ON TO THE NEXT PAGE

MATHEMATICS LEVEL IIC TEST FORM B—Continued

USE THIS SPACE FOR SCRATCHWORK.

7. Which of the following graphs of functions is symmetrical across the line $x = y$?

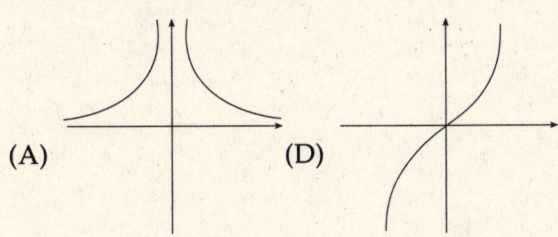
(A)   (D)

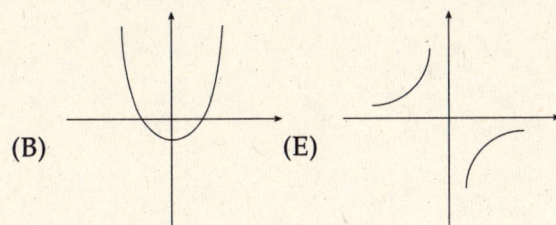
(B)   (E)

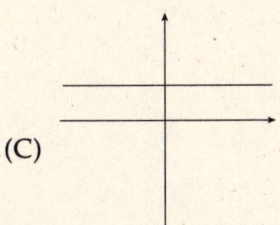
(C)

8. How many distinct triangles may be traced in Figure 1?

(A) 9
(B) 10
(C) 18
(D) 19
(E) 20

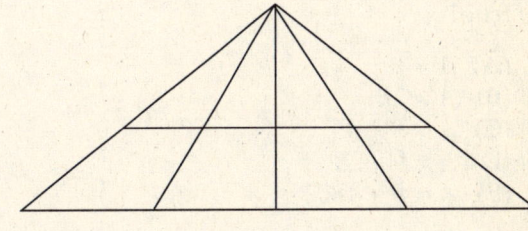

Figure 1

GO ON TO THE NEXT PAGE

MATHEMATICS LEVEL IIC TEST FORM B—Continued

USE THIS SPACE FOR SCRATCHWORK.

9. If $f(x) = \frac{1}{2}x^2 - 6x + 11$, then what is the minimum value of $f(x)$?

   (A) −8.0
   (B) −7.0
   (C) 3.2
   (D) 6.0
   (E) 11.0

10. $|x - y| + |y - x| =$

    (A) 0
    (B) $x - y$
    (C) $y - x$
    (D) $2|x - y|$
    (E) $2|x + y|$

$$0° \leq A \leq 90°$$
$$0° \leq B \leq 90°$$

11. If $\sin A = \cos B$, then which of the following must be true?

    (A) $A = B$
    (B) $A = 2B$
    (C) $A = B + 45$
    (D) $A = 90 - B$
    (E) $A = B + 180$

GO ON TO THE NEXT PAGE

MATHEMATICS LEVEL IIC TEST FORM B—*Continued*

USE THIS SPACE FOR SCRATCHWORK.

12. Each month, some of the automobiles produced at the Carco plant have flawed catalytic converters. According to the chart in Figure 2, what is the probability that a car produced in one of the three months shown will be flawed?

    (A) 0.01
    (B) 0.02
    (C) 0.03
    (D) 0.04
    (E) 0.05

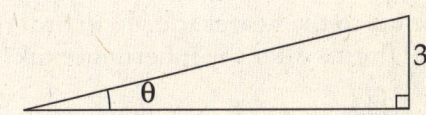

	Total Units Produced	Units Flawed
April	569	15
May	508	18
June	547	16

Figure 2

13. Gartonsville building codes require that a wheelchair ramp must rise at an angle (θ) of no less than 5° and no more than 7° from the horizontal. If a wheelchair ramp rises exactly 3 feet, which of the following could be the length of the ramp?

    (A) 19.0 feet
    (B) 24.0 feet
    (C) 28.0 feet
    (D) 35.0 feet
    (E) 42.0 feet

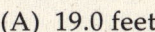

Figure 3

14. Figure 4 represents the graph of the function $y = -x^4 - 4x^3 + 14x^2 + 45x - n$. Which of the following could be the value of n?

    (A) −50
    (B) −18
    (C) 50
    (D) 100
    (E) 150

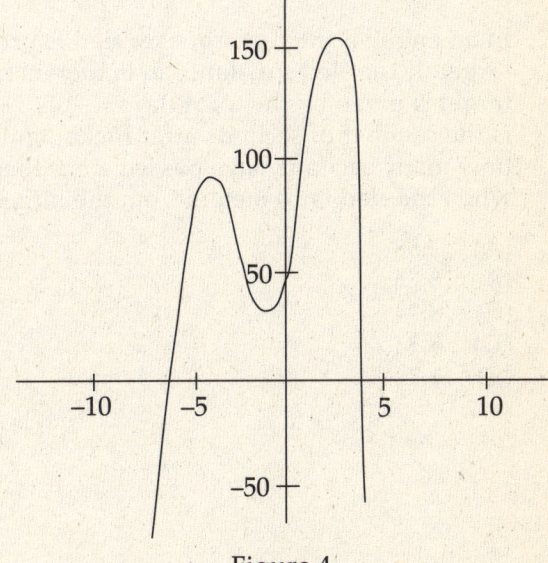

Figure 4

GO ON TO THE NEXT PAGE

MATHEMATICS LEVEL IIC TEST FORM B—Continued

USE THIS SPACE FOR SCRATCHWORK.

15. What value does the expression $\frac{x^2 - x - 6}{3x + 6}$ approach as x approaches -2?

    (A) -1.67
    (B) -0.60
    (C) 0
    (D) 1.00
    (E) 2.33

16. In Titheland, the first 1,000 florins of any inheritance are untaxed. After the first 1,000 florins, inheritances are taxed at a rate of 65%. How large must an inheritance be, to the nearest florin, in order to amount to 2,500 florins after the inheritance tax?

    (A) 7,143
    (B) 5,286
    (C) 4,475
    (D) 3,475
    (E) 3,308

17. In an engineering test, a rocket sled is propelled into a target. If the sled's distance d in meters from the target is given by the formula $d = -1.5t^2 + 120$, where t is the number of seconds after rocket ignition, then how many seconds have passed since rocket ignition when the sled is 10 meters from the target?

    (A) 2.58
    (B) 8.56
    (C) 8.94
    (D) 9.31
    (E) 11.26

GO ON TO THE NEXT PAGE

MATHEMATICS LEVEL IIC TEST FORM B—Continued

USE THIS SPACE FOR SCRATCHWORK.

18. $3 + 1 + \dfrac{1}{3} + \dfrac{1}{9} ... =$

   (A) 4.33
   (B) 4.44
   (C) 4.50
   (D) 6.00
   (E) The sum is not finite.

19. If $e^f = 5$, then $f =$

   (A) 0.23
   (B) 1.61
   (C) 7.76
   (D) 148.41
   (E) It cannot be determined from the information given.

20. If the greatest possible distance between two points within a certain rectangular solid is 12, then which of the following could be the dimensions of this solid?

   (A) $3 \times 3 \times 9$
   (B) $3 \times 6 \times 7$
   (C) $3 \times 8 \times 12$
   (D) $4 \times 7 \times 9$
   (E) $4 \times 8 \times 8$

21. Runner A travels a feet every minute. Runner B travels b feet every second. In one hour, runner A travels how much farther than runner B, in feet?

   (A) $a - 60b$
   (B) $a^2 - 60b^2$
   (C) $360a - b$
   (D) $60(a - b)$
   (E) $60(a - 60b)$

GO ON TO THE NEXT PAGE

22. A right triangle has sides in the ratio of 5:12:13. What is the measure of the triangle's smallest angle, in degrees?

(A) 13.34
(B) 22.62
(C) 34.14
(D) 42.71
(E) 67.38

$$f(x) = \frac{1}{x+1}$$

$$g(x) = \frac{1}{x} + 1$$

23. What is $g(f(x))$?

(A) 2
(B) $x + 2$
(C) $2x + 2$
(D) $\dfrac{x+2}{x+1}$
(E) $\dfrac{2x+1}{x+1}$

24. If $f(x) = (x - \pi)(x - 3)(x - e)$, then what is the difference between the greatest and least roots of $f(x)$?

(A) 0.14
(B) 0.28
(C) 0.36
(D) 0.42
(E) 0.72

MATHEMATICS LEVEL IIC TEST FORM B—Continued

25. $\dfrac{x!}{(x-2)!} =$

   (A) 0.5
   (B) 2.0
   (C) x
   (D) $x^2 - x$
   (E) $x^2 - 2x + 1$

26. What is the volume of the solid created by rotating rectangle $ABCD$ around the y-axis?

   (A) 219.91
   (B) 245.00
   (C) 549.78
   (D) 769.69
   (E) 816.24

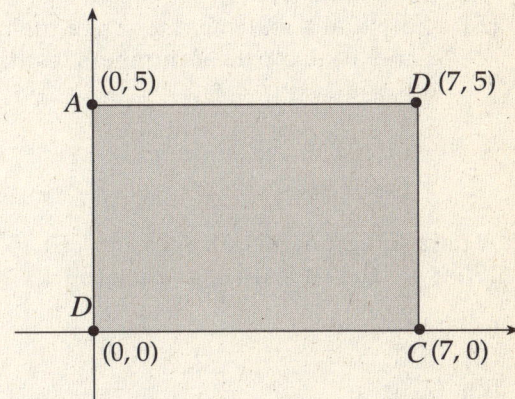

Figure 5

27. If $f(x,y) = \dfrac{x^2 - 2xy + y^2}{x^2 - y^2}$, then $f(-x,-y) =$

   (A) 1
   (B) $\dfrac{1}{x+y}$
   (C) $\dfrac{-x+y}{x+y}$
   (D) $\dfrac{-x+y}{x-y}$
   (E) $\dfrac{x-y}{x+y}$

MATHEMATICS LEVEL IIC TEST FORM B—Continued

USE THIS SPACE FOR SCRATCHWORK.

28. In order to disprove the hypothesis, "No number divisible by 5 is less than 5," it would be necessary to

(A) prove the statement false for all numbers divisible by 5
(B) demonstrate that numbers greater than 5 are often divisible by 5
(C) indicate that infinitely many numbers greater than 5 are divisible by 5
(D) supply one case in which a number divisible by 5 is less than 5
(E) show that a statement true of numbers greater than 5 is also true of numbers less than 5

29. A parallelogram has vertices at (0, 0), (5, 0), and (2, 3). What are the coordinates of the fourth vertex?

(A) (3, –2)
(B) (5, 3)
(C) (7, 3)
(D) (10, 5)
(E) It cannot be determined from the information given.

30. The expression $\dfrac{x^2 + 3x - 4}{2x^2 + 10x + 8}$ is undefined for what values of x?

(A) $x = \{-1, -4\}$
(B) $x = \{-1\}$
(C) $x = \{0\}$
(D) $x = \{1, -4\}$
(E) $x = \{0, 1, 4\}$

GO ON TO THE NEXT PAGE

MATHEMATICS LEVEL IIC TEST FORM B—Continued

USE THIS SPACE FOR SCRATCHWORK.

31. For which of the following functions is $f(x)$ positive for all real values of x?

    I. $f(x) = x^2 + 1$
    II. $f(x) = 1 - \sin x$
    III. $f(x) = \pi(\pi^{x-1})$

    (A) I only
    (B) II only
    (C) I and III only
    (D) II and III only
    (E) I, II, and III.

32. The graph of $y = f(x)$ is shown in Figure 6. Which of the following could be the graph of $y = -f(-x)$?

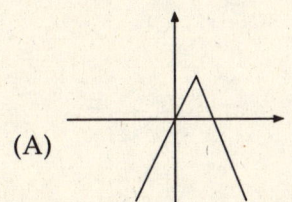
(A)   (B)

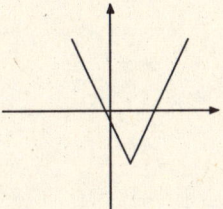

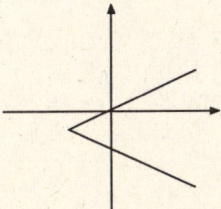
(C)   (D)

(E)

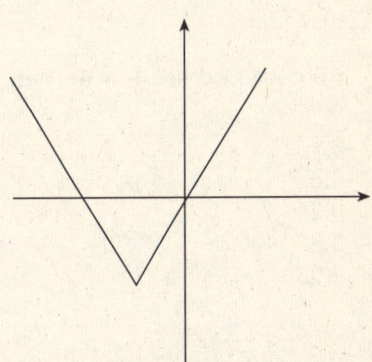

Figure 6

MATHEMATICS LEVEL IIC TEST FORM B—Continued

USE THIS SPACE FOR SCRATCHWORK.

33. A wire is stretched from the top of a two-foot-tall anchor to the top of a 50-foot-tall antenna. If the wire is straight and has a slope of $\frac{2}{5}$, then what is the wire's length in feet?

    (A) 89.18
    (B) 120.00
    (C) 123.26
    (D) 129.24
    (E) 134.63

34. If $\frac{3\pi}{2} < \theta < 2\pi$ and sec $\theta = 4$, then tan $\theta =$

    (A) −3.93
    (B) −3.87
    (C) 0.26
    (D) 3.87
    (E) 3.93

35. Circle O is centered at (−3, 1) and has a radius of 4. Circle P is centered at (4, −4) and has a radius of n. If circle O is externally tangent to circle P, then what is the value of n?

    (A) 4.00
    (B) 4.37
    (C) 4.60
    (D) 5.28
    (E) 6.25

GO ON TO THE NEXT PAGE

36. In triangle ABC, $\frac{\sin A}{\sin B} = \frac{7}{10}$ and $\frac{\sin B}{\sin C} = \frac{5}{2}$. If angles A, B, and C are opposite sides a, b, and c, respectively, and the triangle has a perimeter of 16, then what is the length of a?

   (A) 2.7
   (B) 4.7
   (C) 5.3
   (D) 8.0
   (E) 14.0

x	$h(x)$
−1	0
0	3
1	0
2	3

37. The table of values above shows selected coordinate pairs on $h(x)$. If h is a third-degree polynomial, then which of the following could be $h(x)$?

   (A) $x(x + 1)(x - 1)$
   (B) $(x + 1)^2(x - 1)$
   (C) $(x - 1)(x + 2)^2$
   (D) $(x - 1)^2(x + 3)$
   (E) $(x - 1)(x + 1)(2x - 3)$

$$a + b + 2c = 7$$
$$a - 2b = 8$$
$$3b + 2c = n$$

38. For what values of n does the system of equations above have no real solution?

   (A) $\{n \neq -1\}$
   (B) $\{n \leq 0\}$
   (C) $\{n \geq 1\}$
   (D) $\{n > 7\}$
   (E) $\{n = -15\}$

MATHEMATICS LEVEL IIC TEST FORM B—Continued

USE THIS SPACE FOR SCRATCHWORK.

39. In Figure 7, what is the value of θ in degrees?

    (A) 62.00
    (B) 65.38
    (C) 65.91
    (D) 68.49
    (E) 68.70

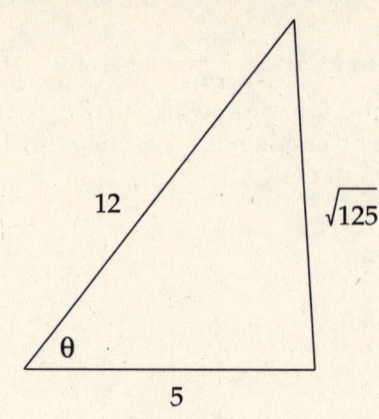

Figure 7

Note: Figure Not Drawn to Scale

$$[a \otimes b] = e^a - ab + 1$$

40. If $[1.08 \otimes x]$ is positive and $[1.09 \otimes x]$ is negative, then which of the following could be the value of x?

    (A) 3.62
    (B) 3.65
    (C) 3.71
    (D) 3.73
    (E) 3.77

41. In the function $g(x) = A[\sin(Bx + C)] + D$, constants are represented by A, B, C, and D. If $g(x)$ is to be altered in such a way that both its period and amplitude are increased, which of the following constants must be increased?

    (A) A only
    (B) B only
    (C) C only
    (D) A and B only
    (E) C and D only

GO ON TO THE NEXT PAGE

MATHEMATICS LEVEL IIC TEST FORM B—Continued

USE THIS SPACE FOR SCRATCHWORK.

42. The elements of set M and set N are arranged in exactly 20 pairs, such that each pair contains exactly one element from set M and one element from set N. If in each such pair, the element from set M is larger than the element from set N, then which of the following statements must be true?

    I. The median of M is greater than the median of N.
    II. The mean of M is greater than the mean of N.
    III. The mode of M is greater than the mode of N.
    IV. The range of M is greater than the range of N.

    (A) I and II only
    (B) II and IV only
    (C) III and IV only
    (D) I, II, and III only
    (E) I, II, and IV only

43. If $\log_{(x-4)} y = 3$, then $y =$

    (A) $4x - 16$
    (B) $4x^2 - 48x + 64$
    (C) $x^3 - 64$
    (D) $x^3 + x^2 + x + 4$
    (E) $x^3 - 16x^2 + 64x - 64$

44. If $0 \leq n \leq \dfrac{\pi}{2}$ and $\cos(\cos n) = 0.8$, then $\tan n =$

    (A) 0.65
    (B) 0.75
    (C) 0.83
    (D) 1.19
    (E) 1.22

GO ON TO THE NEXT PAGE

MATHEMATICS LEVEL IIC TEST FORM B—Continued

USE THIS SPACE FOR SCRATCHWORK.

45. The height of a cylinder is equal to one half of n, where n is equal to one half of the cylinder's diameter. What is the surface area of this cylinder in terms of n?

    (A) $\dfrac{3\pi n^2}{2}$
    (B) $2\pi n^2$
    (C) $3\pi n^2$
    (D) $2\pi n^2 + \dfrac{\pi n}{2}$
    (E) $2\pi n^2 + \pi n$

46. A number is defined as a *trignot* when all of the six trigonometric functions can produce that number. Which of the following numbers is a *trignot*?

    (A) −2.0
    (B) −0.5
    (C) 0.0
    (D) 0.5
    (E) 1.0

47. Which of the following expresses the range of $g(x) = \dfrac{5}{x+4}$?

    (A) $\{y:\ y \neq 0\}$
    (B) $\{y:\ y \neq 1.25\}$
    (C) $\{y:\ y \neq -4.00\}$
    (D) $\{y:\ y > 0\}$
    (E) $\{y:\ y \leq -1 \text{ or } y \geq 1\}$

GO ON TO THE NEXT PAGE

MATHEMATICS LEVEL IIC TEST FORM B—Continued

48. Vectors $\vec{w}$, $\vec{x}$, $\vec{y}$, and $\vec{z}$ have magnitudes of w, x, y, and z, respectively. If $\vec{x} + \vec{y} = \vec{z}$, $\vec{x} - \vec{y} = \vec{w}$, and $w = z$, then which of the following must be true?

(A) $y = 0$
(B) $x = y$
(C) $2y = w$
(D) $z > x$
(E) $x^2 + y^2 = z^2$

49. If $f(x, y) = \dfrac{xy + y}{x + y}$, then which of the following statements must be true?

   I. If $x = 0$ and $y \neq 0$, then $f(x, y) = 1$
   II. If $x = 1$, then $f(x, x) = 1$
   III. If $f(x, y) = f(y, x)$

(A) I only
(B) II only
(C) I and II only
(D) I and III only
(E) I, II, and III

50. A triangle is formed by the x-axis, the y-axis, and the line $y = mx + b$. If $m = -b^3$, then what is the volume of the cone generated by rotating this triangle around the x-axis?

(A) $\dfrac{\pi}{9}$
(B) $\dfrac{\pi}{3}$
(C) 3π
(D) 9π
(E) It cannot be determined from the information given.

STOP

IF YOU FINISH BEFORE TIME IS CALLED, YOU MAY CHECK YOUR WORK ON THIS TEST ONLY.
DO NOT WORK ON ANY OTHER TEST IN THIS BOOK.

HOW TO SCORE THE PRINCETON REVIEW MATH SUBJECT TEST

When you take the real exam, the proctors will collect your text booklet and bubble sheet and send your answer sheet to New Jersey where a computer (yes, a big old-fashioned one that has been around since the 1960s) looks at the pattern of filled-in ovals on your answer sheet and gives you a score. We couldn't include even a small computer with this book, so we are providing this more primitive way of scoring your exam.

DETERMINING YOUR SCORE

STEP 1   Using the answers on the next page, determine how many questions you got right and how many you got wrong on the test. Remember, questions that you do not answer don't count as either right answers or wrong answers.

STEP 2   List the number of right answers here.   (A) _____

STEP 3   List the number of wrong answers here. Now divide that number by 4. (Use a calculator if you're feeling particularly lazy.)   (B) _____ ÷ 4 = (C) _____

STEP 4   Subtract the number of wrong answers divided by 4 from the number of correct answers. Round this score to the nearest whole number. This is your raw score.   (A) _____ − (C) _____ = _____

STEP 5   To determine your real score, take the number from Step 4 above and look it up in the left column of the Score Conversion Table on page 380; the corresponding score on the right is your score on the exam.

MATHEMATICS LEVEL IIC

1. E	11. D	21. E	31. C	41. A
2. B	12. C	22. B	32. A	42. A
3. A	13. C	23. B	33. D	43. E
4. A	14. A	24. D	34. B	44. D
5. D	15. A	25. D	35. C	45. C
6. C	16. B	26. D	36. C	46. C
7. E	17. B	27. E	37. E	47. A
8. E	18. C	28. D	38. A	48. E
9. B	19. B	29. E	39. D	49. C
10. D	20. E	30. A	40. B	50. B

MATHEMATICS LEVEL IIC SUBJECT TEST SCORE CONVERSION TABLE

Raw Score	College Board Scaled Score	Raw Score	College Board Scaled Score
50	800	15	570
49	800	14	560
48	800	13	560
47	800	12	550
46	800	11	530
45	800	10	520
44	800	9	500
43	800	8	480
42	800	7	470
41	790	6	450
40	780	5	430
39	770	4	420
38	760	3	400
37	760	2	380
36	750	1	370
35	740	0	350
34	730	−1	330
33	720	−2	320
32	710	−3	300
31	710	−4	280
30	700	−5	270
29	690	−6	250
28	680	−7	230
27	670	−8	220
26	660	−9	200
25	660	−10	200
24	650	−11	200
23	640	−12	200
22	630		
21	620		
20	610		
19	610		
18	600		
17	590		
16	580		

14
Subject Test Answers and Explanations

LEVEL IC TEST FORM A

1. **(B)** Isolate x by adding x to each side and subtracting 5 from each side to produce the equation $-2 = x$.

   You can also backsolve this question. Just plug in values from the answer choices, starting with (C); the one that makes the equation $3 - x = 5$ true is the correct answer.

2. **(A)** The fraction $\frac{b}{a}$ is the reciprocal of $\frac{a}{b}$; it's just flipped over. To find the numerical value of $\frac{b}{a}$, just flip the numerical value of $\frac{a}{b}$, which is 0.625. Your calculator will tell you that $\frac{1}{0.625} = 1.6$.

3. **(C)** Backsolve. Starting with (C), plug values from the answer choices into the equation $x - 3 = 3(1 - x)$. Use your calculator to see which value makes the equation true. The first one you try, (C), is correct.

   To solve the problem algebraically, isolate x one step at a time. First, multiply through by 3 on the right: $x - 3 = 3 - 3x$. Next, add $3x$ to each side, and then add 3 to each side, to get $4x = 6$. Divide each side by 4 to get $x = \frac{6}{4}$, or 1.5.

4. **(D)** This one's much clearer if you draw it. You can think of segments AC and BD as overlapping segments, where BC is the amount of the overlap. The lengths of AC and BD add up to 27, but it's only a distance of 21 from A to D. The difference, a distance of 6, is the overlap. That's the length of BC.

5. **(A)** See those variables everywhere? Plug in! Choose your own values, ones that will make the math easy. Suppose a group of 5 people buys 3 widgets at 10 dollars each ($z = 5$, $x = 3$, and $y = 10$). Three widgets for $10 each makes a total of $30. If the 5 people split the cost equally, then each pays $6. Find the answer choice that equals 6, and you're home free. Only (A) does the trick.

6. **(A)** The value at which a line intersects the y-axis is called the y-intercept (that's the y-coordinate of the point of intersection; the x-coordinate is zero at every point on the y-axis). If you put the line $3y + 5 = x - 1$ into the form $y = mx + b$, then b will be the y-intercept. The rearranged equation looks like this: $y = \frac{1}{3}x - 2$. The y-intercept is -2, and you know the x-coordinate must be 0, so the point of intersection has the coordinates $(0, -2)$.

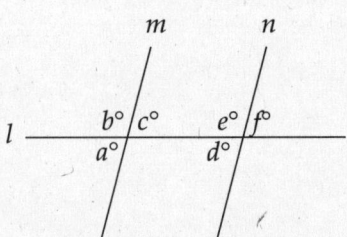

7. **(C)** The fact that m and n are parallel tells you that Fred's Theorem is at work here. That means that all the big angles are equal, all the small angles are equal, and any big angle plus any small angle equals 180°. The big angles measure 125°, so the small ones must measure 55°. Because d and f are both small angles, $d + f = 110°$.

8. **(E)** This is a simple percentage-translation question—if you don't remember how to do it, review the chart on page 236. This question translates directly to the equation $\frac{25}{100} \times 28 = \frac{x}{100} \times 4$, which can be simplified to $\frac{1}{4} \times 28 = \frac{x}{25}$, or $7 = \frac{x}{25}$. Solve, and you find that $x = 175$.

9. **(D)** When you're given two equations extremely similar in form, you're probably looking at classic ETS-style simulta-

neous equations. The best way to solve these? Rack 'em, stack 'em, add and subtract 'em! (Isn't that satisfying?) In this case, adding the two equations cancels out the "b" term, leaving you with this equation: $8a = 16$. This is easily solved, so that $a = 2$.

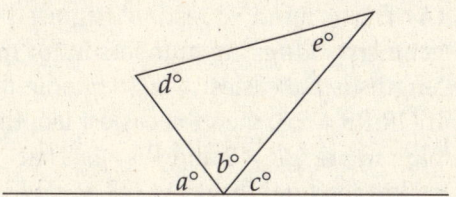

10. **(C)** This problem tests the Rule of 180 for triangles and straight lines. Just remember that both $(a + b + c)$ and $(b + d + e)$ must equal 180°. Keeping that in mind, you can plug in whatever values you want—there's nothing but variables in the question and answer choices. Any number will work as long as you obey the Rule of 180. Suppose for example, that $a = 50$, $b = 60$, and $c = 70$. Then, suppose that $d = 80$ and $e = 40$. The sum of a and b is then 120; the correct answer must also equal 120. Only (C) works.

11. **(E)** Remember that the absolute value of something can be thought of as the distance on a number line between that value and zero. If the absolute value of some value is greater than 3, then that value must be more than 3 away from zero. In this case, you're dealing with y-coordinates and you want all points 3 units or more from the x-axis, where $y = 0$. Only (E) fits the bill. If this kind of reasoning is unclear to you, just try picking points in the shaded region of each graph, and seeing whether they make the equation $|y| \geq 3$ true. Any answer choice whose graph contains "illegal" points can be eliminated.

12. **(E)** Direct variation between two quantities means that they always have the same quotient. In this case, it means that $\frac{m}{n}$ must always equal 5. To find the value of m when $n = 2.2$, set up the equation $\frac{m}{2.2} = 5$, and solve for m. You'll find that $m = 11$.

13. **(B)** To find the slope of the line easily, get its equation into the form $y = mx + b$, where m will be the value of the slope. To express $3y - 5 = 7 - 2x$ in this form, just isolate y. You'll find that $y = -\frac{2}{3}x + 4$. Here, the slope of the line (m) is $-\frac{2}{3}$.

14. **(D)** A quick review of exponent rules: When raising powers to powers, multiply exponents; when multiplying powers of the same base, add exponents; and when dividing powers of the same base, subtract exponents. For this problem, you have to do all three. Take the steps one at a time, following the rule of PEMDAS:

$$\frac{(n^3)^6 \times (n^4)^5}{n^2} = \frac{n^{18} \times n^{20}}{n^2} = \frac{n^{38}}{n^2} = n^{36}.$$

15. **(E)** Be careful here. That's not an absolute value symbol; it's one of the weird symbols ETS sometimes uses to represent functions. It's a basic function question, requiring you to plug the value –3 into the function: $\|-3\| = (-3)^2 - 3(-3) = 9 - (-9) = 18$.

16. **(D)** With few exceptions, logic questions on the Math IC Test you on the *contrapositive*. The idea is that, given a statement like "If A, then B," the only thing you automatically know is the contrapositive: "If not B, then not A." Here, you're told that no doctoral candidates graduate. In Cookie-Monster language, that would read "If doctoral student, then no graduate." The contrapositive would be "If graduate, then not doctoral student." That's almost

exactly what (D) says (since there are only engineers and doctoral candidates at this school, *not* being a doctoral candidate is the same as *being* an engineer). The other answer choices all talk about subjects you don't know anything about, like the number of students or the quality of their scholarly skills. You can eliminate those answer choices.

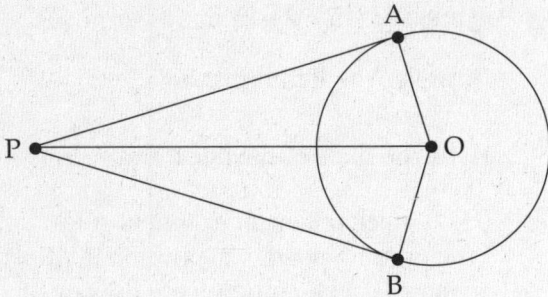

17. **(D)** There are a few simple rules for lines tangent to circles. Most importantly, a tangent line is always perpendicular to the radius it meets. That makes both $\triangle APO$ and $\triangle BPO$ right triangles. In each triangle, segment PO is the hypotenuse, so it's impossible for PB to be longer than PO. Statement I is therefore not true, and (A), (C), and (E) can be eliminated. Both of the remaining answers, (B) and (D), contain statement II, so it must be true; concentrate on statement III. Both $\angle APO$ and $\angle BPO$ are right angles, so they must add up to 180°. Since the other two angles, $\angle APB$ and $\angle AOB$, complete a quadrilateral, they must also add up to 180° (making a total of 360° in the quadrilateral). Statement III must also be true, and (D) is correct.

18. **(E)** The weird decimal answer choices should tip you off that this quadratic polynomial can't be factored using FOIL in reverse. You have to fall back on the quadratic formula $\left(x = \dfrac{-b \pm \sqrt{b^2 - 4ac}}{2a}\right)$, which gives you the roots of any qua-

dratic expression. If you use the quadratic formula on the expression $2x^2 - 5x - 9$, you'll get two possible roots: –1.21 and 3.71. Only one of these appears among the answer choices.

19. **(A)** This is a ratio question disguised as geometry. Plugging numbers in for the lengths is the easiest way to handle this. If $QR:RS$ = 2:3, then the easiest lengths to plug in are QR = 2 and RS = 3. That makes the total length of QS equal 5. Then, if $PQ:QS$ = 1:2, the length of PQ must equal $\dfrac{5}{2}$ or 2.5. The ratio of PQ to RS is then equal to $\dfrac{2.5}{3}$, which equals $\dfrac{5}{6}$.

20. **(C)** The term 3^{x-2} can also be written as $3^x \div 3^2$ (exponents are subtracted when terms are divided). Since $3^x = 54$, then $3^{x-2} = 54 \div 9 = 6$.

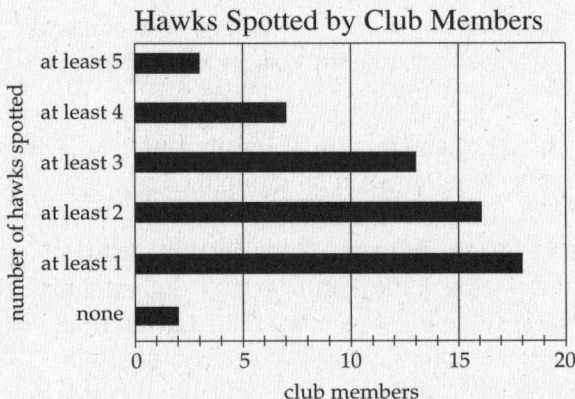

21. **(B)** This is the simplest kind of chart-reading question. It's only difficult if you over-think it. The third bar from the top represents students who saw at least three hawks. According to the bar, 13 students are in this group. Important: You *don't* need to look at any other bar. This one shows you *all* of the students who saw at least three hawks.

22. **(B)** This question is a little odd. The trick to this chart is that each bar (not counting the bar for "none") includes the people in the bars above it. For example, the "at least 3" bar shows everyone who saw 3 hawks, 4 hawks, 5 hawks, or more. People who saw 4 or 5 hawks are counted in the "at least 3" bar. For this reason, the "at least 1" bar includes everybody in the club who saw any hawks at all—18 people. The only people not counted in the "at least 1" bar are those who saw no hawks—the 2 people in the "none" bar. The total membership of the club is therefore 18 + 2, or 20.

23. **(E)** To get from the surface area of a cube to its volume, you need to take the middle step of finding the most basic fact about the cube—the length of its edge. Surface area is given by this formula: $SA = 6s^2$. Given the surface area of the cube ($6x^2 - 36x + 54$), you need to divide by 6 (getting $x^2 - 6x + 9$) and then take the square root (getting $x - 3$). The expression ($x - 3$) represents the length of the cube's edge. Plug this into the volume formula ($V = s^3$), and you've got the answer: $V = (x - 3)^3 = (x^2 - 6x + 9)(x - 3) = x^3 - 6x^2 + 9x - 3x^2 + 18x - 27 = x^3 - 9x^2 + 27x - 27$.

24. **(C)** Draw this! Always make a sketch when a geometrical situation is described but not drawn for you. You're told that triangle ABC must be isosceles, because AB and BC are equal in length. You also know that AC is a diameter (because it bisects the circle). This tells you that $\angle B$ measures 90°, because any inscribed angle that intercepts a diameter is a right angle. This makes triangle ABC an isosceles right triangle—a 45°-45°-90°. Now that you can see the relationships between the triangle and the circle, plug in a value for the radius of the circle. For example, if the radius of the circle is 2, then the triangle is a 45°-45°-90° with a hypotenuse of 4 and legs of $\frac{4}{\sqrt{2}}$. The circle's area is then 12.5664, and the triangle's area is 4. Four is 31.8% of 12.5664.

25. **(B)** To simplify the expression $\frac{(3-i)^2}{2}$, follow PEMDAS and do the exponent first—you'll need to use FOIL to square the binomial on top of the fraction. You should get this: $\frac{9-6i+i^2}{2}$. Since i is the square root of –1, the value of i^2 is –1: $\frac{9-6i-1}{2} = \frac{8-6i}{2}$. Then it's a simple matter to divide by 2 and get the expression in its simplest form: $4 - 3i$.

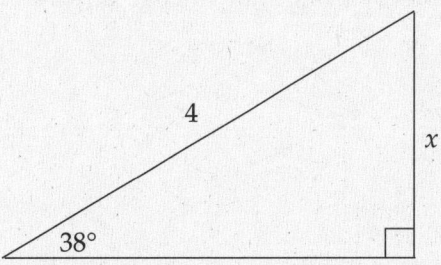

26. **(C)** The diagram of this triangle gives you the length of the hypotenuse and the measure of an angle, and asks for the length of the side opposite that angle. That's enough to set up a simple equation using the SOHCAHTOA definition of the sine: $\sin \theta = \frac{O}{H}$. Plugging the values from this triangle into the equation gets you this: $\sin 38° = \frac{x}{4}$. Use your calculator to find the value of $\sin 38°$, and you get the equation $0.61566 = \frac{x}{4}$. This is easily solved to reveal that $x = 2.4626$.

27. **(A)** The range of a function is the set of values the function can produce; on a graph, the range corresponds to the y-coordinates of the curve. Looking at this graph, you'll see that it seems to continue downward (in the negative y-direction)

forever. The range doesn't seem to have a minimum value; (D) and (E) can therefore be eliminated. The graph has an apparent maximum value of 2. That makes (A) a strong contender. The function's graph is also a continuous curve; that means it occupies a range of values, not just a few specific ones. Only (A) describes such a range of values.

28. **(E)** Not many shortcuts here. The trick is making sure that each triangle will fold around to a separate edge of the square base. In each of the wrong answers, two triangles will overlap a single edge of the base, leaving another edge uncovered.

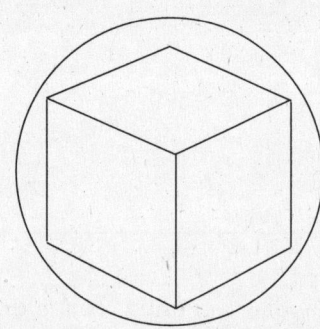

29. **(B)** This one isn't really vulnerable to shortcuts or techniques. You pretty much have to visualize the situation described in each answer-choice and find the one that produces exactly eight points of intersection. When a cube is inscribed in a sphere, each of the cube's 8 corners touches the inside of the sphere.

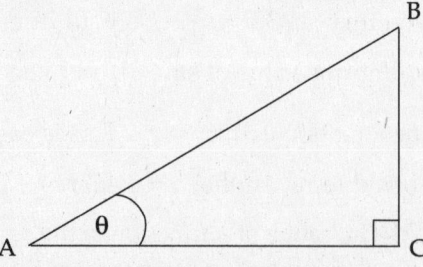

30. **(C)** Use the SOHCAHTOA definition of the sine function $\left(\sin \theta = \frac{O}{H}\right)$ to set up an equation. In this case, $\sin \theta = 0.6 = \frac{BC}{12}$.

Solve this and you'll find that the length of $BC = 7.2$. Now you know the lengths of one leg and the hypotenuse of a right triangle. To find the length of the other leg, use the Pythagorean Theorem: $(7.2)^2 + (AC)^2 = 12^2$. Solving this shows that $AC = 9.6$. Knowing the lengths of both legs, it's no trouble to compute the area of the triangle: $\frac{1}{2}bh = \frac{1}{2}(7.2)(9.6) = 34.56$.

31. **(E)** In questions like this, ETS is trying to trick you into making unjustified assumptions about the information you're given. Here, you don't know what $f(x)$ is. You're given only one coordinate pair from the function, which isn't enough information to figure out *anything* else. The function might be $f(x) = x^2$, or $f(x) = 5x$, or $f(x) = x + 20$. There's an infinite number of possible $f(x)$'s, and no way to know which is right. It's impossible to determine the value of s.

32. **(A)** Backsolve. Plug in values from the answer choices, and see which one makes the equation true. Only (A) does it. When solving this question algebraically, the trick is to isolate x one step at a time. The correct order looks like this:

$$\frac{9(\sqrt{x}-2)^2}{4} = 6.25$$

$$9(\sqrt{x}-2)^2 = 25$$

$$(\sqrt{x}-2)^2 = \frac{25}{9}$$

$$\sqrt{x}-2 = \frac{5}{3} \text{ or } -\frac{5}{3}$$

$$\sqrt{x} = \frac{11}{3} \text{ or } \frac{1}{3}$$

$$x = \frac{121}{9} \text{ or } \frac{1}{9}$$

$$x = 13.44 \text{ or } 0.11$$

Only (A) represents a possible value of x.

33. **(D)** In questions like this, you've got to simplify a complicated trigonometric expression. The best place to start is usually with any tangent functions in the expression, using the fact that $\tan\theta = \frac{\sin\theta}{\cos\theta}$:

$$\left(\frac{1}{\cos\theta} - \frac{\sin\theta}{\tan\theta}\right)(\cos\theta) =$$

$$\left(\frac{1}{\cos\theta} - \frac{\sin\theta}{\frac{\sin\theta}{\cos\theta}}\right)(\cos\theta) =$$

$$\left(\frac{1}{\cos\theta} - \left(\frac{\sin\theta}{1}\right)\left(\frac{\cos\theta}{\sin\theta}\right)\right)(\cos\theta) =$$

$$\left(\frac{1}{\cos\theta} - \cos\theta\right)(\cos\theta) =$$

$$\left(\frac{\cos\theta}{\cos\theta} - \cos^2\theta\right) =$$

$$1 - \cos^2\theta =$$

Once you've simplified it this far, you can use a basic trigonometric identity to simplify the expression one step further: $1 - \cos^2\theta = \sin^2\theta$.

34. **(B)** Plug in! Suppose $x = 2$. Then $f(x) = 64$ and $f(-x) = -64$, and statement I is clearly not true. You can eliminate any answer choice containing I: (A), (C), and (E). Both of the remaining answer choices contain statement II, so it must be true. Go straight to statement III: $\frac{1}{2}f(2) = 32$, and $f\left(\frac{1}{2}(2)\right) = 2$. Statement III is also false; so the correct answer is (B).

35. **(E)** Plugging in works very well here. Suppose $x = 2$. Then $f(-x) = (-2)^3 - 6 = -8 - 6 = -14$. Since you know that $f(-x) = -14$, just flip the signs and you'll see that $-f(-x) = 14$. The correct answer will be the one that equals 14 when you plug in $x = 2$. Only (E) works.

36. **(D)** This is a permutations question. In a 3-digit number containing no zeroes, there are nine possibilities for the first digit (1–9); nine possibilities for the second digit (1–9); and nine possibilities for the third digit (1–9). That makes a total of $9 \times 9 \times 9$ possible 3-digit numbers, or 729.

37. **(B)** At all points on the y-axis of the coordinate plane, $x = 0$. To find the coordinates of the points of intersection of a circle and the y-axis, plug zero in as the x-value and solve for y. In this case, that gets you this equation: $(y - 3)^2 + (0 - 2)^2 = 16$. This equation can be simplified through the following steps:

$$(y - 3)^2 + 4 = 16$$

$$(y - 3)^2 = 12$$

$$y^2 - 6y + 9 = 12$$

$$y^2 - 6y - 3 = 0$$

This quadratic polynomial can't be factored, so use the quadratic formula $\left(x = \frac{-b \pm \sqrt{b^2 - 4ac}}{2a}\right)$ to find the values of y. You'll find that $y = 6.4641$ or -0.4641. That means that the circle intersects the y-axis at the points $(0, -0.46)$ and $(0, 6.46)$.

38. **(A)** Chapter 6 gives you a few basic facts about inscribed solids. One of these is that when a box is inscribed in a sphere, the long diagonal of the box is a diameter of the sphere. To find the length of a box's long diagonal, use the "Super

Pythagorean Theorem": $a^2 + b^2 + c^2 = d^2$, where a, b, and c are the dimensions of the box and d is the length of the diagonal. In this case, you get $16 + 25 + 49 = d^2$. That means that $d = \sqrt{90}$, or 9.4868. Be careful, though; that's the diameter. The question asks you for the radius, half of the diameter: 4.7434.

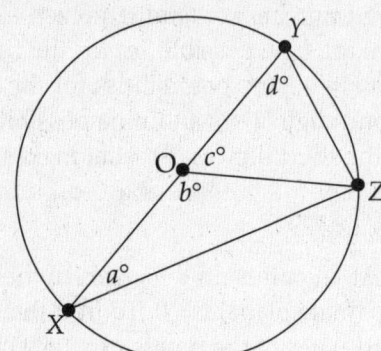

39. **(C)** There are two ways to solve this one. The elegant way is to notice that c is a central angle intersecting the arc YZ. Since a is an inscribed angle that intersects the same arc, it must measure half of angle c. Therefore $2a = c$ (see chapter 5 for central vs. inscribed angles in circles).

If that doesn't jump out at you, then just plug in using triangle rules. Suppose, for example, that $c = 50$. The other two angles in $\triangle OYZ$ would then need to add up to 130; since $\triangle OYZ$ is isosceles (two sides are radii), each of the other angles must equal $65°$. That tells you that $d = 65$. Meanwhile, elsewhere in the triangle, the Rule of 180 tells you that $b = 130$, because $c = 50$. Then, since $\triangle OZX$ is also isosceles (two sides are radii), the other two angles must each equal measure $25°$; that means that $a = 25$. This is just one possible set of numbers; any number would work as long as you obeyed the rules of geometry. Go to your answer choices using $a = 25$, $b = 130$, $c = 50$, and $d = 65$. Using these values, you'll find that the only answer choice that equals c is (C).

40. **(A)** This is a repeated percent-change question, basically the same as a compound-interest question about a bank account. Remember the formula:
Final = Original $\times$ $(1 + $ rate$)^{\text{\# of changes}}$. This colony has an original size of 1,250, and increases by 8% (.08) every month. In two years, it will make 24 of these increases. That's all you need to fill in the formula, which would look like this:
Final = $1{,}250 \times (1 + 0.08)^{24}$. Paying careful attention to the order of operations, run that through your calculator. You should find that Final = 7,926.4759. That's very close to (A), which represents 24 increases of almost exactly 8%.

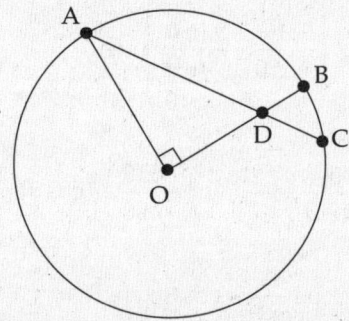

41. **(A)** Both segments OA and OB have lengths of 4, because they're both radii. The ratio you're given, $OD = 3DB$, tells you that the lengths of OD and DB are 3 and 1, respectively. Consequently, you know you're looking at a right triangle with legs of lengths 3 and 4; the length of the hypotenuse must be 5. Using the SOHCAHTOA definition of the sine function, you can determine that
$\sin \angle A = \dfrac{3}{5}$, or 0.6.

42. **(D)** To solve an inequality containing an absolute value, like $|x^3 - 8| \leq 5$, it's necessary to rewrite it as two inequalities, like this: $x^3 - 8 \leq 5$ and $-(x^3 - 8) \leq 5$. Solve each of these, and you'll have the complete solution to the original inequality. Solve $x^3 - 8 \leq 5$ and you get $x \leq \sqrt[3]{13}$,

or $x \leq 2.3513$. Solve $-(x^3 - 8) \leq 5$ and you get $x \geq \sqrt[3]{3}$, or $x \geq 1.4422$. Combine these solutions into one statement, and you get $1.44 \leq x \leq 2.35$.

43. **(C)** The best way to attack this question is just to try it. You'll find that no matter how you arrange the circles and connect them, the resulting polygon always has a perimeter of 24. That's because the polygon is always made up of two radii from in each circle, for a total of 12 radii—each with a length of 2.

44. **(C)** As always, start by putting the equation of the line in slope-intercept form: $y = -\frac{1}{2}x + 3$. This makes the y-intercept easy to find; it's the point (0, 3), because b represents the value at which the line intersects the y-axis. To find the x-intercept, make $y = 0$ and solve for x:

$$0 = -\frac{1}{2}x + 3$$

$$\frac{1}{2}x = 3$$

$$x = 6$$

This tells you that the x-intercept is at the point (6, 0). To find the distance between the two points, sketch their positions on the coordinate plane; you'll see they form a right triangle with legs of lengths 3 and 6, in which the hypotenuse represents the distance between the two points. Use the Pythagorean Theorem to find the length of the hypotenuse: 6.7082. That's the distance between the intercepts.

45. **(A)** The simple, grinding way to do this one is to use the distance formula $\left(d = \sqrt{(x_2 - x_1)^2 + (y_2 - y_1)^2}\right)$ on the answer choices. Any answer choice that produces a distance other than 25 or 26 can be discarded immediately. Only (A) produces distances of 25 and 26 from the two points given.

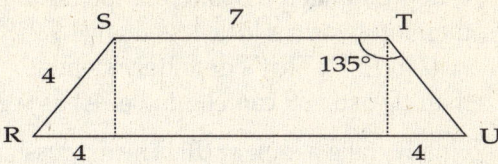

Figure 10

46. **(D)** Because this is an isosceles trapezoid, you know that $\angle S$ is equal to $\angle T$, and so measures 135° as well. The four angles must total 360°, so $\angle R$ and $\angle U$ measure 45° each (you can also use Fred's Theorem to figure that out, since the bases of the trapezoid are parallel lines). Divide the trapezoid into a rectangle and two triangles by drawing altitudes to S and T. Each of the triangles must be a 45-45-90 right triangle. Using the proportions of the 45-45-90, you can find the length of each triangle's legs: $\frac{4}{\sqrt{2}}$, or 2.8284. You can use the lengths of the legs to find that each triangle has an area of 4. The rectangle has an area of $\left(7 \times \frac{4}{\sqrt{2}}\right)$, or 19.7990. The total area of the trapezoid is 4 + 4 + 19.80, or 27.80.

47. **(D)** No ordinary calculator can work with exponents this big, and there's no way to spot the biggest values here by looking at them; you've got to get tricky. The important fact about this question is that it's not necessary to find the exact value of any expression; merely to <u>compare</u> them. The best way to compare these expressions is to get them into similar forms. To start with, rearrange as many answer choices as possible so that they have exponents of 100: (C) can be expressed as $(3^5)^{100}$, or 243^{100}; (D) can be expressed as $(4^4)^{100}$, or 256^{100}; and (E) is already there—250^{100}. Suddenly it's easy to see that (D) is the biggest of the three, and eliminate (C) and (E).

Next, take a look at (A). The exponent 999 is approximately 1,000. The expression is therefore worth a little less than $(1.73^{10})^{100}$, or $(240.12)^{100}$. That's definitely smaller than (D), so you can eliminate (A) as well.

Finally, take a look at (B). The expression 2^{799} is almost equal to 4^{400}. How can you tell? Well, 4^{400} can be written as $(2^2)^{400}$, or 2^{800}. That makes it clear that (D) is bigger than (A). Answer choice (D) reigns supreme.

48. **(E)** You can plug in on this one. Try an easy number like 8 for V, the volume of the cube. This makes the length of an edge of the cube equal 2. To find the length of the long diagonal, use the "Super Pythagorean Theorem:" $a^2 + b^2 + c^2 = d^2$. You should get this: $3(2)^2 = d^2$, so that $d = \sqrt{12}$ or 3.46. The correct answer is the one which gives you 3.46 when you use the value $V = 8$. Only (E) does the trick.

49. **(B)** This question looks a lot scarier than it is. The subscript of any term (the "2" in a_2, for example) just indicates the term's position in the series (a_2 is second). All the statement "$a_n = a_{n-1} + a_{n-2}$" is saying is that any term in the sequence is equal to the sum of the two terms before it. For example, a_8, the eighth term in the series, equals $a_7 + a_6$—the sum of the sixth and seventh terms. To find the value of a_8, just write out the sequence from the beginning, starting with the first and second terms you're given. It will look like this: 1, 2, 3, 5, 8, 13, 21, 34.... If you count carefully, you'll find that the eighth term is 34.

50. **(B)** This is a logarithm comprehension question. If $\log_x(y^x) = z$, then z is the exponent that turns x into y^x. If you think about it that way then it's clear that x raised to the power of z would be y^x. If that doesn't make sense to you, then review chapter 3.

If that doesn't work for you, then it's possible (but a little tricky) to plug in. Do it this way: Plug 10 in for x so that you're working with a common logarithm, the kind your calculator can compute. Plug in 2 for y and you get this: $\log_{10}(2^{10}) = z$. This can be written simply as log 1,024 = z. Your calculator can then compute the value of z: $z = 3.0103$. You can then compute the value of x^z: You get 1,024. The only answer choice that equals 1,024 is (B).

LEVEL IC TEST FORM B

1. **(D)** There are two good approaches to this question. The fastest way to do it is by treating it as a simultaneous-equations question, like this:

   $$\begin{array}{r} \text{Sherry} + \text{Heather} = 240 \\ \text{Rob} + \text{Sherry} = 300 \\ \hline \text{Rob} + \text{Sherry} + \text{Heather} + \text{Sherry} = 540 \end{array}$$

   Because you know that the three people together weigh 410 pounds, you can rewrite that last equation this way:

   410 + Sherry = 540

   Then, by subtracting, you can easily find Sherry's weight:

   Sherry = 130

   If you don't want to handle the question this way, you can simply backsolve. That means starting with (C) and using the answer choices as Sherry's weight. For example, if Sherry's weight is 120 pounds, that makes Robin's weight 180 pounds and Heather's weight 120 pounds. Those three weights don't add up to 410 pounds, so move on and try the next answer choice. Answer choice (D) works.

2. **(C)** Two graphs will intersect at the point where both of their equations are true. That is, where $y = 4$ and $y = \frac{x^3}{2}$. That means the point at which $\frac{x^3}{2} = 4$. Just solve this equation to find the value of x:

   $$\frac{x^3}{2} = 4$$
   $$x^3 = 8$$
   $$x = 2$$

3. **(E)** This is just algebraic manipulation. Plug the value given for r and s into the equation:

   $$\frac{s}{r} + \frac{4}{r^2} =$$

   $$\frac{6}{\frac{2}{3}} + \frac{4}{\left(\frac{2}{3}\right)^2} =$$

   $$\frac{6}{\frac{2}{3}} + \frac{4}{\frac{4}{9}} =$$

   $$\left(6 \times \frac{3}{2}\right) + \left(4 \times \frac{9}{4}\right) =$$

   $$\frac{18}{2} + \frac{36}{4} = 9 + 9 = 18$$

   The correct answer is (E).

4. **(A)** You're given variables and no numbers, so plug in. Suppose $p = 60$. The other two internal angles of the triangle must then add up to 120°. For now, let's plug in 50 and 70, making the angle above n a 50° angle, and the angle above m a 70° angle. Then, because there are 180° in a straight line, you know that $n = 130$ and $m = 110$. Now that you have values for each of the variables, you can go to the answer choices and see which one works. You're looking for the answer choice that is equal to p, or 60. Only (A) works, and that's the correct answer.

5. **(B)** You've got variables in the question and the answer choices, so plug in. Suppose that you're traveling 5 miles ($b = 5$). That means that your fare will include an original $2.50 plus five $.30 charges for the miles traveled, for a total of $4.00. To find the correct answer, plug $b = 5$ into the answer choices and see which one gives you a value of 4. Only (B) does the trick.

6. **(D)** Just make x equal to an even integer, and plug in. Say $x = 4$. Then only one answer choice produces an odd integer: that's (D), which produces a value of 13.

7. **(E)** The easiest way to answer this question is with a little reasoning. The question tells you that a is 30% of something and that b is 6% of the same thing. No matter what their values are, you can see that a must be 5 times as big as b—and that makes b one-fifth of a, or 20% of a. The correct answer is (E).

   If you're not comfortable with that line of reasoning, you can always fall back on plugging in. That will be easiest if you say that $c = 100$. Then $a = 30$ and $b = 6$. After that, it's easy to see that 6 is one-fifth of 30, meaning that b is 20% of a.

8. **(C)** The solution set of a system of equations is the set of values that makes all the equations in the system true. Individually, these equations aren't hard to solve. Here's the solution for the first one:

   $$x(x - 1) = 6$$
   $$x^2 - x = 6$$
   $$x^2 - x - 6 = 0$$
   $$(x - 3)(x + 2) = 0$$
   $$x = -2, 3$$

   And here's the solution for the second equation:

   $$x^2 + 1 = 5$$
   $$x^2 = 4$$
   $$x = -2, 2$$

   The solution set for the system of equations is the set of values that makes *both* equations true. And as you can see, there's only one value that appears in the solution sets of both equations, and that's −2. The solution set of the system is therefore $x = -2$. The correct answer is (C).

9. **(B)** A friendly reminder: The perimeter is what you get when you add up all a polygon's sides. The most common careless mistake on perimeter questions is to calculate the area instead; notice answer choice (E).

   In the end, this is a simple addition question. The polygon can be seen as a 7×10 rectangle with two notches cut into it. The notches add to the polygon's perimeter, but figuring out how much they add is the tricky part. A plain 7×10 rectangle has a perimeter of 34. The notch cut into the bottom of this rectangle adds 4 units to this perimeter, not 7. The three horizontal segments on the bottom of the figure must still add up to 10, the length of the rectangle; only the vertical segments add length. In the same way, the notch cut into the right side of the rectangle adds 4 units of length, not 6, because the vertical sides must still add up to 7, the rectangle's height; only the horizontal segments add length. The total perimeter is $34 + 4 + 4$, or 42. The correct answer is (B).

10. **(A)** The thing to be careful about here is taking the absolute value at the right time. To find $|x + y|$, you have to add x and y together first, and then take the absolute value. If you take the absolute value of each quantity before adding, you're likely to get a wrong answer. The only pair of coordinates whose sum has an absolute value greater than 5 is (A), because $-4 + -2 = -6$, and $|-6| = 6$.

11. **(B)** Keri sold more than 300 units in only four months in 1996: April, August, September, and December. That means four bonuses, for a total of $4,000.00.

12. **(D)** The trick here is to remember the bonuses. In the 3-month periods shown in answer choices (A) and (B), Keri sold 1,000 units, earning $1,000.00. Each 3-month period also includes one bonus, raising her

income for that period to $11,000.00. In the 3-month period shown in answer choice (D), however, Keri earns more. She sells 950 units, earning $9,500.00, and also receives two bonuses, for a total of $11,500.00. The correct answer is (D).

13. **(B)** When a varies directly as b, then $\frac{a}{b}$ will always come out to the same value. You start out with $a = 14$ and $b = 8$. That makes $\frac{a}{b} = \frac{14}{8}$, or $\frac{7}{4}$. That's the constant value of $\frac{a}{b}$. When $b = 6$, then, you can set up this equation:

$$\frac{a}{6} = \frac{7}{4}$$

Then, just solve to find the value of a:

$$a = 6\left(\frac{7}{4}\right)$$

$$a = \frac{42}{4}$$

$$a = 10.5$$

The correct answer is (B).

14. **(C)** It's algebra time. Whenever two fractions are equal, you can cross-multiply:

$$\frac{1}{x} = \frac{4}{5}$$

$$4x = 5$$

$$x = \frac{5}{4}$$

You're not done when you've solved for x, though. The question asks not for x but for $\frac{x}{3}$. To produce the right answer, you have to divide by 3:

$$x = \frac{5}{4}$$

$$\frac{x}{3} = \frac{5}{12} = 0.41666...$$

The correct answer is (C).

15. **(A)** The four right triangles inside the quadrilateral are congruent; they have legs and hypotenuses of equal length. The exact lengths of the segments don't matter. For convenience's sake, you can plug in values for the lengths. Suppose that the right triangles are all 3-4-5 triangles. That makes the hypotenuses 5. Segments RT then has a total length of 6 (two legs of length 3) and SU has a total length of 8 (two legs of length 4).

The value of $\sin \angle RSU$, according to SOHCAHTOA, will now be $\frac{3}{5}$. The correct answer choice will be the one that also equals $\frac{3}{5}$. Answer choice (A) is correct.

16. **(D)** There are two common sorts of non-real numbers that are important on the Math Subject Tests. These are the square roots of negative numbers (imaginary numbers) and numbers divided by zero (undefined). The expression $\sqrt{x} + \frac{1}{x-3}$ will therefore be real only if x is zero or greater (so that the quantity under the radical isn't negative), and only if x doesn't equal 3 (so that the fraction's denominator isn't zero). You can see that x doesn't have to be greater than 1 ($x = 0$ is allowed, for example), so statement I is out. This allows you to eliminate (A), (C), and (E). Statement II is definitely true. And statement III must also be true, since x must be greater than or equal to zero. The correct answer is (D).

17. **(E)** Always sketch situations that are described but not shown. Here's the situation described in the question:

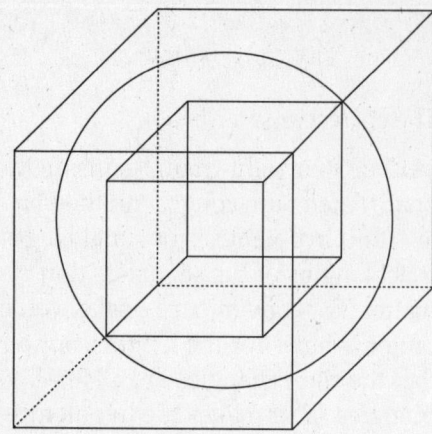

Cube A is the outermost shape. Sphere O is inscribed inside it, and cube B is then inscribed within the sphere. You can use this simple sketch to check your answer choices. As you can see from your sketch, one edge of A is equal in length to the diagonal of B. The correct answer is (E).

18. **(A)** This is a fancy simultaneous-equations question. It's impossible to solve for the value of x, y, or z, but you can find the value of their sum. It's done by adding the equations, like this:

$$a - x = 12$$
$$b - y = 7$$
$$c - z = 15$$
$$\overline{a + b + c - x - y - z = 34}$$

This equation can be written like this: $(a + b + c) - (x + y + z) = 34$. And since the question tells you that $a + b + c = 50$, you can simplify the equation even further:

$$50 - (x + y + z) = 34$$
$$-(x + y + z) = -16$$
$$x + y + z = 16$$

The correct answer is (A).

19. **(B)** Since Cassie's stuck in the driver's seat, the number of permutations of the people in the car is simply determined by the possible arrangements of passengers in the 3 passenger's seats. The number of permutations of 3 items in 3 spaces is given by 3!, which equals $3 \times 2 \times 1$, or 6. The correct answer is (B).

20. **(E)** It's algebra time again. Here's how to solve this equation:

$$\frac{1}{2}x - 3 = 2\left(\frac{x-1}{5}\right)$$
$$\frac{1}{2}x - 3 = \frac{2x-2}{5}$$
$$\frac{1}{2}x = \frac{2x-2}{5} + 3$$
$$x = \frac{4x-4}{5} + 6$$
$$5x = 4x - 4 + 30$$
$$5x - 4x = -4 + 30$$
$$x = 26$$

The correct answer is (E).

21. **(C)** Remember that slope is equal to "rise" over "run". A line that has a slope greater than 1 is changing vertically (along the y-axis) faster than it's changing horizontally (along the x-axis). Line l might look like this:

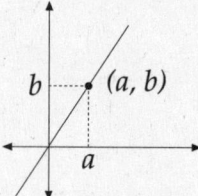

As you can see, a and b don't have to be equal—in fact, they can't be. You can also see that a doesn't have to be less than b—a will be less than b when they're both positive, but b will be less than a when they're negative. Of all the statements in the answer choices, only (C) must be true. Since b is always farther from zero than a, its square will always be greater than the square of a.

22. **(C)** This one is a little tricky. The best way to approach it is by experimenting with sketches. If you examine the possibilities carefully, you'll see that there are three possible locations of the missing vertex:

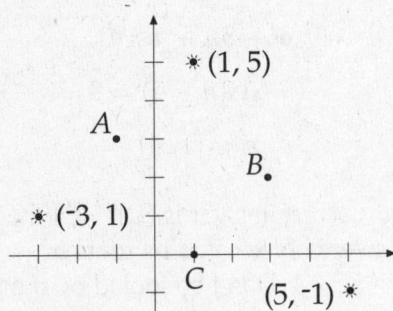

The correct answer is (C).

23. **(E)** Backsolving works well on this one. Just take the values of x from the answer choices, fill in the values from the other angles using the formulas $y = \frac{2}{3}x$ and $w = 2z$. You'll know you've got the right answer when the values produced obey the rule of 180°. If $x = 72$, then $y = \frac{2}{3}(72)$, or 48. The sum of x and y is then 120, which means that $w = 120$, since it's a vertical angle. And that makes $z = 60$. If you check all those numbers against the Rule of 180°, you'll see that (E) is the correct answer.

24. **(D)** The new radius of circle O is $(r + t)$. To find the circle's area, just plug this quantity into the formula for the area of a circle, $A = \pi r^2$. To do this, you'll need to use FOIL:

$A = \pi(r + t)^2$
$A = \pi(r^2 + rt + rt + t^2)$
$A = \pi(r^2 + 2rt + t^2)$

The correct answer is (D).

25. **(C)** You can find the radius of the circle from the area you're given, just by plugging the area into the formula and solving:

$$A = \pi r^2$$
$$\pi r^2 = 9\pi$$
$$r^2 = 9$$
$$r = 3$$

Because OA and OB are both radii, each must have a length of 3. And because they're perpendicular, they must be legs of an isosceles right triangle. That means that $\triangle ABO$ has angles measuring 45°, 45°, and 90°, and must therefore have sides in the ratio $1:1:\sqrt{2}$. The length of the hypotenuse, AB, is simply $3\sqrt{2}$, or 4.24. The correct answer is (C).

26. **(A)** Solving $x^2 + 2x - 3 = 0$ is just a matter of factoring:

$$x^2 + 2x - 3 = 0$$
$$(x - 1)(x + 3) = 0$$
$$x = -3, 1$$

The funny looking equation $x < |x|$, however, tells you that x is negative. That means that x can only equal -3. To find the answer, just plug $x = -3$ into the expression $2x + 4$:

$$2(-3) + 4 = -6 + 4 = -2$$

The correct answer is (A).

27. **(B)** The corresponding parts of similar triangles are proportional. To figure out a value in one triangle from the corresponding value in the other, you need to know what the proportion is. Luckily, this question gives you the hypotenuse of both triangles: 5 and 4, respectively. This allows you to set up this proportion in your mind:

$$\frac{\text{any part of } \triangle ABC}{\text{the matching part of } \triangle CBD} = \frac{5}{4}$$

To find the area of $\triangle CBD$, you'll need to know the lengths of both of its legs. Use the proportion to find these lengths from the legs of the larger triangle. First, find the longer leg, BD:

$$\frac{4}{BD} = \frac{5}{4}$$

$$5BD = 16$$

$$BD = \frac{16}{5} = 3.2$$

Then, find CD:

$$\frac{3}{CD} = \frac{5}{4}$$

$$5CD = 12$$

$$CD = \frac{12}{5} = 2.4$$

Once you know that $\triangle CBD$ is a right triangle with legs of 2.4 and 3.2, all you have to do is plug these lengths into the area formula for triangles:

$$A = \frac{1}{2}bh$$

$$A = \frac{1}{2}(2.4)(3.2)$$

$$A = 3.84$$

The correct answer is (B).

28. **(D)** For this question, you need to remember how FOIL works, and that $i^2 = -1$. If you've got that, the rest is easy:

$$(5 - 3i)(4 + 2i) =$$
$$20 + 10i - 12i - 6i^2 =$$
$$20 - 2i - 6i^2 =$$
$$20 - 2i - 6(-1) =$$
$$20 + 6 - 2i =$$
$$26 - 2i$$

The correct answer is (D).

29. **(D)** Well, you know from the definition of the function that $f(n) = n^2 - 5n$. You're also told that $f(n) = -4$. These two statements allow you to set up an equation and solve for n:

$$n^2 - 5n = -4$$
$$n^2 - 5n + 4 = 0$$
$$(n - 1)(n - 4) = 0$$
$$n = \{1, 4\}$$

The correct answer is (D). No other answer choice contains roots of $n^2 - 5n + 4$. Don't be fooled by the negative answers (A), (B), and (C).

30. **(B)** It's a good idea to sketch the situation that's described in this question:

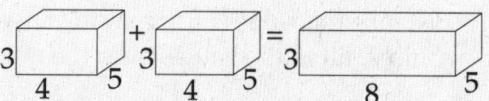

As you can see, only one dimension of this solid is doubled. It's now a rectangular solid of dimensions $3 \times 5 \times 8$. The longest line that can be drawn within this solid is the long diagonal, which can be found using the "Super Pythagorean Theorem:"

$$d^2 = a^2 + b^2 + c^2$$
$$d^2 = 3^2 + 5^2 + 8^2$$
$$d^2 = 9 + 25 + 64$$
$$d^2 = 98$$
$$d = \sqrt{98} = 9.899$$

The correct answer is (B).

31. **(E)** *Don't* use FOIL on this expression! FOIL is for squaring binomials (pairs of terms that are being added together). The two terms in parentheses here are being multiplied together, so FOIL isn't necessary. You can write this multiplication this way:

$$(5.5 \times 10^4)^2 =$$
$$(5.5 \times 10^4)(5.5 \times 10^4) =$$
$$5.5 \times 5.5 \times 10^4 \times 10^4 =$$

The tricky part is remembering how to combine the terms. The "5.5" terms can be multiplied normally. For the "10^4" terms, you have to remember that exponents are added when you multiply terms with identical bases:

$$5.5 \times 5.5 \times 10^4 \times 10^4 =$$
$$30.25 \times 10^8 =$$

Finally, you need to put the result in proper scientific notation by moving the decimal point over one place to the left. That's the same as dividing by 10, so to keep the whole expression from changing value, you need to multiply by 10 at the same time:

$$3.025 \times 10 \times 10^8 =$$
$$3.025 \times 10^9 \approx 3.0 \times 10^9$$

The correct answer is (E).

32. **(A)** Use the slope formula, which is $m = \dfrac{y_2 - y_1}{x_2 - x_1}$, to find the slope of the line:

$$m = \frac{y_2 - y_1}{x_2 - x_1}$$
$$m = \frac{-3 - 2}{2 - (-2)}$$
$$m = \frac{-3 - 2}{2 + 2}$$
$$m = \frac{-5}{4}$$

The correct answer is (A).

33. **(B)** Don't let all the scientific talk scare you. All you're being asked to do is find the value of t for which $n = 500$. To do that, just set the equation for n ($698 - 2t - 0.5t^2$) equal to 500, and solve:

$$698 - 2t - 0.5t^2 = 500$$
$$198 - 2t - 0.5t^2 = 0$$
$$-0.5t^2 - 2t + 98 = 0$$

As you can see, this equation doesn't look like a very good bet for factoring. To solve for t, you have to use the quadratic formula:

$$x = \frac{-b \pm \sqrt{b^2 - 4ac}}{2a}$$

$$n = \frac{-(-2) \pm \sqrt{(-2)^2 - 4(-0.5)(198)}}{2(-0.5)}$$

$$n = \frac{2 \pm \sqrt{4 + (2 \times 198)}}{-1}$$

$$n = \frac{2 \pm \sqrt{400}}{-1}$$

$$n = \frac{2 \pm 20}{-1} = -1(2 \pm 20)$$

$$n = -1(22), -1(-18)$$

$$n = -22, 18$$

A length of time can't be negative, so you can throw out the negative value. The sample will reach 500° C after 18 seconds. The correct answer is (B).

34. **(E)** Perpendicular lines have negative reciprocal slopes. That means that if line m has a slope of $-\dfrac{1}{2}$, line l, which is perpendicular to it, must have a slope of 2. That alone lets you eliminate the equations in (A), (B) and (C), all of which have the wrong slope. The only difference between (D) and (E) is the y-intercept—the point at which the line crosses the y-axis, given by b in an equation of the

form $y = mx + b$. To find the y-intercept of line l, take what you know about the line ($y = 2x + b$) and plug a point on the line. In this case, you know that l passes through $(4, 5)$, allowing you to set up this equation:

$$y = 2x + b$$
$$5 = 2(4) + b$$
$$5 = 8 + b$$
$$b = -3$$

The y-intercept is -3, which means that the line's equation is $y = 2x - 3$. The correct answer is (E).

35. **(B)** For this question, it's important to remember the difference between central angles and inscribed angles. Central angles, like $\angle AOB$ and $\angle BOC$, are like pie-slices that start at the circle's center. The arcs they intercept are equal in measure to the angles themselves. Inscribed angles, on the other hand, start on the edge of the circle, like $\angle BDA$ and $\angle CAD$. The arcs they intercept are twice as great in measure as the angles themselves.

Since $\angle BDA$ is an inscribed angle measuring $25°$, the arc it intercepts, AB, must measure $50°$. For the same reason, the arc intercepted by $\angle BDA$ must measure $64°$. That's a total of $114°$ out of the semicircle $ABCD$, which leaves $66°$ out of the $180°$ half-circle. That's the measure of arc BC. Since arc BC measures $66°$, the central angle BOC must also measure $66°$. The correct answer is (B).

36. **(A)** This question tests your understanding of exponent rules. You're told that $4^{x+2} = 48$, but how do you solve for 4^x? Remember that when you multiply exponential terms that have a common base, you add the exponents. In the same way, you can express addition in an exponent as multiplication:

$$4^{x+2} = 48$$
$$4^x \times 4^2 = 48$$

Once you've taken this step, solving is easy:

$$4^x \times 16 = 48$$
$$4^x = 3$$

The correct answer is (A).

37. **(C)** In this question, you know that $s(x)$ is the function that turns $r(x)$ into $2x - 1$. In other words, it turns $6x + 5$ into $2x - 1$. What's involved in doing this? Well, you can get there in two simple steps. First, subtract 8:

$$6x + 5 - 8 = 6x - 3$$

Then, divide by 3:

$$(6x - 3) \div 3 = 2x - 1$$

And you're there. Function $s(x)$ must subtract 8 and then divide by 3. Expressed algebraically, that looks like this:

$$s(x) = (x - 8) \div 3 = \frac{x - 8}{3}$$

The correct answer is (C).

38. **(A)** You've got to make a sketch when you're adding vectors. To add vectors, put the tail of one vector to the tip of the other, and see where they lead you. Here's a sketch of the addition of vectors a and b:

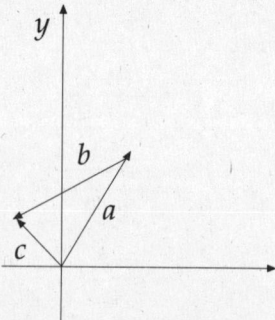

The little vector, $\vec{c}$, is the sum of $\vec{a}$ and $\vec{b}$. The only answer choice that looks anything like it is (A), and that's the correct answer.

39. **(C)** There are two steps to solving a logarithm with a strange base, like this one. First, use the definition of a logarithm to put the equation in exponential form:

$$\log_3 6 = n$$
$$3^n = 6$$

Once the equation is in exponential form, take the logarithm of both sides:

$$\log 3^n = \log 6$$

Then, use the exponent rule for logarithms to pull the exponent out and make it a coefficient:

$$n \log 3 = \log 6$$

And finally, isolate n and use your calculator to find a numerical value:

$$n = \frac{\log 6}{\log 3} = \frac{0.778}{0.477} = 1.63$$

The correct answer is (C).

40. **(A)** The cylindrical cup has a radius of 2 and a height of 3. Its volume, given by $V = \pi r^2 h$, is $\pi (2)^2 (3)$, or 12π. The volume of the rectangular tank, given by $V = lwh$, is $6 \times 7 \times 8$, or 336. To find out how many times the cup can be filled completely from the tank, just divide 336 by 12π using your calculator:

$$\frac{336}{12\pi} = 8.913$$

As you can see, the cup can be filled almost 9 times, but can be filled <u>completely</u> only 8 times. The correct answer is (A).

41. **(D)** Always sketch situations that are described but not shown. The situation described in this question would look something like this:

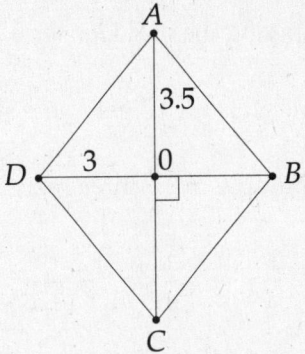

Segments AC and BD form right angles, and each cuts the other exactly in half. Segments BO and DO have lengths of 3, and segments AO and CO have lengths of 3.5. If you sketch segments AB, BC, CD, and DA, then you have four identical right triangles. You can use the Pythagorean Theorem to find the length of each hypotenuse:

$$a^2 + b^2 = c^2$$
$$3^2 + 3.5^2 = c^2$$
$$9 + 12.25 = c^2$$
$$21.25 = c^2$$
$$c = 4.61$$

When you know the lengths of each side, you can use SOHCAHTOA to figure out the value of $\sin \angle ADO$. The side opposite $\angle ADO$ is AO, and the hypotenuse is AD. You can use the lengths of these segments to find the value of the trig function:

$$\sin x = \frac{\text{opposite}}{\text{hypotenuse}}$$

$$\sin \angle ADO = \frac{AO}{AD}$$

$$\sin \angle ADO = \frac{3.5}{4.61} = 0.76$$

The correct answer is (D).

42. **(A)** Each term in this geometric sequence is one-fourth of the previous term. In other words, each term is produced by

multiplying the previous term by $\frac{1}{4}$. The third term is $\frac{3}{2}$. To get to the sixth term, just multiply by $\frac{1}{4}$ three more times:

$$\frac{3}{2} \times \frac{1}{4} \times \frac{1}{4} \times \frac{1}{4} = \frac{3}{128}$$

The correct answer is (A).

43. **(A)** Plugging in can work wonders for you on this one. The constant k can be anything; let's say it equals 3. If that case $f(x) = 3x$, and $g(x) = x + 3$. Using these values makes it easy to plug into the function to test each of the three statements.

Suppose $x = 2$, and take a look at statement I. Using our values, we can plug into the statement to see whether it must be true:

$$f(2x) = 2f(x)$$
$$f(4) = 2f(2)$$
$$3 \times 4 = 2(3 \times 2)$$
$$12 = 12$$

Statement I works so far, so let's keep it and move on. Here's how statement II looks with our values plugged into it:

$$f(x + 2) = f(x) + 2$$
$$f(4) = f(2) + 2$$
$$3 \times 4 = (3 \times 2) + 2$$
$$12 = 8$$

Statement II is definitely NOT true, so you can eliminate it. That gets rid of answer choices (B), (C), and (E), leaving only (A) and (D). Since statement I is present in both answers, you can forget about it. To pick the right answer, you need to check out statement III:

$$f(g(x)) = g(f(x))$$
$$f(g(2)) = g(f(2))$$
$$f(2 + 3) = g(3 \times 2)$$

$$f(5) = g(6)$$
$$3 \times 5 = 6 + 3$$
$$15 = 9$$

Statement III definitely isn't true either, so you can get rid of it. That eliminates answer choice (D), leaving only (A).

44. **(E)** This is a Pythagorean Theorem question disguised by a complicated physical description. Making a sketch of the wall in question can help clear this up:

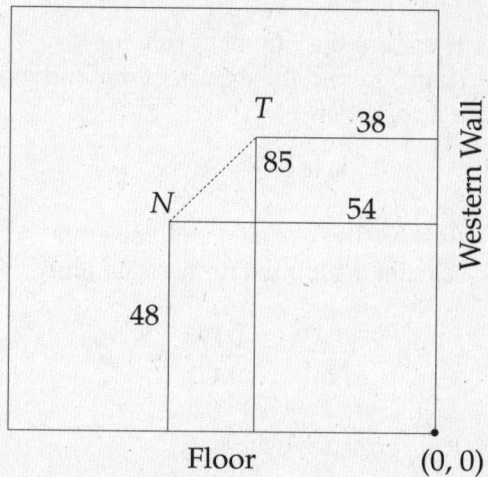

As you can see, the nail and tack can be viewed as points on a coordinate plane. The coordinates of the points are given by their distances from the floor and the western wall. The coordinates of the tack are (–38, 85), and the coordinates of the nail are (–54, 48). These are larger coordinates than you're used to working with, but you can work with them normally. To find the distance between the tack and nail, use the distance formula:

$$d = \sqrt{(y_2 - y_1)^2 + (x_2 - x_1)^2}$$

$$d = \sqrt{(48 - 85)^2 + (-54 - (-38))^2}$$

$$d = \sqrt{(-37)^2 + (-16)^2}$$

$$d = \sqrt{1369 + 256}$$

$d = \sqrt{1625}$

$d = 40.31$

The correct answer is (E).

45. **(C)** The domain of a function $s(x)$ is the set of values that you're allowed to plug into the function in the x position. The only numbers not in a function's domain are those that make the function produce non-real numbers—that is, fractions with zero in the denominator and square roots of negative numbers. There are no fractions in this function, so any values not in the domain of s will be those that make the square root negative.

For $\sqrt{12-x^2}$ to be a real number, the quantity under the radical, $12-x^2$, must be zero or positive. To find the values of x that make $s(x)$ real, just write that as an inequality, and solve:

$$12 - x^2 \geq 0$$
$$-x^2 \geq -12$$
$$x^2 \leq 12$$
$$x \leq \sqrt{12} \text{ and } x \geq -\sqrt{12}$$
$$-\sqrt{12} \leq x \leq \sqrt{12}$$

This is the domain of $s(x)$. The correct answer is (C).

46. **(C)** The basic rule of logic most commonly tested on the Math Subject Tests is the contrapositive. When you see any statement in the form "If A, then B," then you automatically know that the statement "If not B, then not A" is also true. That's the contrapositive statement. In this case, you're given the statement "If a tree falls in the forest, a sound is heard." The contrapositive of this statement is, "If no sound is heard, a tree doesn't fall in the forest."

The question asks you to pick the logically impossible answer choice. That will be a statement that contradicts either the original statement or the contrapositive of the original statement. Answer choice (C) is directly opposed to the contrapositive. The correct answer is (C).

47. **(D)** An exclamation point following a number indicates a factorial. The factorial of a number is the product of every integer from 1 to the number itself, inclusive. The fraction $\dfrac{8!}{5!3!}$ can therefore be rewritten like this:

$$\frac{8 \times 7 \times 6 \times 5 \times 4 \times 3 \times 2 \times 1}{(5 \times 4 \times 3 \times 2 \times 1)(3 \times 2 \times 1)} =$$

And once the fraction is written out, there's a lot of canceling you can do:

$$\frac{8 \times 7 \times \cancel{6} \times \cancel{5} \times \cancel{4} \times \cancel{3} \times \cancel{2} \times \cancel{1}}{(\cancel{5} \times \cancel{4} \times \cancel{3} \times \cancel{2} \times \cancel{1})(\cancel{3} \times \cancel{2} \times \cancel{1})} =$$

$$\frac{8 \times 7}{1} = 56$$

The correct answer is (D).

48. **(C)**

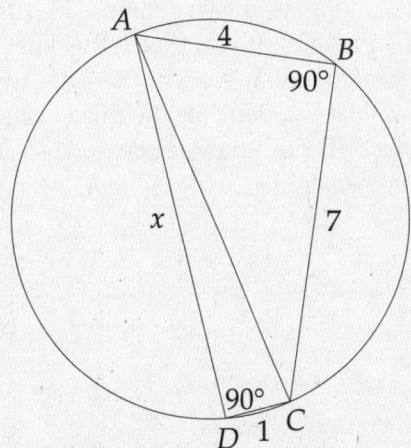

If you draw diameter AC, you divide the quadrilateral $ABCD$ into two triangles. Both $\angle ADC$ and $\angle ABC$ are inscribed angles that intercept a half-circle, or an arc of $180°$. Since inscribed angles intercept arcs that have twice the measure of the angles themselves, that means that

∠ADC and ∠ABC are both 90° angles, and that therefore △ADC and △ABC are right triangles that share a hypotenuse. This hypotenuse is the key finding the length of segment AD.

Because you know the lengths of the legs of right triangle ABC (AB = 4 and BC = 7) you can use the Pythagorean Theorem to find the hypotenuse's length: $4^2 + 7^2 = c^2$, and so $c = \sqrt{65}$. Once you know that the hypotenuse of both triangles has a length of $\sqrt{65}$, you can use the Pythagorean Theorem again to find the length of the missing leg of the other right triangle:

$$1^2 + x^2 = \sqrt{65}^2$$
$$1 + x^2 = 65$$
$$x^2 = 64$$
$$x = 8$$

The segment AD must therefore have a length of 8. The correct answer is (C).

49. **(E)** The segment from (0, 0) to (–2, 2) has a length of $2\sqrt{2}$, or 2.83 (you can think of it as the diagonal of a square with sides 2 units long, or the hypotenuse of a 45°-45°-90° triangle). If you sketch this line segment, you'll see that there are two possible locations for the third vertex of an equilateral triangle constructed from that segment:

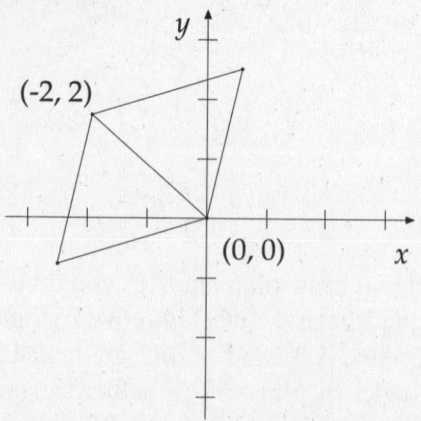

Before you start doing more complicated calculations, make the best sketch you can of the situation that's described. You may find that you can eliminate a few answer choices just by looking at your sketch and guesstimating.

There are a couple of ways to find the coordinates of these vertices, but the simplest way is to use trigonometry. Let's start with the upper vertex. Draw an altitude from this point to the x-axis to produce a right triangle:

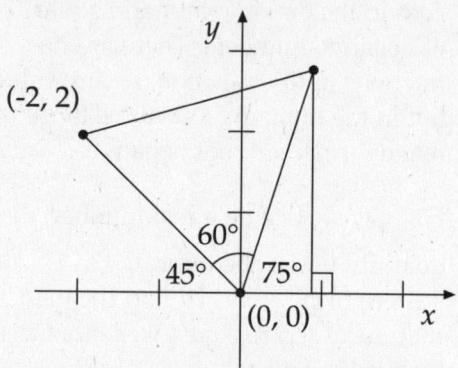

Since the hypotenuse of this triangle is one side of the equilateral triangle, you know it has a length of $2\sqrt{2}$, or 2.83. The rule of 180° also lets you figure out the measure of the lower acute angle in this right triangle; it must measure 75°, since it forms a straight line with the 45° and 60° angles. All you need to find the coordinates of the third vertex are the legs of this right triangle. Trigonometry is the easiest way to find them. The length of the horizontal leg, adjacent to the 75° angle, will be (cos 75°)($2\sqrt{2}$) = (0.26)(2.83) = 0.73. The length of the vertical leg will be (sin 75°)($2\sqrt{2}$) = (0.97)(2.83) = 2.73. The coordinates of the third vertex in this position will therefore be (0.73, 2.73). That's one of the answer choices, so you're done; the correct answer is (E).

50. **(B)** Remember, in compound-function questions you should always start with the inner function. In this case, that's f(x). The graph of f(x) is a fairly simple line that looks like this (it may make things

simpler to think of f(x) as –x + 1 rather than 1 – x):

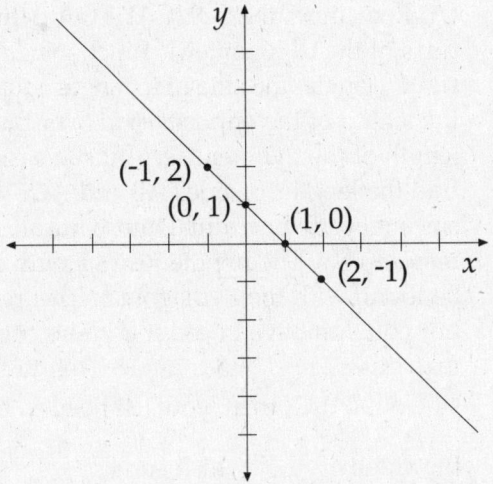

The results of the function f(x) are then going to be run through the function g(x). The function g(x) is pretty simple; it just subtracts 1 from part of the function and adds 1 to the rest of it. But this is the tricky part. The output of f(x) is being put into g(x), so the definition of g(x) now looks like this:

$$g(x) = \begin{cases} f(x) - 1, f(x) \geq 2 \\ f(x) + 1, f(x) < 2 \end{cases}$$

That means that you add 1 to f(x) wherever f(x) is less than 2—NOT where x is less than 2. As long as you're clear on that part, then modifying the graph is easy. Just slide the graph up one unit where f(x) < 2, which is to the right of the point (–1, 2), and slide the graph down one unit where f(x) ≥ 2, which is at and to the left of point (–1, 2). The correct answer is (B).

LEVEL IIC TEST FORM A

1. **(B)** Algebraic manipulation is the easiest way to solve this one. Start by adding s to each side, producing the inequality $r > r + 2s$. Then subtract r from each side to get $0 > 2s$. If $2s < 0$, then $s < 0$ (just divide both sides by 2). This can also be solved by plugging in, but since only certain values will make the original inequality true, it can take some time to plug in enough different values to eliminate all the wrong answers. Algebraic manipulation is faster here.

2. **(C)** You'll want to use the process of elimination here to knock out some answer choices. The best numbers to plug into the function are the ones in the answer choices: 10 and –10. Start by finding out whether the statement $f(x) = f(-x)$ is true when $x = 10$. $f(10) = |10| + 10 = 20$. $f(-10) = |-10| + 10 = 10 + 10 = 20$. Since $f(10)$ and $f(-10)$ are equal, $f(x) = f(-x)$ is true when $x = 10$, and every answer choice that does not include 10 can be eliminated. That gets rid of (A), (D), and (E). To choose between the remaining answers (B) and (C), plug in a number included in (C) but not in (B). A simple number like zero works best. $f(0) = |0| + 10 = 10$. $f(-0) = |-0| + 10 = 0 + 10 = 10$. You can see that $f(0) = f(-0)$, so zero must be part of the correct answer. (B) does not include zero, so you can eliminate it. (C) is correct.

3. **(D)** Remember how factorials cancel out in fractions. If you expand the factorials in the fraction $\frac{15!}{13!\,2!}$, then you get

$$\frac{15 \cdot 14 \cdot 13 \cdot 12 \cdot 11 \cdot 10 \cdot 9 \cdot 8 \cdot 7 \cdot 6 \cdot 5 \cdot 4 \cdot 3 \cdot 2 \cdot 1}{(13 \cdot 12 \cdot 11 \cdot 10 \cdot 9 \cdot 8 \cdot 7 \cdot 6 \cdot 5 \cdot 4 \cdot 3 \cdot 2 \cdot 1)(2 \cdot 1)}.$$

You can see that every factor from 1 to 13 in the numerator is also in the denominator; these factors cancel out, leaving you with $\frac{15 \cdot 14}{2 \cdot 1}$, or $\frac{210}{2}$, which equals 105.

4. **(A)** Remember the SOHCAHTOA definitions of the trigonometric functions. In a right triangle, the sine of an angle equals the length of the opposite side over the length of the hypotenuse. You can easily find the lengths of sides AB and AC, since they are horizontal and vertical, respectively. The hypotenuse's length can be found with the Pythagorean Theorem, but you don't even need it if you recognize this as a 5-12-13 triangle. To find the sine of $\angle ABC$, then, you just need to find the value of $\frac{AC}{BC}$, which equals $\frac{5}{13}$.

5. **(D)** A "system" is another way of describing a set of simultaneous equations. To find the complete solution set of a system of equations, you need to solve the equations simultaneously. Instead of doing the algebra here, however, you should notice that the answer choices all have numbers in them. That means you can backsolve. As always, start with (C), which in this case is the coordinate pair $(-4, -3)$, making $x = -4$ and $y = -3$. To backsolve, put these numbers into both equations. These numbers work in the first equation, because $(-4)^2 + (-3)^2 = 16 + 9 = 25$. They don't work in the second equation, however, because $-4 \neq -3 + 1$. That means that both (C) and (D) can be eliminated, since both contain the coordinate pair $(-4, -3)$. Both (A) and (B) can be eliminated quickly, because each contains a single coordinate pair that works in neither equation. The correct answer is (E). You can check by plugging these numbers into the equations, but it's not necessary. Once the other answers are clearly eliminated, (E) must be correct.

6. **(E)** When there are variables in the question and the answer choices, plug in. Remember to select numbers that make your math easy. In this case, it's important to choose numbers that make the fraction $\frac{j}{k}$ work out conveniently. Making $j = 4$ and $k = 2$ turns out well, because it makes $\frac{j}{k} = 2$. The expression $\dfrac{jk - \frac{j}{k}}{\frac{j}{k}}$ then works out to $\dfrac{(4)(2) - 2}{2} = \dfrac{8-2}{2} = \dfrac{6}{2} = 3$. To find the correct answer, just go quickly through the answer choices to find the one that also equals 3 when $j = 4$ and $k = 2$. Only (E) works out equal to 3. (E) is correct.

7. **(E)** This is a visual perception question, and there's no hard and fast technique to follow to solve it. The best plan is to experiment with sketches in your test booklet and use common sense. Remember that this is an EXCEPT question, so you're looking for a shape that *can't* be made. Any time you find a way to make one of the shapes in the answer choices, that choice can be eliminated.

The intersection of a cube and a plane can be a triangle:

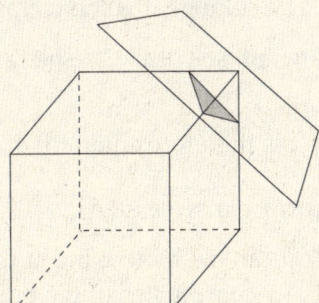

a point:

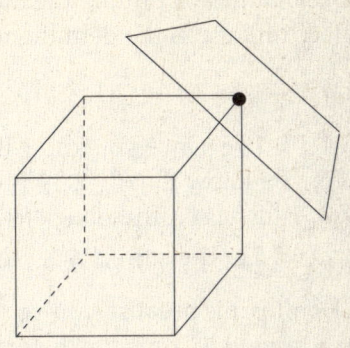

a rectangle:

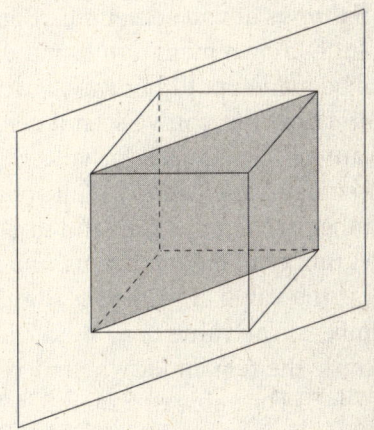

or a line segment:

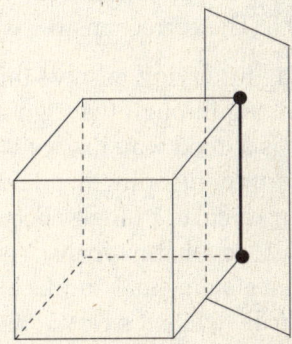

but there's no way to produce a circle, since none of a cube's edges or faces is curved. The correct answer is (E).

8. **(B)** This is a compound-function question, in which you'll have to apply two functions in combination. The most important rule to remember is that you have to

work from the inside out. Start by finding $g(2.3)$. That means putting the number 2.3 in place of the x in the definition of $g(x)$:
$g(2.3) = \frac{1}{2}\sqrt{2.3} + 1 = \frac{1}{2}(1.52) + 1 =$
$0.76 + 1 = 1.76$. Once you know that $g(2.3) = 1.76$, you know that $f(g(2.3))$ is equal to $f(1.76)$, which is easily solved: $f(1.76) = \sqrt[3]{1.76} = 1.21$. The correct answer is (B).

9. **(D)** Don't panic because you've never seen a term like "x mod y" before. This isn't something you slept through in math class. It's just one of those terms ETS throws at you sometimes. Some will be little-known math terms, and others will be made up. Either way, it doesn't matter whether you've seen it before, because ETS defines it for you. To find the value of any "x mod y", just take the number in the x position and divide it by the number in the y position. The remainder is the value of "x mod y" for those numbers. The value of 61 mod 7 is 5, because the remainder when 61 is divided by 7 is 5. The value of 5 mod 5 is 0, because the remainder when 5 is divided by 5 is zero. The expression (61 mod 7) – (5 mod 5) can be rewritten 5 – 0, which equals 5. The correct answer is (D).

10. **(D)** Plug in! Trying to solve this question by thinking through the trigonometric theory is a good way to give yourself a brain cramp. Just plug in a few numbers on your calculator (angles between 0° and 90°) and see what happens. You'll soon find that statements I and III always prove true, while it's easy to find an exception to statement II. The correct answer is (D).

11. **(B)** The statement $f(x) = \begin{cases} 2, x \neq 13 \\ 4, x = 13 \end{cases}$ can be read, "f of x equals 2 when x does not equal 13, and f of x equals 4 when x equals 13." Since none of the value's you're given equals 13, the function will always come out to 2: $f(15) - f(14) = 2 - 2 = 0$. The correct answer is (B).

12. **(D)** This question can be backsolved if you get completely stuck, but algebraic manipulation is simpler and faster. Just isolate x. Start by multiplying both sides by 25, so that $\frac{x^5}{25} = 25$ becomes $x^5 = 625$. Then take the fifth root of each side to get x alone. On most calculators, you take the fifth root of a value by raising that value to the power of one-fifth, or 0.2: $x = \sqrt[5]{625} = 3.62$. The correct answer is (D).

13. **(A)** On the Math IIC Test, most trigonometry questions like this one are solved by using trigonometric identities to change the form of equations. Most often, the most successful strategy is to start by getting everything in terms of sine and cosine. The ratio you're given can be written in fractional form, like this: $\frac{\sec x}{\csc x} = \frac{1}{4}$. The secant and cosecant can also be expressed in terms of sine and cosine: $\frac{\frac{1}{\cos x}}{\frac{1}{\sin x}} = \frac{1}{4}$. This fraction simplifies to $\frac{\sin x}{\cos x} = \frac{1}{4}$, and since $\tan x = \frac{\sin x}{\cos x}$, this can be written as $\tan x = \frac{1}{4}$. The cotangent is the reciprocal of the tangent, so $\cot x = 4$. The ratio of $\tan x$ to $\cot x$ is therefore equal to $\frac{\frac{1}{4}}{4}$, or $\frac{1}{16}$. The correct answer is (A).

14. **(D)** You can tell by looking at the figure that the x-coordinates of the points in region J include everything from $x = 0$ to $x = 6$. The y coordinates of the points in the region include everything from $y = 0$ to $y = 3$. A rectangular region containing

all points $(2x, y-1)$ would therefore stretch from $x = 0$ to $x = 12$ (doubling both values) and from $y = -1$ to $y = 2$ (subtracting 1 from both values). The resulting region would look like this:

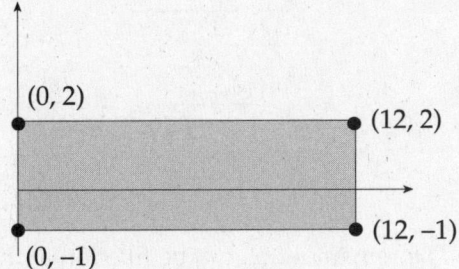

And would have a length of 12, a width of 3, and an area of 36. The correct answer is (D).

15. **(D)** Draw it! It's always a good idea to sketch any figure that is described but not shown. The triangle described here would look something like this:

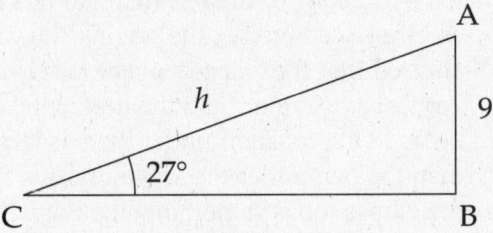

where h is the unknown length of the hypotenuse. As you can see, you're given the measure of an angle and the length of the opposite side in a right triangle, and you're trying to find the hypotenuse. The easiest way to find it is to use a trigonometric function that relates all these quantities. Since $\sin\theta = \frac{opposite}{hypotenuse}$, the sine is the function you want. Just plug the information you have into the formula, and solve for the missing piece:

$\sin 27° = \frac{9}{h}$. To solve, start by isolating h:

$h = \frac{9}{\sin 27°}$. Then use your calculator to find a numerical value: $h = \frac{9}{\sin 27°} = \frac{9}{0.454} = 19.82$. The correct answer is (D).

16. **(B)** A glance at the answer choices tells you that you can't factor this quadratic expression, because none of the answers are integers. Since you can't factor it, you'll have to use the quadratic formula:

$x = \frac{-b \pm \sqrt{b^2 - 4ac}}{2a}$, for equations in the form $y = ax^2 + bx + c$. The function $f(x) = x^2 + 6x - 12$ has coefficients $a = 1$, $b = 6$, and $c = -12$. Plug these coefficients into the quadratic formula to find the function's "zeroes," or roots:

$x = \frac{-6 \pm \sqrt{6^2 - 4(1)(-12)}}{2(1)} = \frac{-6 \pm \sqrt{84}}{2} =$

$\frac{-6 \pm 9.165}{2} = -3 \pm 4.58 = \{1.58, -7.58\}$.

Only one of these roots appears in the answer choices: -7.58. The correct answer is (B).

17. **(B)** There are two basic trigonometric identities being used here: $\tan x = \frac{\sin x}{\cos x}$ and $\sin^2 x + \cos^2 x = 1$. To find the value of $\tan x$ in terms of m, you first need to know the values of $\sin x$ and $\cos x$ in terms of m. The first step is easy: It's given that $\sin x = m$. The second step is trickier. To find the value of $\cos x$, plug the value of $\sin x$ into the identity $\sin^2 x + \cos^2 x = 1$, producing the equation $m^2 + \cos^2 x = 1$. Then isolate $\cos x$:

$$\cos^2 x = 1 - m^2$$

$$\cos x = \sqrt{1 - m^2}.$$

Once you can express $\sin x$ and $\cos x$ in terms of m, then finding the value of $\tan x$ is easy: $\tan x = \frac{\sin x}{\cos x} = \frac{m}{\sqrt{1-m^2}}$. The correct answer is (B).

18. **(C)** Use the definition of a logarithm to rewrite the equation $\log_y 2 = 8$ in exponential form: $y^8 = 2$. To find the value of y, take the eighth root of both sides. This can be done on your calculator by raising 2 to the power of one-eighth, or 0.125. You'll find that $y = 1.09$. The correct answer is (C).

19. **(D)** An angle of $\frac{\pi}{4}$ radians is equivalent to an angle of 45°. You're therefore looking for a an angle between 45° and –45° whose sine equals $\frac{1}{3}$. The easiest way to find this number is to enter the quantity $\frac{1}{3}$ into your calculator and take its inverse sine. This will show you the smallest positive angle whose sine is $\frac{1}{3}$. This angle, θ, is 19.47 degrees, or 0.34 radians. To find $\cos(2\theta)$, just double this value and take its cosine. You'll get 0.777..., or $\frac{7}{9}$. The correct answer is (D).

20. **(A)** The best way to solve inverse-function questions on the Math IIC Test is to plug in numbers. Be sure to pick a number that will work out neatly. For example, for the function $f(x) = \sqrt{x} - 1$, a good value for x would be 4. Then, $f(4) = \sqrt{4} - 1 = 2 - 1 = 1$. You can see that $f(x)$ turns 4 into 1. The inverse of $f(x)$ will therefore turn 1 into 4. To find the correct answer, plug 1 into all of the answer choices. The correct answer will produce a value of 4. Only (A) comes out to 4, since $(1 + 1)^2 = 4$, and so (A) is the correct answer.

21. **(A)** This is a job for polynomial division. Remember that you can divide polynomials using long division just as if they were numbers. This is the division problem you're being asked to figure out:

    $x+1 \overline{)4x^2 + 6x + L}$. If you solve the problem as far as you can using the information you have, this is what you get:

    $$\begin{array}{r} 4x+2 \\ x+1 \overline{)4x^2 + 6x + L} \\ \underline{4x^2 + 4x} \\ 2x + L \\ \underline{2x + 2} \\ 2 \end{array}$$

    The remainder, 2, can be filled in at the bottom of the problem because it's given by the question. You don't know the value of L to begin with, but by the time you've completed the problem, you can see that $L - 2 = 2$. And that tells you that $L = 4$. The correct answer is (A).

22. **(C)** If you've studied the section of this book dealing with ellipses, then you can tell by looking at the formula that this is an ellipse centered at the origin. You can also tell that the ellipse's major axis is the vertical axis rather than the horizontal, because the constant under the y is larger than the one under the x. That means that the ellipse looks something like this:

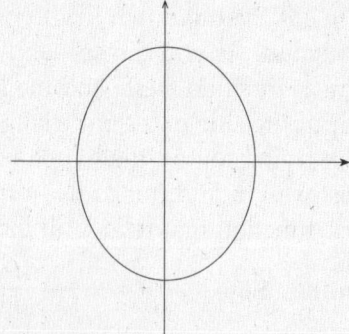

The major axis is the vertical axis of the ellipse. The endpoints of the major axis are the two points at which the ellipse intersects the y-axis. Since every point on the y-axis has an x-coordinate of 0, then all you have to do to find the coordinates of the endpoints is plug $x = 0$ into the equation of the ellipse, and see what y-coordinates that produces:

$$\frac{0^2}{10} + \frac{y^2}{20} = 1$$

The zero causes the entire first term to equal zero, effectively eliminating it:

$$0 + \frac{y^2}{20} = 1$$

$$\frac{y^2}{20} = 1$$

$$y^2 = 20$$

$$y = \pm\sqrt{20}$$

$$y = 4.472, -4.472$$

The endpoints of the ellipse's major axis are therefore (0, –4.472) and (0, 4.472). Because the points lie on a vertical line, it's not even necessary to use the distance formula to find the length of the segment between them. The distance between the points is equal to 4.472 – (–4.472), or 8.944. This is the length of the ellipse's major axis. The correct answer is (C).

23. **(B)** The function in this question is a *definitional* function. That is, it defines the function's operation verbally. In this case, [x] means the greatest integer less than or equal to x. In simple terms, that means that [x] equals x if x is an integer, and if x isn't an integer, then [x] is the next smallest integer. The number 2.75 becomes 2.0, –3.54 becomes –4.0, and so on. To find the graph of the function $f\left(\frac{x}{2}\right) - 1$, just plug in numbers. Start with something very easy, like zero: $f\left(\frac{0}{2}\right) - 1 =$ [0] – 1 = 0 – 1 = –1. The function $f\left(\frac{x}{2}\right) - 1$ must contain the point (0, –1). Only (B) and (D) contain that point; answer choices (A), (C), and (E) may be eliminated (remember, a function does not include points marked by open circles). To choose between (B) and (D), plug in another easy number, like 2: $f\left(\frac{2}{2}\right) - 1 =$ [1] – 1 = 1 – 1 = 0. The function $f\left(\frac{x}{2}\right) - 1$ must therefore also contain the point (2, 0). Only (B) contains that point—(D) contains the point (2, 1) instead. The correct answer is (B).

24. **(D)** If you remember that the secant is the reciprocal of the cosine, then you can solve this one quickly. If the cosine of an angle is 0.3527, then the secant of that angle is $\frac{1}{0.3527}$, or 2.8353. The expression "arccos[0.3527]" represents the angle between 0° and 180° whose cosine equals 0.3527. To find arccos[0.3527], just type 0.3527 into your calculator and find its inverse cosine: it comes out to 69.347°. Then, just find the secant of 69.347°, which is 2.835. If your calculator doesn't have a secant function, just remember that $\sec x = \frac{1}{\cos x}$. You can find the secant of 69.347° by finding the cosine of the angle and then taking its reciprocal using the 1/x key. The correct answer is (D).

25. **(B)** The function $f(x) = x^2 + 5x + 6$ is a quadratic function, which means that its graph will be a parabola. Since the coefficient of the x^2 term is positive, you know the parabola opens upward. The parabola therefore has a minimum value—its vertex. To find the value of x at which $f(x)$ reaches its minimum value, just find the x-coordinate of the parabola's vertex. You can do this by using the vertex formula: For any parabola in the form $y = ax^2 + bx + c$, the x-coordinate of the vertex equals $\frac{-b}{2a}$. For

the parabola $f(x) = x^2 + 5x + 6$, that places the vertex at $\frac{-5}{2}$. The correct answer is (B).

26. **(B)** An arithmetic sequence is one that increases by adding a constant amount again and again. The most important information to have when you're working with an arithmetic series is the size of the interval between any two consecutive terms in the series. Since the 20th term in the series is 20 and the 50th term is 100, you know that 30 steps in the series produce an increase of 80. That makes each step worth $\frac{80}{30}$, or $2.66\bar{6}$. To find the first term in the series, you can use the formula for the nth term in an arithmetic series: $a_n = a_1 + (n-1)d$, where n is the number of the term and d is the interval between any two consecutive terms. In this case, you know that $a_{20} = 20$, and that $d = 2.66\bar{6}$ or $\frac{8}{3}$. You can then fill those values into the formula, like this:

$$20 = a_1 + (20-1)(2.66\bar{6})$$

You can then solve for the value of a_1:

$$20 = a_1 + 19 \cdot 2.66\bar{6}$$
$$20 = a_1 + 50.66\bar{6}$$
$$a_1 = -30.66\bar{6}$$

The correct answer is (B).

27. **(A)** This question is a lot simpler than it looks. ETS will never require you to do complex calculations with polar coordinates; you only need to know the basics. As you've seen in this book, polar coordinates can be converted to rectangular coordinates very simply, with these two equations: $x = r \cos \theta$ and $y = r \sin \theta$. That means that the equation $r \sin \theta = 1$ simply translates to $y = 1$ in rectangular coordinates. And that's the equation of a horizontal line. The correct answer is (A).

28. **(B)** Remember, a function is a relation in which every element in the domain corresponds to only one element in the range. Simply put, that means that there's only one y for every x. If you look at simple tables of values for the functions you're given, you can see that they obey the rule:

I. $f(x) = x^2$ II. $f(x) = x^3$ III. $f(x) = |x|$

x	$f(x)$	x	$f(x)$	x	$f(x)$
−2	4	−2	−8	−2	2
−1	1	−1	−1	−1	1
0	0	0	0	0	0
1	1	1	1	1	1
2	4	2	8	2	2

But the inverses of these functions will *reverse* these tables of values, switching the x and $f(x)$ values. After all, the inverse undoes the original function, turning all the $f(x)$ values back into the x values. If you reverse the tables of values for these three functions, you'll find that only one of them remains a function:

I. $f(x) = \pm\sqrt{x}$ II. $f(x) = \sqrt[3]{x}$ III. $f(x) = \pm x$

x	$f(x)$	x	$f(x)$	x	$f(x)$
0	0	−8	−2	0	0
1	1, −1	−1	−1	1	1, −1
4	2, −2	0	0	2	2, −2
		1	1		
		8	2		

The inverses of functions I and III are not functions themselves: in each case, some x values correspond to more than one $f(x)$ value. That is, there's more than one y for each x. The inverse of function II, however, *is* a function, because there's still only one y for every x. The correct answer is (B).

29. **(D)** This is a classic ETS limit question. Ordinarily, to find the limit of an expression as x approaches a certain value, you just plug that value in for x, and *voila*, there's your answer. This, however, is one of those annoying expressions that become undefined (zero in the denominator) when you plug in the value. That could mean that the limit does not exist; but before you decide that, you've got to factor the top and bottom and see whether anything cancels out. In this case, the expression is factorable:

$$\frac{x^3 - x}{x+1} =$$

$$\frac{x(x^2 - 1)}{x+1} =$$

$$\frac{x(x+1)(x-1)}{x+1} =$$

$$\frac{x(x-1)}{1} = x^2 - x$$

The whole expression reduces to $x^2 - x$. In order to find the value of $\lim_{x \to -1} x^2 - x$, just plug in -1 for x: $(-1)^2 - (-1) = 1 + 1 = 2$. The correct answer is (D).

30. **(A)** The statement $g(f(x)) = x$ means that $g(x)$ and $f(x)$ are inverse functions: each function reverses the operation of the other. You're given the function $f(x) = \frac{e^{7x} + \sqrt{3}}{2}$. To find the inverse function, just isolate x:

$$y = \frac{e^{7x} + \sqrt{3}}{2}$$

$$2y = e^{7x} + \sqrt{3}$$

$$2y - \sqrt{3} = e^{7x}$$

Continue by using the definition of a natural logarithm to put the equation into logarithmic form:

$$\log_e(2y - \sqrt{3}) = 7x$$

$$\ln(2y - \sqrt{3}) = 7x$$

$$\frac{\ln(2y - \sqrt{3})}{7} = x$$

When you've isolated x, then the thing on the other side of the equal sign is the inverse function, $g(x)$:

$$g(x) = \frac{\ln(2x - \sqrt{3})}{7}$$

The correct answer is (A).

31. **(D)** When a figure is described but not shown, always sketch it:

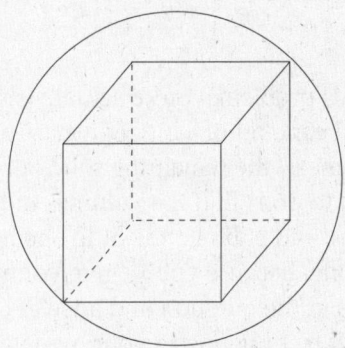

When a cube is inscribed in a sphere, the long diagonal of the cube is a diameter of the sphere. Since the sphere has a radius of 6 and a diameter of 12, you know that the sphere has a long diagonal of 12. The long diagonal of a cube is related to a side of the cube using the formula $d = s\sqrt{3}$. You can use that formula to find the length of a side: $s = \frac{d}{\sqrt{3}}$. In this case $s = \frac{12}{\sqrt{3}} = \frac{12}{1.732} = 6.93$. Once you know the length of one side of the cube, you can easily find the cube's volume using the formula $V = s^3$. You'll find that the cube has a volume of 332.55. Answers (A), (B), and (C) can immediately be eliminated. To choose between (D) and (E), just calculate the decimal value of one of them. Since $192\sqrt{3} = 332.55$, (D) is the correct answer.

32. **(E)** When a figure is described but not shown, always sketch it:

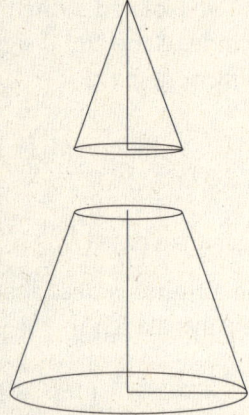

In this problem, you're actually slicing a small cone off of a bigger cone. To find the volume of the remaining solid, all you need to do is find the volumes of the two cones, and subtract the little one from the big one. Because you're given only variables in the question and answer choices, you'll be plugging in your own values. Let's say the larger cone has a height of $h = 4$ and a radius of $r = 2$. Then we can find its volume using the formula $V = \frac{1}{3}\pi r^2 h$. In this case, you get:

$V = \frac{1}{3}\pi r^2 h = \frac{1}{3}\pi r^2 h = \frac{1}{3}\pi r^2 h$. To find the volume of the smaller cone, you first need to figure out its base and height. The height is easy: since the problem says the larger cone is cut "midway," you know that the smaller cone's height equals 2, or half of the larger cone's height. Because one cone is a piece of the other, they are proportional. The radius of the smaller cone must equal 1, or half of the larger cone's radius. Now you can find the volume of the smaller cone:

$V = \frac{1}{3}\pi(1)^2(2) = \frac{1}{3}\pi(2) = \frac{2\pi}{3}$. Once you know the volumes of both cones, it's easy to subtract to find the volume of the leftover solid: $\frac{16\pi}{3} - \frac{2\pi}{3} = \frac{14\pi}{3}$. That's your target number. To find the right answer just plug $h = 4$ and $r = 2$ into the answer choices, and see which answer produces a value of $\frac{14\pi}{3}$. Only (E) does the trick.

33. **(B)** The equation $e^{x^2} = \frac{1}{\sqrt{3}^x}$ looks very difficult to solve. The presence of the constant e, however, should tip you off that you'll be solving this problem using natural logarithms. Many exponential equations can also be expressed as logarithms. Before you put this equation into natural log form, though, express everything in terms of exponents. Raising something to the power of $\frac{1}{2}$ is the same as taking a square root; and raising something to the power of -1 is the same as putting it on the bottom of a fraction. That means that $\frac{1}{\sqrt{3}^x}$ can also be written as $((3^{\frac{1}{2}})^x)^{-1}$, or $3^{-\frac{x}{2}}$. The original equation can then be rewritten like this: $e^{x^2} = 3^{-\frac{x}{2}}$. Using the definition of the natural logarithm, you can convert this equation to logarithmic form: $\log_e 3^{-\frac{x}{2}} = x^2$, or $\ln 3^{-\frac{x}{2}} = x^2$. You can then pull the exponent out of the logarithm and make it a coefficient, like this: $-\frac{x}{2}\ln 3 = x^2$. Finally, divide each side by x, and you've got the value of x: $-\frac{1}{2}\ln 3 = x$. Use your calculator to figure out the value of $-\frac{1}{2}\ln 3$. You should get -0.549. The correct answer is (B).

34. **(D)** The equation $y = 2x^2 - 6x + c$ is in quadratic form, which means that its graph will be a parabola. If it's tangent to the x-axis, that means the vertex of the parabola lies on the x-axis (sketch it and you'll see that's the only way to get a point of tangency). You know that the y-coordinate of the vertex must therefore be zero. The x-coordinate can be found using the vertex formula, $x = -\frac{b}{2a}$. In this case, $x = -\frac{-6}{2(2)} = \frac{6}{4} = 1.5$. You can then plug the coordinates (1.5, 0) into the equation $y = 2x^2 - 6x + c$ and solve for c:

$$0 = 2(1.5)^2 - 6(1.5) + c$$
$$0 = 4.5 - 9.0 + c$$
$$4.5 = c$$

The correct answer is (D).

35. **(C)** You're dealing with imaginary numbers here, so your calculator won't be much use. To solve this problem, just plug $(i - 1)$ into the expression in place of x and do the math:

$$x^2 + 2x + 2 =$$
$$(i - 1)^2 + 2(i - 1) + 2 =$$
$$(i - 1)(i - 1) + 2(i - 1) + 2 =$$
$$i^2 - 2i + 1 + 2i - 2 + 2 =$$
$$i^2 + 1 =$$
$$-1 + 1 = 0$$

The correct answer is (C).

36. **(E)** With a graphing calculator, this is easy. Just type in the simple equations in the answer choices and see what their graphs look like. If you haven't got a graphing calculator, then you've got to proceed by process of elimination. The easiest way to eliminate answers is to observe that this curve does not appear to cross the x-axis or the y-axis. Any answer choice whose graph *does* intercept an axis can therefore be eliminated. That means plugging $x = 0$ and $y = 0$ into the answer choices to see what happens.

You should know roughly what the graph of $y = e^x$ looks like (it appears earlier in this book). If you remember it, you'll recall that it crosses the y-axis. Even if you don't remember the shape of the graph, it's easy to discover that the curve crosses the axis just by plugging in $x = 0$ and seeing that the equations in both (A) and (B) contain the point (0, 1). That's a y-intercept, so you can eliminate both choices.

Answer choice (C) can be eliminated because it's the equation of a line—no exponents.

Answer choice (D) is the equation of a parabola, which might fit the graph you're given. The equation factors, however, into $f(x) = (x - 1)(x - 2)$, which means that it has roots at $x = 1$ and $x = 2$. Those are x-intercepts, so you can eliminate (D) as well.

That leaves only (E), which is the equation of a hyperbola. You can easily tell from the equation that $x \neq 0$ and $y \neq 0$. That makes (E) the correct answer.

37. **(C)** Remember that when you calculate the probability of multiple events, you've got to find the probabilities of the individual events and multiply them together. On the first drawing, the odds of getting a dime are $\frac{8}{11}$, because there are 8 dimes out of a total of 11 coins. On the second drawing, the odds of getting a dime are $\frac{7}{10}$, because there are 7 dimes remaining out of a total of ten remaining coins. The total probability of drawing two dimes, then, is $\frac{8}{11} \times \frac{7}{10} = \frac{56}{110}$ which reduces to $\frac{28}{55}$. The correct answer is (C).

38. **(C)** You can recognize the graph of an even function because it's symmetrical across the *y*-axis. The graphs in (A), (C), and (D) are symmetrical across the *y*-axis, making them even functions. Graphs (B) and (E) are not even, so you can eliminate them.

    You can recognize the graph of an odd function because it looks the same when rotated 180°. The only remaining graph that looks just the same when the page is turned upside-down is (C).

39. **(D)** Always draw figures that are described but not shown. The diagram described in this question would look about like this:

    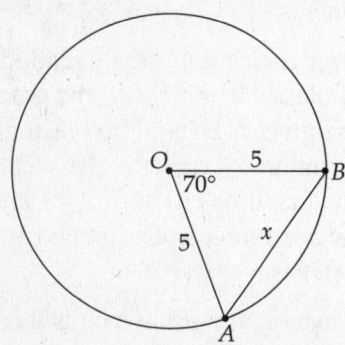

    The circle is just window-dressing here. The important thing in this problem is the triangle. You've got an isosceles triangle with two sides of length 5 that meet at a 70° angle. Your mission is to find the length *x* of the mystery side opposite the 70° angle. Estimation is the fastest way to choose the right answer. This is almost an equilateral triangle, but one angle is opened wider than 60°, making the other two angles smaller. Side *x*, opposite the big angle, must be bigger than 5 but much smaller than 10. Only (D) gives you such a number. Of course, it's also possible to find the exact value of *x*. For this, the Law of Cosines is your best tool: $c^2 = a^2 + b^2 - 2ab \cos C$. This formula contains the lengths of all three sides of a triangle (*a*, *b*, and *c*) and the measure of one angle (*C*). To solve the problem, just plug the two sides and angle that you know into the formula, and solve for the unknown side. Remember that the unknown side in this case is opposite the 70° angle:

    $$c^2 = (5)^2 + (5)^2 - 2(5)(5)\cos 70°$$
    $$c^2 = 25 + 25 - 50 \cos 70°$$
    $$c^2 = 50 - 50(0.342)$$
    $$c^2 = 50 - 17.10$$
    $$c^2 = 32.90$$
    $$c = \pm 5.74$$

    All lengths must be positive, so the length of the missing side must be 5.74. The correct answer is (D).

40. **(D)** The basic tangent and cotangent curves are very similar. The tangent curve, however, always has a positive slope, while the cotangent curve's slope is always negative. The slope of this graph is negative, so you can eliminate choices (A), (C), and (E). To choose between (B) and (D), plug in. You know from the graph that the function contains the point (0, 1). Just plug zero into one of the remaining functions and see which one produces a value of 1 (remember that $\frac{\pi}{4}$ radians is equivalent to 45°). You'll find that (B) produces a value of –1, while (D) produces a value of 1. The correct answer is (D).

41. **(A)** This is a vector-addition question with a minor twist: The negative sign. To add vectors *v* and *w*, look at each vector in terms of its *x*- component and *y*- component. This is particularly easy in this case, because both vectors start at the origin. The coordinates of the vectors' endpoints give you their *x*- and *y*- components. Vector *v* goes over –3 and up 4; vector *w* goes over 12 and up 5. To find the sum of *v* and *w*, just add these components. This produces a vector (*v* + *w*) that goes over 9 and up 9, ending at point (9, 9). Because *z* = –(*v* + *w*), you

must flip the signs of the x- and y-coordinates to get z. Vector z will therefore end at point $(-9, -9)$.

42. **(A)** This question looks frighteningly complicated, but it's not actually that bad. Just remember that, on the Math IIC Test, most equations containing the constant e will be solved using natural logarithms. You're asked to find the value of x when $f(x) = 0.33$, so start by setting up this equation:

$$\frac{1}{\sqrt{2\pi}} e^{-\frac{x^2}{2}} = 0.33$$

The next step to simplifying this equation is to get rid of that negative exponent:

$$\frac{1}{\sqrt{2\pi}} \times \frac{1}{e^{\frac{x^2}{2}}} = 0.33$$

Then, isolate the term containing x, and simplify:

$$\frac{1}{\sqrt{2\pi} \, e^{\frac{x^2}{2}}} = 0.33$$

$$\sqrt{2\pi} \, e^{\frac{x^2}{2}} = \frac{1}{0.33} = 3.03$$

$$e^{\frac{x^2}{2}} = \frac{3.03}{\sqrt{2\pi}}$$

$$e^{\frac{x^2}{2}} = 1.2089$$

Once you've simplified the equation this much, it's simple to convert it into logarithmic form using the definition of a natural logarithm:

$$\ln 1.2089 = \frac{x^2}{2}$$

$$0.1897 = \frac{x^2}{2}$$

$$x^2 = 0.3794$$

$$x = 0.616, -0.616$$

The correct answer is (A).

43. **(A)** In this figure, minor arc ST is intercepted by both central angle $\angle SOT$ and inscribed angle $\angle SVT$. Remember that a *central* angle of a certain degree measure intercepts an arc with an equal degree measure. An *inscribed* angle intercepts an arc of twice its own degree measure. That tells you that minor arc ST has a measure of $120°$, and that angle $\angle SVT$ has a measure of $60°$. Knowing that, you can see that statement I must be true, while statement III is definitely *not* true. This allows you to eliminate answer choices (B), (D), and (E).

To choose between (A) and (C), you've got to decide whether statement II is true. Statement II will be true only if point V is in the center of arc SVT. If you try redrawing the figure, however, you'll see that V doesn't have to be in the center of the arc. Statement II doesn't have to be true, and answer choice (C) can be eliminated. The correct answer is (A).

44. **(C)** The statements $f(a, b) = 15$, $f(b, c) = 20$, and $f(a, c) = 10$ can be rewritten as equations by inserting the definition of the function $f(x, y)$, like this:

$$\frac{ab}{3} = 15, \quad \frac{bc}{3} = 20, \text{ and } \frac{ac}{3} = 10$$

These equations can be simplified easily by getting the 3 out of the denominator:

$$ab = 45, \, bc = 60, \text{ and } ac = 30$$

The easiest way to find the product of a, b, and c, is to multiply these three equations together, like so:

$$ab \times bc \times ac = 45 \times 60 \times 30$$

$$aabbcc = 81{,}000$$

$$a^2 b^2 c^2 = 81{,}000$$

$$abc = \sqrt{81{,}000}$$

$$abc = 284.60$$

The correct answer is (C).

45. **(B)** You've got variables in the question and variables in the answer choices. That means it's time to plug in. Pick numbers that make the calculation easy. In this case, let's say that $x = 3$ and $y = 81$. The expression $\log_{x^2} y$ then becomes $\log_{3^2} 81$, or $\log_9 81$, and is equal to 2—the exponent that turns 9 into 81. Since $\log_{3^2} 81 = 2$, the correct answer will be the one that also equals 2 when $x = 3$ and $y = 81$. Quickly going through statements I, II and III, you'll find that:

   Statement I isn't necessarily true, because $\log_3 81^2 \neq 2$ (you don't need to compute the value of $\log_3 81^2$, it's obviously not 2, because $3^2 \neq 81^2$). You can therefore eliminate (A) and (D).

   Statement II is true, because $\log_3 9 = 2$. You can therefore eliminate (C), because it doesn't contain statement II.

   Statement III isn't true, because $\log_3\left(\dfrac{81}{2}\right) \neq 2$ (again, you don't have to compute the exact value of $\log_3\left(\dfrac{81}{2}\right)$, it's clearly not 2). You can therefore eliminate answer choice (E).

   The correct answer is (B).

46. **(D)** It's important to draw this one. First, realize that diagonal GD is not equal in length to diagonal BD, because GD cuts across *two* corners of the polygon, while BD only cuts across one corner. Diagonal GD will be longer. This lets you eliminate (A), (B), and (C), because only (D) and (E) are larger than 9. To find its exact length, you have to turn this into a triangle question by drawing diagonal BG.

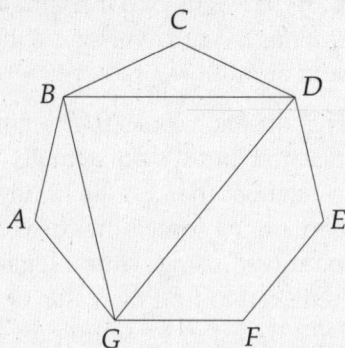

Since BG also cuts across one corner of this regular polygon, you know it has a length of 9, just like BD. You now know two sides of triangle BDG. All you need to solve the third side is the measure of any angle in the triangle.

To find the measure of an angle, start with the internal angles of the seven sided polygon $ABCDEFG$. It's a regular polygon, so all of its angles are equal, and the sum of its internal angle measures will equal $(n - 2) \times 180°$, where n, the number of sides, equals 7. That means that the total measure of the internal angles is 900°, and the measure of each angle is $\dfrac{900°}{7}$, or 128.57°. You can use this fact to solve any other angle in the figure. For example, the small triangle $\triangle BCD$ contains angle $\angle BCD$, which must measure 128.57°. Angle $\angle CBD$ and angle $\angle CDB$ are equal, and must therefore each measure 25.72° to make the angles of the triangle total 180°. Since $\triangle GAB$ is identical to $\triangle BCD$, its acute angles must also measure 25.72° each. Knowing that, you can find the measure of $\angle GBD$ by subtracting $\angle ABG$ and $\angle CBD$ from $\angle ABC$, which must measure 128.57°. You'll find that the measure of $\angle GBD$ equals 77.13°. Finally, you can use the Law of Cosines to find the length of GD:

$$c^2 = a^2 + b^2 - 2ab \cos C$$
$$c^2 = 9^2 + 9^2 - 2(9)(9) \cos 77.13°$$

$$c^2 = 81 + 81 - (162)(0.223)$$
$$c^2 = 125.92$$
$$c = \sqrt{125.92} = 11.22$$

The correct answer is (D).

47. **(C)** This is a complex number, because it has both real and imaginary components. To find its absolute value, visualize the term $6 - 3i$ on the complex plane, where one axis represents real values and the other represents imaginary values:

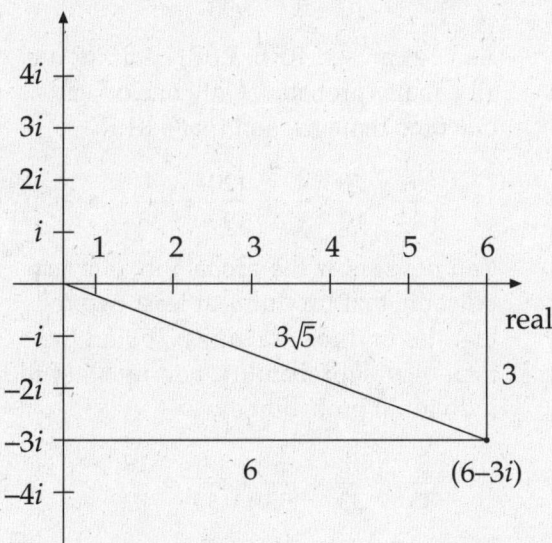

The point representing $6 - 3i$ would be six steps along the real axis and 3 steps along the imaginary axis, as shown. The point's absolute value is its distance from the origin, which is simply the hypotenuse of a right triangle with legs of length 3 and 6, respectively. Just plug those values into the Pythagorean Theorem, and you've got the absolute value:

$$h^2 = 3^2 + 6^2$$
$$h^2 = 45$$
$$h = 3\sqrt{5}$$

The correct answer is (C).

48. **(D)** This is a combinations question, since rearranging the order of the dishes doesn't change the dinner. Since there the dinner's being ordered in two parts (the part from column A and the part from column B), you should calculate the number of combinations in two parts. Start with the five dishes from column A: the number of permutations for five items selected from a group of ten is given by $10 \times 9 \times 8 \times 7 \times 6 = 30,240$. To find the number of combinations, divide that number by 5!: $\dfrac{10 \times 9 \times 8 \times 7 \times 6}{5 \times 4 \times 3 \times 2 \times 1} = 252$. There are 252 possible combinations of five dishes from column A. You're also selecting five dishes from column B, which has twenty selections. The number of combinations for column B will therefore be given by:

$\dfrac{20 \times 19 \times 18 \times 17 \times 16}{5 \times 4 \times 3 \times 2 \times 1} = 15,504$. So there are 15,504 combinations of 5 dishes from column B. To find the total number of possible combinations, multiply these figures together: $15,504 \times 252 = 3,907,008$. The correct answer is (D).

49. **(A)** This is a glorified Law of Cosines question. Sketch the information you're given, and you'll see that the question describes a triangle:

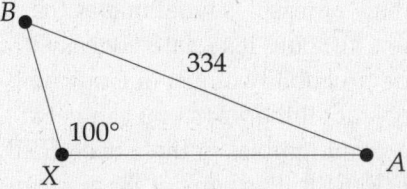

The paths of trucks A and B are 100° apart. Because the trucks begin traveling at the same time, and truck A travels twice as fast as B, truck A will always be twice as far as truck B from the point of origin. Four hours after they start, the trucks are 334 miles apart; that's the longest side of the triangle. You can think of the distances traveled by the trucks as d and $2d$ for trucks B and A, respectively. This allows you to relate all of these quantities with the law of cosines in an equation in one variable:

$$c^2 = a^2 + b^2 - 2ab\cos C$$
$$(334)^2 = d^2 + (2d)^2 - 2(d)(2d)\cos 100°$$
$$111{,}556 = 5d^2 - (4d^2)(-0.1736)$$
$$111{,}556 = 5d^2 + 0.69d^2$$
$$111{,}556 = 5.69d^2$$
$$d^2 = 19{,}605.62$$
$$d = \sqrt{19{,}605.62} = 140.02$$

After four hours, then, truck B has traveled 140.02 miles, which reflects a constant speed in miles per hour of $\dfrac{140.02}{4}$, or 35.01 miles per hour. The correct answer is (A).

50. **(D)** This probability question starts out with a trick. Since it *tells* you what happens in the first two drawings, they don't even enter into your calculations. Their outcome is certain. The real question starts after that: Given a container that holds 5 blue marbles and 6 red ones, what is the probability that three drawings will produce *at least* two red marbles?

That "at least" is what makes the question difficult. It's relatively easy to find the probability of just one outcome—but look at all the ways you can get *at least* two red marbles in three tries: RRB, RBR, BRR, RRR. Each one of these is a separate outcome. To compute the total probability of getting at least two red marbles, you need to find the probability of each outcome that does the trick, and then add them up.

Here's the probability of drawing RRB:

$$\frac{6}{11} \times \frac{5}{10} \times \frac{5}{9} = \frac{150}{990} = \frac{5}{33}$$

There are 6 red marbles out of a total of 11 on the first drawing; 5 red marbles out of a total of 10 on the second; and 5 blue marbles out of a total of 9 on the third. The total probability of this outcome is $\dfrac{5}{33}$.

Here's the probability of drawing RBR:

$$\frac{6}{11} \times \frac{5}{10} \times \frac{5}{9} = \frac{150}{990} = \frac{5}{33}$$

And here's the probability of BRR:

$$\frac{5}{11} \times \frac{6}{10} \times \frac{5}{9} = \frac{150}{990} = \frac{5}{33}$$

As you can see, RRB, RBR, and BRR are all equally probable. Only one other outcome remains, and that's RRR:

$$\frac{6}{11} \times \frac{5}{10} \times \frac{4}{9} = \frac{120}{990} = \frac{4}{33}$$

You now know the probability of every outcome that produces at least two red marbles in three drawings. To find the total overall probability, add up all of the individual probabilities:

$$\frac{5}{33} + \frac{5}{33} + \frac{5}{33} + \frac{4}{33} = \frac{19}{33}$$

The correct answer is (D).

LEVEL IIC TEST FORM B

1. **(E)** The statement $xy \neq 0$ means that neither x nor y is zero. To find the value of $\frac{y}{x}$, rearrange the equation so that $\frac{y}{x}$ is isolated on one side of the equal sign; whatever's on the other side will be the answer. In this case, the easiest way to isolate $\frac{y}{x}$ is to divide each side of the equation by x, getting $3 = \frac{0.3y}{x}$, and then divide both sides by 0.3, getting $\frac{3}{0.3} = \frac{y}{x}$. Your calculator will tell you that $\frac{3}{0.3}$ is equal to 10.

2. **(B)** To find the increase in $f(x)$ as x goes from 2 to 3, calculate $f(2)$ and $f(3)$ by plugging those numbers into the definition of the function. You'll find that $f(2) = 0.0589$ and $f(3) = 1.4308$. The increase in $f(x)$ is the difference between these two numbers: 1.3719.

3. **(A)** The fact that the line has a y-intercept of 3 and an x-intercept of -5 tells you that it includes the points $(0, 3)$ and $(-5, 0)$. This lets you find the slope of the line using the slope formula: $m = \frac{3-0}{0-(-5)}$.

    This gives you a slope of $\frac{3}{5}$, or 0.6. That's enough to pick out the correct answer already. To find the whole equation of the line, simply plug the value given for the y-intercept (3) in as b in the formula $y = mx + b$.

4. **(A)** This question just requires you to solve an equation for a. The trick is the big radical in the equation. The easiest way to get rid of it is to get the radical by itself on one side of the equal sign: $a + 4 = \sqrt{5a + 18}$. Then just square both sides, getting rid of the radical: $a^2 + 8a + 16 = 5a + 18$. Finally, combine like terms: $a^2 + 3a - 2 = 0$ and solve using the quadratic formula: $a = 0.56$ or $a = -3.56$. The positive value is the correct answer.

    You can also check your answer by backsolving: Plug 0.56 in for a in the original expression and see whether it makes the equation true. If you're having trouble with the algebra, you can also backsolve the entire question. Starting with (C), plug the values in the answer choices in for a in the original expression. If the value makes the equation true, you've got the right answer. If not, then pick another answer choice and try again.

5. **(D)** An arithmetic series is one that increases by a constant added amount. From 4 to 15 is a total increase of 11, which happens from the second term in the arithmetic series to the tenth, taking 8 steps. This means that the each step is worth one-eighth of 11: $\frac{11}{8}$, or 1.375. That's the constant amount added to each term in the series to get the next term. To find the first term in the series, just take one step backward from the second term, that is, subtract 1.375 from 4. You get 2.625.

    You can also solve this equation by using the formula for the nth term of an arithmetic sequence: $a_n = a_1 + (n-1)d$. You'd start by figuring out d (the difference between any two consecutive terms) just as you did above, finding that $d = 1.375$. Then, take one of the terms you're given (for example, $a_2 = 4$, in which case $n = 2$) and use these values to fill in the formula: $4 = a_1 + (2-1)(1.375)$. Then just solve for a_1. Once again you'll find that $a_1 = 2.625$.

6. **(C)** This simple function question just requires you to plug 6 into $g(x)$. You can start by eliminating (A) and (B), because the entire function is contained within an

absolute value sign, so it can't produce negative values. To find the exact value of $g(6)$, plug the number in. You get $\left|5(6)^2 - (6)^3\right|$, or $|-36|$, which equals 36.

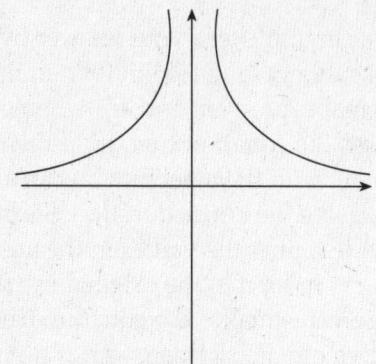

7. **(E)** The line $x = y$ is shown above. A graph symmetrical across this line will look like it is reflected in this line as though it were a mirror. Another way to think about it is that the two halves of a curve symmetrical across the line $x = y$ would meet perfectly if you folded the paper along that line. Of the five choices, only (E) has this kind of symmetry.

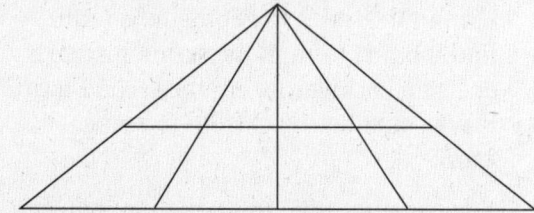

8. **(E)** Tackle visual-perception questions like these systematically. One good way is to start with the smallest triangles and work outward. In the top half of the figure, there are 4 "little" triangles; 3 triangles each formed by pairs of little triangles; 2 triangles formed by triplets of little triangles; and, finally, 1 triangle formed by all four little triangles together. That's a total of 10 triangles in the top half of the figure. Looking at the whole figure, you find the same pattern repeated: 4 tall skinny triangles; 3 triangles formed by pairs of skinny triangles; 2 triangles formed by triplets of skinny triangles; and one big triangle formed by all the skinny triangles together. That's another 10 triangles, for a total of 20. And that's all there are.

9. **(B)** This is the equation of a parabola which opens upward. The minimum value will be the y-value of the vertex, which you can find using the vertex formula. The x-coordinate of the vertex is given by $x = \dfrac{-b}{2a}$, which gives you $x = 6$ in this case. Plug this value back into the equation to get the y-coordinate of the vertex: $\dfrac{1}{2}(6)^2 - 6(6) + 11 = -7$. The function's minimum value is -7.

10. **(D)** This is a great question for plugging in. Plug in a couple of simple values, like $x = 3$ and $y = 5$, and solve: $|3 - 5| + |5 - 3| = |-2| + |2| = 2 + 2 = 4$. The correct answer must also equal 4. Using $x = 3$ and $y = 5$, you'll find that only (D) gives you the correct value.

11. **(D)** Plugging in works very well here. Find simple values that make the original equation true, like $A = 90°$ and $B = 0°$: $\sin 90° = 1$ and $\cos 0° = 1$. Then go through the answer choices to find the one that is also true when $A = 90°$ and $B = 0°$. In this case, only (D) works. In fact, (D) states a basic trigonometric identity: $\sin A = \cos(90 - A)$.

12. **(C)** To figure out a probability, divide the number of things you're looking for (in this case, flawed automobiles) by the total (all automobiles produced).

	Total Units Produced	Units Flawed
April	569	15
May	508	18
June	547	16

Since you're calculating the odds for the entire three-month period, you have to add up the two columns of the chart. You find that 49 flawed automobiles were produced out of a total of 1,624. Divide 49 by 1,624 to find the probability: 0.03.

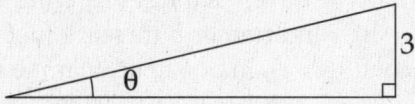

13. **(C)** The easiest way to find the possible ramp lengths is to find the shortest and longest legal lengths. The length of the ramp is the hypotenuse of a right triangle. Using the SOHCAHTOA definition of sine $\left(\sin = \frac{O}{H}\right)$, you can set up equations to find the lengths of a 5° ramp and a 7° ramp: $\sin 5° = \frac{3}{H}$ and $\sin 7° = \frac{3}{H}$. Solve for H in each case, and you'll find that the shortest possible hypotenuse has a length of 24.62, while the longest has a length of 34.42. Only (C) is between these limits.

14. **(A)** This question is simpler than it looks. The constant term $-n$ represents the y-intercept of the whole complicated function. Just look to see where the function crosses the y-axis. It does so at $y = 50$. That means that $-n$, the y-intercept, equals 50. Just plain n, however, equals -50. Watch out for (C), a trap answer!

15. **(A)** Here's a classic limit question. You can't just plug the x-value in question into the function, because it makes the denominator equal zero, which means the function is undefined at that point. To find the limit of the function approaching that point (assuming the limit exists), try to cancel out the troublesome term. First, factor the top and bottom of the function: $\frac{(x+2)(x-3)}{3(x+2)}$. As you can see, the term $(x + 2)$ occurs in the numerator and the denominator, so you can cancel it out. You're left with $\frac{(x-3)}{3}$. Now you can plug $x = -2$ into the function without producing an undefined quantity. The result, -1.67, is the limit of the function as x approaches -2.

16. **(B)** This is a great backsolving question. When you receive an inheritance in Titheland, you get the first 1,000 florins free and clear. After that, you only get 35% of the remaining amount; the government keeps the other 65%. To find the right answer, take the numbers from the answer choices and see which one would give you 2,500 florins after taxes.

    Answer choice (C) is 4,475 florins. Starting with that amount, you'd get 1,000 florins free and clear, and 3,475 would be taxed. The government would take 65% of 3,475, or 2,258.75. That would leave you with 1,216.25 plus the first 1,000, for a total of 2,216.25 florins after taxes—not enough. Your next step is to select a the next larger answer choice and try again. Answer choice (B) is 5,286 florins, which will give you 1,000 untaxed and 4,286 taxed. That means the government takes 65% of 4,286, or 2,785.9 florins, leaving you with 1,500.1. Add the untaxed 1,000, and you've got a total of 2,500.1 florins after taxes—right on the money.

17. **(B)** This question simply tests your understanding of the parts of a formula. To find the answer, just plug in the distance you're given, 10, and solve for the corresponding time: $-1.5t^2 + 120 = 10$. Solve and you'll find that $t^2 = 73.333$, which means that t equals either 8.56 or -8.56. Only a positive time makes sense, so the answer is 8.56.

18. **(C)** Formula time. The "dot dot dot..." at the end of the series indicates that it goes on forever. It's an infinite geometric

series. The formula for the sum of an infinite geometric series is $s = \frac{a_1}{1-r}$, where s is the sum, a_1 is the starting term (3 in this case), and r is the factor by which each term is multiplied to produce the following term (in this case, $\frac{1}{3}$). Just plug the values into the formula to get the sum: $\frac{9}{2}$, or 4.50.

19. **(B)** It may not be obvious at first, but this is a natural logarithm question. The equation $e^f = 5$ can be rewritten in logarithmic form, like this: $\log_e 5 = f$. A logarithm to the base e is called a natural logarithm, which can be written like this: $\ln 5 = f$. And that's something you can just punch into your calculator. The natural logarithm of 5 is 1.60944, which rounds to 1.61.

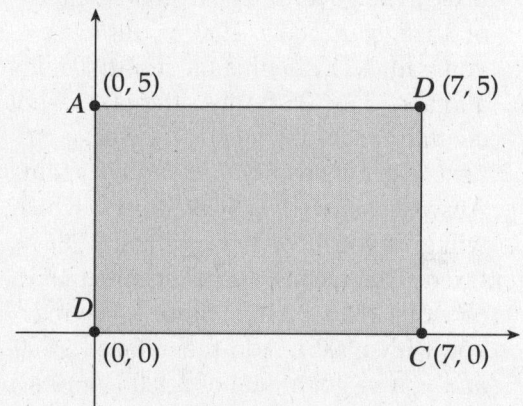

20. **(E)** The greatest distance within a rectangular solid is the length of the long diagonal—the line between diagonally opposite corners, through the center of the solid (shown above). The length of this line can be determined using the "Super Pythagorean Theorem:" $a^2 + b^2 + c^2 = d^2$, where a, b, and c are the dimensions of the solid, and d is the length of the diagonal. To find the possible coordinates of the box, use Super Pythagoras on each of the answer choices, and find the one that gives a diagonal of

12. Only (E) produces the right number: $4^2 + 8^2 + 8^2 = d^2$, and so $d = \sqrt{144}$, or 12.

21. **(E)** The variables in the answer choices tell you that this is a perfect plugging in question. Pick a rate for each runner. Say Runner A travels 100 feet every minute ($a = 100$) and runner B travels 1 foot every second ($b = 1$). In one hour, that means A travels 60×100 feet, or 6,000 feet, while B travels $3,600 \times 1$ feet, or 3,600 feet. In this case, A travels 2,400 feet farther than B. This makes 2,400 your target number—the number the correct answer will equal. Only (E) equals 2,400 using these values.

22. **(B)** This question is easier when you draw the triangle. This right triangle has a hypotenuse of 13 and legs of 5 and 12. The smallest angle will clearly be the angle opposite the side of length 5. Knowing this, you can use SOHCAHTOA to figure out the angle (let's call it θ): $\sin \theta = \frac{5}{13} = 0.3846$. Take the inverse sine of both sides, and you get $\theta = 22.62°$.

23. **(B)** Plugging in makes this question easy. Suppose, for example, that $x = 2$. To evaluate the expression $g(f(2))$, start on the inside: $f(2) = \frac{1}{2+1}$, or $\frac{1}{3}$. Then, work with the outside function: $g\left(\frac{1}{3}\right) = \frac{1}{\frac{1}{3}} + 1$, or $3 + 1$, which equals 4. That's the value of $g(f(2))$. The correct answer choice will be the one that gives the same value (4) when you plug in $x = 2$. Answer choice (B) is the only one that does.

24. **(D)** The roots of an expression are the values that make that expression equal to zero. In this case, there are three roots: π, 3, and e. To figure out which are the greatest and least roots, use your calculator to find the exact value of π (3.14159...) and e (2.71828...). These are the greatest

and least roots, so subtract them to find their difference: 0.42331. [Hint: if your calculator doesn't have an *e* key, just take the inverse ln*x* of 1, and you'll get 2.71828. If your calculator doesn't have a ln*x* key, get one that does.]

25. **(D)** The exclamation points indicate factorials. Remember that a factorial equals the product of every integer from 1 to the indicated number, inclusive. The value of $x!$ therefore equals $x \cdot (x-1) \cdot (x-2) \cdot (x-3) \cdot (x-4)...$, and the value of $(x-2)!$ equals $(x-2) \cdot (x-3) \cdot (x-4)....$ When you put these expanded values in fraction form, most of the terms cancel out:

$$\frac{x \cdot (x-1) \cdot (x-2) \cdot (x-3) \cdot (x-4)...}{(x-2) \cdot (x-3) \cdot (x-4)...}$$ leaving

you with $x(x-1)$, or $x^2 - x$.

If this bit of algebra seems slippery to you, then try plugging in. Suppose that $x = 4$, for example. Then $\frac{x!}{(x-2)!} = \frac{4!}{2!} = \frac{4 \times 3 \times 2 \times 1}{2 \times 1} = 12$. The correct answer will be the one that equals 12 when $x = 4$.

26. **(D)** A rectangle rotated around one edge generates a cylinder. This rectangle is being rotated around the vertical axis, so the cylinder will have a radius of 7 and a height of 5. Just plug these values into the formula for the volume of a cylinder: $V = \pi r^2 h$.

$$V = \pi(7)^2(5)$$
$$= \pi \times 49 \times 5$$
$$= 245$$
$$= 769.69$$

27. **(E)** To make this problem easier, simplify the original function. You can do this by factoring the top and bottom of the fraction: $f(x,y) = \frac{(x-y)(x-y)}{(x-y)(x+y)}$. You can cancel out an $(x-y)$ term on the top and bottom, producing the simplified function $f(x,y) = \frac{(x-y)}{(x+y)}$. Then, to answer the question, just plug $(-x)$ in for x and $(-y)$ in for y: $f(-x,-y) = \frac{(-x)-(-y)}{(-x)+(-y)} = \frac{-x+y}{-x-y}$.

Then, just multiply by $\left(\frac{-1}{-1}\right)$ to flip the signs on the top and bottom of this fraction: $\left(\frac{-x+y}{-x-y}\right) \times \left(\frac{-1}{-1}\right) = \frac{x-y}{x+y}$.

28. **(D)** In order to disprove a rule, it's only necessary to find one exception. That's what (D) is saying. Even if you didn't know that, however, there are still ways to eliminate wrong answers here. Use POE and some common sense. Answer choices (A), (C), and (E) are all pretty much impossible—they each require you to accomplish an enormous or even endless task. Answer choice (B) is more reasonable, but it doesn't relate to the question, which asks about numbers *less* than 5. Answer choice (D) is the only one which is both possible and relevant to the question.

29. **(E)** Careful with this one. Given three corners of a rectangle, you know for sure where the fourth one is. But this is just a parallelogram—capable of having many different shapes. You could place the fourth vertex at (–3, 3), (3, –3), or (7, 3), and still have a parallelogram. You don't know for sure where the fourth vertex is.

30. **(A)** An expression is undefined when its denominator equals zero. To find out what values might do that wicked deed, factor the expression. You'll find it equals $\frac{(x+4)(x-1)}{2(x+4)(x+1)}$. Two values will make this expression undefined: $x = -1$ and $x = -4$,

SUBJECT TEST ANSWERS AND EXPLANATIONS ◆ 383

both of which make the denominator equal to zero. (A) is correct. [Note: it's true that the term $(x + 4)$ cancels out of the factored expression. That doesn't mean that the original expression is defined at $x = -4$, however. It just means that you can calculate the limit of the expression as x approaches -4. Don't confuse the existence of a limit with the defined/undefined status of an expression.]

31. **(C)** In a I, II, III question, tackle the statements one at a time and remember the process of elimination. The expression in statement I is always positive, because x^2 must be positive and adding 1 can only increase it. That means that (B) and (D) can be eliminated, because they don't include statement I. Statement II is trickier: The sine of an expression can be anywhere from -1 to 1, inclusive. That means that most values of $(1 - \sin x)$ will be positive. If $\sin x = 1$, however, then the expression equals zero—not a positive value. Statement II is out, and you can eliminate (E). Finally, statement III can be simplified: $\pi(\pi^{x-1}) = \pi^{x-1+1} = \pi^x$. Since π is positive, and no exponent can change the sign of a base, π^x is always positive. (A) is out, and (C) is the correct answer.

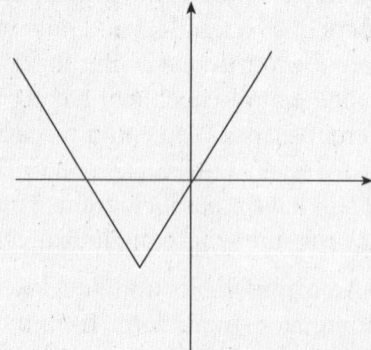

32. **(A)** Changing the sign of x in the expression $f(x)$ will flip the function's graph around the x-axis. Changing the sign of the whole function to produce $y = -f(-x)$ will flip the graph over the y-axis as well. The graph in answer choice (A) represents the original graph flipped both horizontally and vertically.

33. **(D)** It's helpful to draw this one. The wire has a slope of $\frac{2}{5}$, meaning that it rises 2 feet for every 5 feet it runs. Since its total rise is 48 (2×24), its total run must be 120 (5×24). Don't pick answer choice (B), though. The question asks for the length of the wire, not the distance between the anchor and antenna. The wire's length is the hypotenuse of a right triangle with legs of 48 and 120. The Pythagorean Theorem will tell you that its length is 129.24.

34. **(B)** To start with, you can do some useful elimination. The statement $\frac{3\pi}{2} < \theta < 2\pi$ tells you that you're working in the fourth quadrant of the unit circle, where the tangent is negative. You can immediately eliminate (C), (D), and (E). If $\sec \theta = 4$, then $\cos \theta = 0.25$, because by definition $\sec \theta = \frac{1}{\cos \theta}$. Here you have to be careful. If you take the inverse cosine of 0.25, your calculator will display a value whose cosine equals 0.25—in radians, you should get 1.3181; but remember that different angles can produce the same cosine. In this case, you know that $\frac{3\pi}{2} < \theta < 2\pi$, which means that the angle in question is in the fourth quadrant of the unit circle. The angle in that quadrant with an equivalent cosine can be expressed in radians as -1.3181, or as $2\pi - 1.3181 = 4.9651$. Take the tangent of either of these values, and you'll get -3.8730.

35. **(C)** Drawing this one is helpful. You'll find that the radii of the two circles have to add up to the distance between the circles' centers. You can find that distance using the distance formula: $d = \sqrt{(x_2 - x_1)^2 + (y_2 - y_1)^2}$. You'll find

that the two centers are separated by a distance of 8.6023. Since one circle has a radius of 4, the other must have a radius of 4.6023.

36. **(C)** The equations in the beginning of this question can be rearranged into the Law of Sines. A little algebraic manipulation gets you $\dfrac{\sin A}{7} = \dfrac{\sin B}{10}$ and $\dfrac{\sin B}{5} = \dfrac{\sin C}{2}$. These equations can be combined into $\dfrac{\sin A}{7} = \dfrac{\sin B}{10} = \dfrac{\sin C}{4}$, which is the Law of Sines. This tells you that the lengths of the triangle's sides are in a ratio of 7:10:4. A ratio box would look like this:

	Side a	Side b	Side c	Total
ratio	7	10	4	21
multiplier	0.7619	0.7619	0.7619	0.7619
real #s	5.33	7.62	3.05	16.00

You can complete the ratio box by looking at the "Total" column, because you know that the ratio has a total of 21 parts, and the triangle has a perimeter of 16. You then know that there's a multiplier, x, for the ratio and that $21x = 16$. Solving this gives you the multiplier: $x = 0.7619$. You can then use this multiplier in the ratio box to find the real length of any side.

x	$h(x)$
–1	0
0	3
1	0
2	3

37. **(E)** Test each expression with the values from the table. The easiest one to use is 0; when you make $x = 0$, the function should equal 3. Only (D) and (E) equal 3 when $x = 0$. Then, notice that the function is equal to zero when $x = 1$ or -1; these values must be roots of the function. Only (E) contains both 1 and -1 as roots. It's the right answer.

38. **(A)** This is a tricky simultaneous equations problem. After some experimentation, you might notice that the second and third equations can be added together: $a + b + 2c = n + 8$, which is very similar to the first equation: $a + b + 2c = 7$. From this, you can determine that $n + 8 = 7$. If n is any value other than -1, this is impossible—no values of a, b, and c can make this system of equations true. There is no solution for this system if $n \neq -1$.

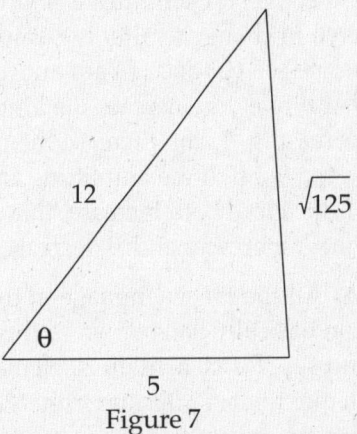

Figure 7

Note: Figure Not Drawn to Scale

39. **(D)** When you know all three sides of a triangle, and you need to determine the measures of the angles, it's time to use the Law of Cosines: $c^2 = a^2 + b^2 - 2ab \cos C$. Just plug in the lengths of the sides, making sure that you make c the side opposite θ:

$\sqrt{125}^2 = (12)^2 + (5)^2 - 2(12)(5)\cos \theta$.

Simplifying this gives you this equation: $\cos \theta = 0.3667$. Taking the inverse cosine of both sides shows that $\theta = 68.4898$.

40. **(B)** In this type of question, ETS is asking you not to solve for an exact value, but to pick out one possible value. Usually, there's a range of possible values. The

best way to determine what value lies within that range is to find the endpoints of the range. In this case, the question tells you that $[1.08 \otimes x] > 0$ and that $[1.09 \otimes x] < 0$. To see what that tells you about x, expand the functions: $e^{1.08} - 1.08x + 1 > 0$ and $e^{1.09} - 1.09x + 1 < 0$. Isolating x in each of these inequalities tells you that $x < 3.65248$ and $x > 3.64612$. Only (B) is between these values.

41. **(A)** This is a little tricky; you've got to pay attention to that underlined word, "increased." The values in the equation $g(x) = A[\sin(Bx + C)] + D$ that determine amplitude and period are A and B, respectively. Quantities C and D do not have to change to alter the amplitude or period, so (C) and (E) are out. The trick is that while you increase the amplitude by increasing A, you increase the period by *decreasing* B. If the amplitude and period of the curve both increase, that means that A increases and B decreases.

42. **(A)** All the fancy language in this question basically boils down to this: Set M and set N each contain 20 elements; each element in set M is larger than the corresponding element in set N.

    Once you have a clear idea of what that means, tackle the statements one at a time. Since the middle two values of M are bigger than the middle two values of N, the median of M must be greater; (B) and (C) can be eliminated because they do not include statement I. Since the total of the elements in M must be greater than the total of those in N, the mean of M must also be greater; II must also be in the correct answer. However, you don't know anything about the modes of the two sets; if the mode of M is its lowest value and the mode of N is its highest value, then the mode of N might be greater. Similarly, if M contains twenty values of 2 and N contains twenty values of 1, then the two sets have the same range: 0. Only statements I and II must be true.

43. **(E)** Start by rewriting the logarithm in exponential form: $(x-4)^3 = y$. Then, simply cube the binomial $x - 4$ using FOIL. Answer choice (E) represents the cube of $x - 4$.

44. **(D)** The statement $0 \leq n \leq \frac{\pi}{2}$ tells you that you're working in the first quadrant of the unit circle where both sine and cosine are never negative. The unit $\frac{\pi}{2}$ also tells you that you'll be working with angles in radians, not degrees; make sure your calculator is in the correct mode. The question tells you that the cosine of the cosine of n is 0.8. To find n, just take the inverse cosine of 0.8, and then take the inverse cosine of the result. You should get 0.8717—that's n. If you get an error, your calculator is probably in "degrees" mode. Finally, take the tangent of 0.8717. You should get 1.1895.

45. **(C)** This cylinder has a radius of n (because n is half the diameter) and a height of $\frac{n}{2}$. Just plug these values into the formula for the surface area of a cylinder: $SA = 2\pi r^2 + 2\pi rh$. You get $2\pi n^2 + \pi n^2$, or $3\pi n^2$.

46. **(C)** The sine and cosine functions can both produce any value from –1 to 1. The tangent and cotangent functions can both produce any real number. The secant and cosecant functions can both produce any number greater than or equal to 1, or less than or equal to –1. The only values that can be produced by all six trig functions are 1 and –1. Only 1 is an answer choice.

47. **(A)** The only way to tackle this one is by trying to disprove each of the answer choices. If you start with (A), you're lucky. There's no way to divide 5 by another quantity and get zero; it's the right answer. Even if you weren't sure, the other answer choices are pretty easy to disprove. Just set $\frac{5}{x+4}$ equal to a quantity prohibited by each answer

choice, and solve for x. If there's a real value of x that solves the equation, then the value *is* in the range after all, and the answer choice is incorrect.

48. **(E)** If $\vec{x} + \vec{y}$ is equal in magnitude to $\vec{x} - \vec{y}$, there are two possible explanations. Either $\vec{y}$ has a magnitude of zero, or $\vec{x}$ and $\vec{y}$ are at right angles to one another. The addition and subtraction of vectors either produces several identical vectors (if y = 0) or a couple of right triangles in which w and z, the sum and difference of the vectors, are the hypotenuses. In either case, statement (E) must be true, either because the vectors form right triangles, or because y = 0. All of the other answer choices are possible, but none of them *must* be true.

49. **(C)** There's no sophisticated math here. It's just an annoying function question with a lot of steps. As usual with function questions, tackle the statements one at a time and remember the process of elimination. Statement I must be true; if x = 0, then the function comes out to $\frac{y}{y}$, which equals 1 no matter what y is (since y can't be zero). Answer choice (B) can be eliminated since it doesn't contain statement I. Statement II must be true; since the value 1 is being plugged into the function in the x and y positions, the function will always equal $\frac{2}{2}$, or 1. Answer choices (A) and (D) can be eliminated because they don't include statement II. Finally, statement III is not necessarily true; $f(x, y) = \frac{xy+y}{x+y}$, and $f(y, x) = \frac{xy+x}{x+y}$. If x and y have different values, then these expressions will not be equal. Answer choice (E) can also be eliminated, because statement III is false. That leaves only (C).

50. **(B)** Plug in! Since no numerical values are assigned to m and b, you can plug in whatever you want. For example, say m = –8 and b = 2. The line's equation is then y = –8x + 2. The x-intercept is $\frac{1}{4}$ and the y-intercept is 2. The triangle formed by this line and the axes has a base of length $\frac{1}{4}$ and a height of length 2 (it's helpful to sketch this). If you rotate this triangle around the x-axis to generate a cone, the cone will have a radius of 2 and a height of $\frac{1}{4}$. Plug those values into the formula for the volume of a cone: $V = \frac{1}{3}\pi r^2 h$. You'll find that the volume equals $\frac{\pi}{3}$. This immediately eliminates answers (A), (C), and (D). Things look pretty good for (B), but you can't be sure it's not (E) until you've tried another set of numbers. If you try m = –1 and b = 1, you get a line with the equation y = –x + 1. This line has an x-intercept of 1 and a y-intercept of 1. The cone generated by rotation would then have a radius of 1 and a height of 1. Plug those numbers into the volume formula and you will once again get $\frac{\pi}{3}$. That's proof enough, the answer's clearly (B). That's the power of plugging in.

[Note: On questions late in the test that seem to have no solution, "It cannot be determined from the information given" is almost NEVER correct. Eliminate it first if you have to guess.]

Index

Symbols

30-60-90 triangle 86
45-45-90 Triangle 85

A

Absolute value 16
Acute angle 145
Adding ranges 60
Adding vectors 223
Altitude 78
Amplitude 172
Arc 78, 98, 145
Area 90
Area of a circle 97
Area of a parallelogram 94
Area of a rectangle 93
Area of a square 93
Area of a trapezoid 94
Area of a triangle 90
Area of an equilateral triangle 90
Arithmetic mean 16, 201
Arithmetic series 217
Arrangements 208
Asymptote 172, 190
Average 27, 201, 202
Average Speed 65
Average wheel 28
Axis of symmetry 134, 135

B

Backsolving 56
Base 30
Binomial 48, 69
Bisector 78

C

Calculator 21, 147
Center of an ellipse 138
Central angle 98
Chord 78
Circle 97, 136
Circumference 97
Circumscribed 78
Coefficient 30, 31
Combinations 202, 208
Complementary angles 78
Complex plane 228
Compound functions 175
Cones 109, 117
Consecutive numbers 16
Constants 47, 70
Contrapositive 225
Coordinate geometry 123
Coordinate plane 123, 124, 162
Coordinate space 140
Coordinates 124
Cosecant 153
Cosine 146, 150, 157
Cotangent 153
Cubes 107
Cylinders 108, 114, 117

D

Decimals 22
Decrease 24
Degree 172
Degree of a function 197
Degrees and radians 160
Denominator of zero 180
Diagonals 92
Diameter 97
Direct variation 61
Distance formula 131
Distance in a Three-Dimensional Space 140
Distributing 49
Dividing exponents 31
Domain 172, 180, 184

E

e 42
Ellipse 137
Equation of a circle with center at origin 136
Equation of a hyperbola with center at origin 139
Equation of a line 125
Equation of a parabola 134
Equation of an ellipse with center at origin 138
Equilateral triangles 82, 90
Even and odd numbers 18
Even functions 172, 194
Even power 30
Exponents 30

F

Face 106
Factoring 49
Factoring quadratics 70
Factorizations 16

First-degree function   198
Foci   137
Focus   137
FOIL   69
Fourth-degree function   198
Fractional exponents   33
Fractions   22
Fred's Theorem   80
Frequency   172
Functions   51, 171, 180

G

General equation of a hyperbola   139
General equation of an ellipse   137
General form of the equation of a parabola   135
Geometric series   219
Graphing functions   185
Graphing trigonometric functions   155
Group questions   211

H

Heptagon   96
Hexagon   96
Hyperbola   139
Hypotenuse   83, 146

I

Imaginary axis   228
Imaginary numbers   16, 30, 227
Increase   24
Indirect variation   62
Inequalities   58
Inequality signs   58
Inscribed   78
Inscribed angle   99
Inscribed solids   114
Integers   15
Interest   35
Intersection   202, 213
Interval   184
Inverse Functions   178
Inverse proportion   62
Inversely proportional   61
Irrational Numbers   15
Isosceles right triangle   85
Isosceles triangles   82

L

Lateral area   109
Law of cosines   164
Law of sines   163
Legs   83
Limits   221
Line   77

Line segments   77, 130
Linear inequalities   132
Logarithms   38, 39
Logic   225
Long diagonal   106
Long diagonal of a cube   107

M

Major arc   98
Mean   201, 202
Measuring angles   160
Median   16, 201, 202
Midpoint   78
Midpoint of a line segment   132
Minor arc   98
Mode   16, 201, 202
Multiple-average question   28
Multiplying exponents   31
Multiplying ranges   61

N

Natural Logarithms   42
Negative exponents   33
Negative root   32
Negative slopes   128
nth term of a geometric series   219
nth term of an arithmetic series   218

O

Obtuse angle   145
Octagon   96
Odd function   172, 195
Odd power   30
Opposite angles   79
Order of operations   20
Origin   124
Origin symmetry   195

P

Parabola   133
Parallel lines   78, 80, 130
Parallelograms   94
PEMDAS   20
Pentagon   96
Percent change   24, 25, 35
Percentages   22
Perimeter   78
Period   162, 172
Periodic functions   155, 162
Permutation   202, 208
Perpendicular   78, 79
Perpendicular lines   130
Plane   78
Plane geometry   77

Plugging in   53
Point-slope form of the equation of a line   126
Points of tangency   99
Polar Coordinates   166
Polygon   78
Polynomial   48
Polynomial division   230
Positive and negative numbers   19
Positive difference   16
Positive root   30, 32
Positive slopes   128
Power rule   40
Prime factorization   17
Prime numbers   15
Principal root   30, 32
Probability   203
Probability of multiple events   204
Product rule   39
Proportion   61
Proportionality of triangles   82
Pyramids   110
Pythagorean Theorem
   83, 106, 107, 108, 111, 131, 140, 164
Pythagorean triplets   84, 147

Q

Quadrants   124
Quadratic   48
Quadratic equation   70
Quadratic formula   73
Quadratic functions   133
Quadratic identities   71
Quadratic polynomial   70
Quadrilateral   78, 92
Quotient rule   39

R

Radian   145
Radical   32
Radius   78, 97
Range   59, 172, 180, 184, 201, 202
Rational numbers   15
Real numbers   16
Real roots   30
Rectangles   92
Rectangular prism   106
Rectangular solids   106, 111, 114
Regular polygons   96
Right angle   79
Right triangles   83, 84, 146
Rise   126
Roots   32, 48, 172, 193
Rotation   116
Rule of 180°   81
Run   126

S

Scientific notation   38
Secant   153
Second-degree function   198
Sector   78, 98
Semicircle   99
Sets   213
Short axis   137
Similar triangles   87
Simultaneous equations   66
Sine   146, 150, 157
Slant height   109
Slope   123, 128
Slope formula   129
Slope-intercept form of the equation of a line   125
Slope-intercept formula   125
SOHCAHTOA   146, 150, 153
Solid geometry   105
Spheres   110, 114, 116
Split function   173
Square root   32
Square root of a negative number   180
Squares   93
Standard deviation   202
Standard form of the equation of a circle   136
Standard form of the equation of a parabola   134
Statistics   201
Straight angle   79
Subtracting ranges   60
Subtracting vectors   223
Sum of a geometric series   219
Sum of an arithmetic series   218
Sum of an infinite geometric series   220
Supplementary angles   78
Surface area   106
Surface area of a cone   109
Surface area of a cube   107
Surface area of a cylinder   108
Surface area of a sphere   110
Symmetry   194
Symmetry in Functions   194

T

Tangent   78, 110, 146, 150, 158
Tangent line   99
Term   47
Third side rule   82
Third-degree function   198
Three-dimensional coordinate system   140
Trapezoids   94
Travel   63, 65
Triangles   81
Trigonometric functions   146, 150, 153, 155
Trigonometric graphs   162
Trigonometric identities   150

Trigonometry   145

U

Union   202, 213
Unit circle   155

V

Variable   47
Vectors   223
Vertex   134
Vertical angles   79
Vertical-line test   186
Vertices   137
Volume   106, 112
Volume of a cone   109
Volume of a cube   107
Volume of a cylinder   108
Volume of a pyramid   111
Volume of a sphere   110

W

Word-problem translation   23
Work   63

X

x-axis   124
x-intercept   124

Y

y–intercept   135
y-axis   124
y-intercept   124

Z

z-axis   140
Zero slope   129

ABOUT THE AUTHOR

Jonathan Spaihts was born in 1970. He is a graduate of Princeton University, and by pure coincidence works for The Princeton Review as a teacher, researcher, and writer. In that capacity he has helped to develop Princeton Review courses for the SAT I, SAT II, and a number of other standardized tests. He may also be seen in thrilling full-motion video on The Princeton Review's *Inside the SAT* and *Inside the GRE* CD-ROMs. When not working for The Princeton Review, Jonathan pursues various arcane writings of his own.

The Princeton Review
Diagnostic Test Form ○ Side 1

Completely darken bubbles with a No. 2 pencil. If you make a mistake, be sure to erase mark completely. Erase all stray marks.

1.
YOUR NAME: _____ Last _____ First _____ M.I. _____
(Print)
SIGNATURE: _____ DATE: ___/___/___
HOME ADDRESS: _____ Number and Street
(Print)
_____ City _____ State _____ Zip Code
PHONE NO.: _____
(Print)

IMPORTANT: Please fill in these boxes exactly as shown on the back cover of your test book.

2. TEST FORM

3. TEST CODE

4. REGISTRATION NUMBER

5. YOUR NAME — First 4 letters of last name | FIRST INIT | MID INIT

6. DATE OF BIRTH

MONTH	DAY	YEAR
○ JAN		
○ FEB		
○ MAR		
○ APR		
○ MAY		
○ JUN		
○ JUL		
○ AUG		
○ SEP		
○ OCT		
○ NOV		
○ DEC		

7. SEX
○ MALE
○ FEMALE

SCANTRON® FORM NO. F-592-KIN
© SCANTRON CORPORATION 1989 3289-C553-5
ALL RIGHTS RESERVED.

Begin with number 1 for each new section of the test. Leave blank any extra answer spaces.

SECTION 1

1–100. Ⓐ Ⓑ Ⓒ Ⓓ Ⓔ

The Princeton Review
Diagnostic Test Form ○ Side 2

Completely darken bubbles with a No. 2 pencil. If you make a mistake, be sure to erase mark completely. Erase all stray marks.

Begin with number 1 for each new section of the test. Leave blank any extra answer spaces.

SECTION 2

1	A	B	C	D	E	26	A	B	C	D	E	51	A	B	C	D	E	76	A	B	C	D	E
2	A	B	C	D	E	27	A	B	C	D	E	52	A	B	C	D	E	77	A	B	C	D	E
3	A	B	C	D	E	28	A	B	C	D	E	53	A	B	C	D	E	78	A	B	C	D	E
4	A	B	C	D	E	29	A	B	C	D	E	54	A	B	C	D	E	79	A	B	C	D	E
5	A	B	C	D	E	30	A	B	C	D	E	55	A	B	C	D	E	80	A	B	C	D	E
6	A	B	C	D	E	31	A	B	C	D	E	56	A	B	C	D	E	81	A	B	C	D	E
7	A	B	C	D	E	32	A	B	C	D	E	57	A	B	C	D	E	82	A	B	C	D	E
8	A	B	C	D	E	33	A	B	C	D	E	58	A	B	C	D	E	83	A	B	C	D	E
9	A	B	C	D	E	34	A	B	C	D	E	59	A	B	C	D	E	84	A	B	C	D	E
10	A	B	C	D	E	35	A	B	C	D	E	60	A	B	C	D	E	85	A	B	C	D	E
11	A	B	C	D	E	36	A	B	C	D	E	61	A	B	C	D	E	86	A	B	C	D	E
12	A	B	C	D	E	37	A	B	C	D	E	62	A	B	C	D	E	87	A	B	C	D	E
13	A	B	C	D	E	38	A	B	C	D	E	63	A	B	C	D	E	88	A	B	C	D	E
14	A	B	C	D	E	39	A	B	C	D	E	64	A	B	C	D	E	89	A	B	C	D	E
15	A	B	C	D	E	40	A	B	C	D	E	65	A	B	C	D	E	90	A	B	C	D	E
16	A	B	C	D	E	41	A	B	C	D	E	66	A	B	C	D	E	91	A	B	C	D	E
17	A	B	C	D	E	42	A	B	C	D	E	67	A	B	C	D	E	92	A	B	C	D	E
18	A	B	C	D	E	43	A	B	C	D	E	68	A	B	C	D	E	93	A	B	C	D	E
19	A	B	C	D	E	44	A	B	C	D	E	69	A	B	C	D	E	94	A	B	C	D	E
20	A	B	C	D	E	45	A	B	C	D	E	70	A	B	C	D	E	95	A	B	C	D	E
21	A	B	C	D	E	46	A	B	C	D	E	71	A	B	C	D	E	96	A	B	C	D	E
22	A	B	C	D	E	47	A	B	C	D	E	72	A	B	C	D	E	97	A	B	C	D	E
23	A	B	C	D	E	48	A	B	C	D	E	73	A	B	C	D	E	98	A	B	C	D	E
24	A	B	C	D	E	49	A	B	C	D	E	74	A	B	C	D	E	99	A	B	C	D	E
25	A	B	C	D	E	50	A	B	C	D	E	75	A	B	C	D	E	100	A	B	C	D	E

SECTION 3

1	A	B	C	D	E	26	A	B	C	D	E	51	A	B	C	D	E	76	A	B	C	D	E
2	A	B	C	D	E	27	A	B	C	D	E	52	A	B	C	D	E	77	A	B	C	D	E
3	A	B	C	D	E	28	A	B	C	D	E	53	A	B	C	D	E	78	A	B	C	D	E
4	A	B	C	D	E	29	A	B	C	D	E	54	A	B	C	D	E	79	A	B	C	D	E
5	A	B	C	D	E	30	A	B	C	D	E	55	A	B	C	D	E	80	A	B	C	D	E
6	A	B	C	D	E	31	A	B	C	D	E	56	A	B	C	D	E	81	A	B	C	D	E
7	A	B	C	D	E	32	A	B	C	D	E	57	A	B	C	D	E	82	A	B	C	D	E
8	A	B	C	D	E	33	A	B	C	D	E	58	A	B	C	D	E	83	A	B	C	D	E
9	A	B	C	D	E	34	A	B	C	D	E	59	A	B	C	D	E	84	A	B	C	D	E
10	A	B	C	D	E	35	A	B	C	D	E	60	A	B	C	D	E	85	A	B	C	D	E
11	A	B	C	D	E	36	A	B	C	D	E	61	A	B	C	D	E	86	A	B	C	D	E
12	A	B	C	D	E	37	A	B	C	D	E	62	A	B	C	D	E	87	A	B	C	D	E
13	A	B	C	D	E	38	A	B	C	D	E	63	A	B	C	D	E	88	A	B	C	D	E
14	A	B	C	D	E	39	A	B	C	D	E	64	A	B	C	D	E	89	A	B	C	D	E
15	A	B	C	D	E	40	A	B	C	D	E	65	A	B	C	D	E	90	A	B	C	D	E
16	A	B	C	D	E	41	A	B	C	D	E	66	A	B	C	D	E	91	A	B	C	D	E
17	A	B	C	D	E	42	A	B	C	D	E	67	A	B	C	D	E	92	A	B	C	D	E
18	A	B	C	D	E	43	A	B	C	D	E	68	A	B	C	D	E	93	A	B	C	D	E
19	A	B	C	D	E	44	A	B	C	D	E	69	A	B	C	D	E	94	A	B	C	D	E
20	A	B	C	D	E	45	A	B	C	D	E	70	A	B	C	D	E	95	A	B	C	D	E
21	A	B	C	D	E	46	A	B	C	D	E	71	A	B	C	D	E	96	A	B	C	D	E
22	A	B	C	D	E	47	A	B	C	D	E	72	A	B	C	D	E	97	A	B	C	D	E
23	A	B	C	D	E	48	A	B	C	D	E	73	A	B	C	D	E	98	A	B	C	D	E
24	A	B	C	D	E	49	A	B	C	D	E	74	A	B	C	D	E	99	A	B	C	D	E
25	A	B	C	D	E	50	A	B	C	D	E	75	A	B	C	D	E	100	A	B	C	D	E

FOR TPR USE ONLY	V1	V2	V3	V4	M1	M2	M3	M4	M5	M6	M7	M8

The Princeton Review
Diagnostic Test Form ○ Side 1

Completely darken bubbles with a No. 2 pencil. If you make a mistake, be sure to erase mark completely. Erase all stray marks.

1.
YOUR NAME: _____
(Print)   Last   First   M.I.

SIGNATURE: _____ DATE: ___/___/___

HOME ADDRESS: _____
(Print)   Number and Street

_____
City   State   Zip Code

PHONE NO.: _____
(Print)

IMPORTANT: Please fill in these boxes exactly as shown on the back cover of your test book.

2. TEST FORM

3. TEST CODE

4. REGISTRATION NUMBER

5. YOUR NAME — First 4 letters of last name | FIRST INIT | MID INIT

6. DATE OF BIRTH
- MONTH: JAN, FEB, MAR, APR, MAY, JUN, JUL, AUG, SEP, OCT, NOV, DEC
- DAY
- YEAR

7. SEX
- MALE
- FEMALE

SCANTRON® FORM NO. F-592-KIN
© SCANTRON CORPORATION 1989   3289-C553-5
ALL RIGHTS RESERVED.

Begin with number 1 for each new section of the test. Leave blank any extra answer spaces.

SECTION 1

Questions 1–100, each with answer choices A B C D E.

The Princeton Review
Diagnostic Test Form ○ Side 2

Completely darken bubbles with a No. 2 pencil. If you make a mistake, be sure to erase mark completely. Erase all stray marks.

Begin with number 1 for each new section of the test. Leave blank any extra answer spaces.

SECTION 2

1	A	B	C	D	E	26	A	B	C	D	E	51	A	B	C	D	E	76	A	B	C	D	E
2	A	B	C	D	E	27	A	B	C	D	E	52	A	B	C	D	E	77	A	B	C	D	E
3	A	B	C	D	E	28	A	B	C	D	E	53	A	B	C	D	E	78	A	B	C	D	E
4	A	B	C	D	E	29	A	B	C	D	E	54	A	B	C	D	E	79	A	B	C	D	E
5	A	B	C	D	E	30	A	B	C	D	E	55	A	B	C	D	E	80	A	B	C	D	E
6	A	B	C	D	E	31	A	B	C	D	E	56	A	B	C	D	E	81	A	B	C	D	E
7	A	B	C	D	E	32	A	B	C	D	E	57	A	B	C	D	E	82	A	B	C	D	E
8	A	B	C	D	E	33	A	B	C	D	E	58	A	B	C	D	E	83	A	B	C	D	E
9	A	B	C	D	E	34	A	B	C	D	E	59	A	B	C	D	E	84	A	B	C	D	E
10	A	B	C	D	E	35	A	B	C	D	E	60	A	B	C	D	E	85	A	B	C	D	E
11	A	B	C	D	E	36	A	B	C	D	E	61	A	B	C	D	E	86	A	B	C	D	E
12	A	B	C	D	E	37	A	B	C	D	E	62	A	B	C	D	E	87	A	B	C	D	E
13	A	B	C	D	E	38	A	B	C	D	E	63	A	B	C	D	E	88	A	B	C	D	E
14	A	B	C	D	E	39	A	B	C	D	E	64	A	B	C	D	E	89	A	B	C	D	E
15	A	B	C	D	E	40	A	B	C	D	E	65	A	B	C	D	E	90	A	B	C	D	E
16	A	B	C	D	E	41	A	B	C	D	E	66	A	B	C	D	E	91	A	B	C	D	E
17	A	B	C	D	E	42	A	B	C	D	E	67	A	B	C	D	E	92	A	B	C	D	E
18	A	B	C	D	E	43	A	B	C	D	E	68	A	B	C	D	E	93	A	B	C	D	E
19	A	B	C	D	E	44	A	B	C	D	E	69	A	B	C	D	E	94	A	B	C	D	E
20	A	B	C	D	E	45	A	B	C	D	E	70	A	B	C	D	E	95	A	B	C	D	E
21	A	B	C	D	E	46	A	B	C	D	E	71	A	B	C	D	E	96	A	B	C	D	E
22	A	B	C	D	E	47	A	B	C	D	E	72	A	B	C	D	E	97	A	B	C	D	E
23	A	B	C	D	E	48	A	B	C	D	E	73	A	B	C	D	E	98	A	B	C	D	E
24	A	B	C	D	E	49	A	B	C	D	E	74	A	B	C	D	E	99	A	B	C	D	E
25	A	B	C	D	E	50	A	B	C	D	E	75	A	B	C	D	E	100	A	B	C	D	E

SECTION 3

1	A	B	C	D	E	26	A	B	C	D	E	51	A	B	C	D	E	76	A	B	C	D	E
2	A	B	C	D	E	27	A	B	C	D	E	52	A	B	C	D	E	77	A	B	C	D	E
3	A	B	C	D	E	28	A	B	C	D	E	53	A	B	C	D	E	78	A	B	C	D	E
4	A	B	C	D	E	29	A	B	C	D	E	54	A	B	C	D	E	79	A	B	C	D	E
5	A	B	C	D	E	30	A	B	C	D	E	55	A	B	C	D	E	80	A	B	C	D	E
6	A	B	C	D	E	31	A	B	C	D	E	56	A	B	C	D	E	81	A	B	C	D	E
7	A	B	C	D	E	32	A	B	C	D	E	57	A	B	C	D	E	82	A	B	C	D	E
8	A	B	C	D	E	33	A	B	C	D	E	58	A	B	C	D	E	83	A	B	C	D	E
9	A	B	C	D	E	34	A	B	C	D	E	59	A	B	C	D	E	84	A	B	C	D	E
10	A	B	C	D	E	35	A	B	C	D	E	60	A	B	C	D	E	85	A	B	C	D	E
11	A	B	C	D	E	36	A	B	C	D	E	61	A	B	C	D	E	86	A	B	C	D	E
12	A	B	C	D	E	37	A	B	C	D	E	62	A	B	C	D	E	87	A	B	C	D	E
13	A	B	C	D	E	38	A	B	C	D	E	63	A	B	C	D	E	88	A	B	C	D	E
14	A	B	C	D	E	39	A	B	C	D	E	64	A	B	C	D	E	89	A	B	C	D	E
15	A	B	C	D	E	40	A	B	C	D	E	65	A	B	C	D	E	90	A	B	C	D	E
16	A	B	C	D	E	41	A	B	C	D	E	66	A	B	C	D	E	91	A	B	C	D	E
17	A	B	C	D	E	42	A	B	C	D	E	67	A	B	C	D	E	92	A	B	C	D	E
18	A	B	C	D	E	43	A	B	C	D	E	68	A	B	C	D	E	93	A	B	C	D	E
19	A	B	C	D	E	44	A	B	C	D	E	69	A	B	C	D	E	94	A	B	C	D	E
20	A	B	C	D	E	45	A	B	C	D	E	70	A	B	C	D	E	95	A	B	C	D	E
21	A	B	C	D	E	46	A	B	C	D	E	71	A	B	C	D	E	96	A	B	C	D	E
22	A	B	C	D	E	47	A	B	C	D	E	72	A	B	C	D	E	97	A	B	C	D	E
23	A	B	C	D	E	48	A	B	C	D	E	73	A	B	C	D	E	98	A	B	C	D	E
24	A	B	C	D	E	49	A	B	C	D	E	74	A	B	C	D	E	99	A	B	C	D	E
25	A	B	C	D	E	50	A	B	C	D	E	75	A	B	C	D	E	100	A	B	C	D	E

FOR TPR USE ONLY	V1	V2	V3	V4	M1	M2	M3	M4	M5	M6	M7	M8

The Princeton Review
Diagnostic Test Form ○ Side 1

Completely darken bubbles with a No. 2 pencil. If you make a mistake, be sure to erase mark completely. Erase all stray marks.

1.
YOUR NAME: _____
(Print)   Last   First   M.I.

SIGNATURE: _____ DATE: __/__/__

HOME ADDRESS: _____
(Print)   Number and Street

_____
City   State   Zip Code

PHONE NO.: _____
(Print)

IMPORTANT: Please fill in these boxes exactly as shown on the back cover of your test book.

2. TEST FORM

3. TEST CODE

4. REGISTRATION NUMBER

5. YOUR NAME — First 4 letters of last name | FIRST INIT | MID INIT

6. DATE OF BIRTH

MONTH	DAY	YEAR
○ JAN		
○ FEB		
○ MAR		
○ APR		
○ MAY		
○ JUN		
○ JUL		
○ AUG		
○ SEP		
○ OCT		
○ NOV		
○ DEC		

7. SEX
○ MALE
○ FEMALE

SCANTRON® FORM NO. F-592-KIN
© SCANTRON CORPORATION 1989 3289-C553-5
ALL RIGHTS RESERVED.

Begin with number 1 for each new section of the test. Leave blank any extra answer spaces.

SECTION 1

(Answer bubbles 1–100, each with options A B C D E)

The Princeton Review
Diagnostic Test Form ○ Side 2

Completely darken bubbles with a No. 2 pencil. If you make a mistake, be sure to erase mark completely. Erase all stray marks.

Begin with number 1 for each new section of the test. Leave blank any extra answer spaces.

SECTION 2

1	A	B	C	D	E	26	A	B	C	D	E	51	A	B	C	D	E	76	A	B	C	D	E
2	A	B	C	D	E	27	A	B	C	D	E	52	A	B	C	D	E	77	A	B	C	D	E
3	A	B	C	D	E	28	A	B	C	D	E	53	A	B	C	D	E	78	A	B	C	D	E
4	A	B	C	D	E	29	A	B	C	D	E	54	A	B	C	D	E	79	A	B	C	D	E
5	A	B	C	D	E	30	A	B	C	D	E	55	A	B	C	D	E	80	A	B	C	D	E
6	A	B	C	D	E	31	A	B	C	D	E	56	A	B	C	D	E	81	A	B	C	D	E
7	A	B	C	D	E	32	A	B	C	D	E	57	A	B	C	D	E	82	A	B	C	D	E
8	A	B	C	D	E	33	A	B	C	D	E	58	A	B	C	D	E	83	A	B	C	D	E
9	A	B	C	D	E	34	A	B	C	D	E	59	A	B	C	D	E	84	A	B	C	D	E
10	A	B	C	D	E	35	A	B	C	D	E	60	A	B	C	D	E	85	A	B	C	D	E
11	A	B	C	D	E	36	A	B	C	D	E	61	A	B	C	D	E	86	A	B	C	D	E
12	A	B	C	D	E	37	A	B	C	D	E	62	A	B	C	D	E	87	A	B	C	D	E
13	A	B	C	D	E	38	A	B	C	D	E	63	A	B	C	D	E	88	A	B	C	D	E
14	A	B	C	D	E	39	A	B	C	D	E	64	A	B	C	D	E	89	A	B	C	D	E
15	A	B	C	D	E	40	A	B	C	D	E	65	A	B	C	D	E	90	A	B	C	D	E
16	A	B	C	D	E	41	A	B	C	D	E	66	A	B	C	D	E	91	A	B	C	D	E
17	A	B	C	D	E	42	A	B	C	D	E	67	A	B	C	D	E	92	A	B	C	D	E
18	A	B	C	D	E	43	A	B	C	D	E	68	A	B	C	D	E	93	A	B	C	D	E
19	A	B	C	D	E	44	A	B	C	D	E	69	A	B	C	D	E	94	A	B	C	D	E
20	A	B	C	D	E	45	A	B	C	D	E	70	A	B	C	D	E	95	A	B	C	D	E
21	A	B	C	D	E	46	A	B	C	D	E	71	A	B	C	D	E	96	A	B	C	D	E
22	A	B	C	D	E	47	A	B	C	D	E	72	A	B	C	D	E	97	A	B	C	D	E
23	A	B	C	D	E	48	A	B	C	D	E	73	A	B	C	D	E	98	A	B	C	D	E
24	A	B	C	D	E	49	A	B	C	D	E	74	A	B	C	D	E	99	A	B	C	D	E
25	A	B	C	D	E	50	A	B	C	D	E	75	A	B	C	D	E	100	A	B	C	D	E

SECTION 3

1	A	B	C	D	E	26	A	B	C	D	E	51	A	B	C	D	E	76	A	B	C	D	E
2	A	B	C	D	E	27	A	B	C	D	E	52	A	B	C	D	E	77	A	B	C	D	E
3	A	B	C	D	E	28	A	B	C	D	E	53	A	B	C	D	E	78	A	B	C	D	E
4	A	B	C	D	E	29	A	B	C	D	E	54	A	B	C	D	E	79	A	B	C	D	E
5	A	B	C	D	E	30	A	B	C	D	E	55	A	B	C	D	E	80	A	B	C	D	E
6	A	B	C	D	E	31	A	B	C	D	E	56	A	B	C	D	E	81	A	B	C	D	E
7	A	B	C	D	E	32	A	B	C	D	E	57	A	B	C	D	E	82	A	B	C	D	E
8	A	B	C	D	E	33	A	B	C	D	E	58	A	B	C	D	E	83	A	B	C	D	E
9	A	B	C	D	E	34	A	B	C	D	E	59	A	B	C	D	E	84	A	B	C	D	E
10	A	B	C	D	E	35	A	B	C	D	E	60	A	B	C	D	E	85	A	B	C	D	E
11	A	B	C	D	E	36	A	B	C	D	E	61	A	B	C	D	E	86	A	B	C	D	E
12	A	B	C	D	E	37	A	B	C	D	E	62	A	B	C	D	E	87	A	B	C	D	E
13	A	B	C	D	E	38	A	B	C	D	E	63	A	B	C	D	E	88	A	B	C	D	E
14	A	B	C	D	E	39	A	B	C	D	E	64	A	B	C	D	E	89	A	B	C	D	E
15	A	B	C	D	E	40	A	B	C	D	E	65	A	B	C	D	E	90	A	B	C	D	E
16	A	B	C	D	E	41	A	B	C	D	E	66	A	B	C	D	E	91	A	B	C	D	E
17	A	B	C	D	E	42	A	B	C	D	E	67	A	B	C	D	E	92	A	B	C	D	E
18	A	B	C	D	E	43	A	B	C	D	E	68	A	B	C	D	E	93	A	B	C	D	E
19	A	B	C	D	E	44	A	B	C	D	E	69	A	B	C	D	E	94	A	B	C	D	E
20	A	B	C	D	E	45	A	B	C	D	E	70	A	B	C	D	E	95	A	B	C	D	E
21	A	B	C	D	E	46	A	B	C	D	E	71	A	B	C	D	E	96	A	B	C	D	E
22	A	B	C	D	E	47	A	B	C	D	E	72	A	B	C	D	E	97	A	B	C	D	E
23	A	B	C	D	E	48	A	B	C	D	E	73	A	B	C	D	E	98	A	B	C	D	E
24	A	B	C	D	E	49	A	B	C	D	E	74	A	B	C	D	E	99	A	B	C	D	E
25	A	B	C	D	E	50	A	B	C	D	E	75	A	B	C	D	E	100	A	B	C	D	E

FOR TPR USE ONLY	V1	V2	V3	V4	M1	M2	M3	M4	M5	M6	M7	M8

The Princeton Review
Diagnostic Test Form ○ Side 1

Completely darken bubbles with a No. 2 pencil. If you make a mistake, be sure to erase mark completely. Erase all stray marks.

1. YOUR NAME: (Print) Last / First / M.I.
SIGNATURE: ___ **DATE:** __/__/__
HOME ADDRESS: (Print) Number and Street / City / State / Zip Code
PHONE NO.: (Print)

IMPORTANT: Please fill in these boxes exactly as shown on the back cover of your test book.

2. TEST FORM

3. TEST CODE

4. REGISTRATION NUMBER

5. YOUR NAME — First 4 letters of last name / FIRST INIT / MID INIT

6. DATE OF BIRTH — MONTH / DAY / YEAR
JAN, FEB, MAR, APR, MAY, JUN, JUL, AUG, SEP, OCT, NOV, DEC

7. SEX — ○ MALE ○ FEMALE

SCANTRON® FORM NO. F-592-KIN
© SCANTRON CORPORATION 1989 3289-C553-5
ALL RIGHTS RESERVED.

Begin with number 1 for each new section of the test. Leave blank any extra answer spaces.

SECTION 1

(Answer bubbles A–E for questions 1–100)

The Princeton Review
Diagnostic Test Form ○ Side 2

Completely darken bubbles with a No. 2 pencil. If you make a mistake, be sure to erase mark completely. Erase all stray marks.

Begin with number 1 for each new section of the test. Leave blank any extra answer spaces.

SECTION 2

1	A	B	C	D	E	26	A	B	C	D	E	51	A	B	C	D	E	76	A	B	C	D	E
2	A	B	C	D	E	27	A	B	C	D	E	52	A	B	C	D	E	77	A	B	C	D	E
3	A	B	C	D	E	28	A	B	C	D	E	53	A	B	C	D	E	78	A	B	C	D	E
4	A	B	C	D	E	29	A	B	C	D	E	54	A	B	C	D	E	79	A	B	C	D	E
5	A	B	C	D	E	30	A	B	C	D	E	55	A	B	C	D	E	80	A	B	C	D	E
6	A	B	C	D	E	31	A	B	C	D	E	56	A	B	C	D	E	81	A	B	C	D	E
7	A	B	C	D	E	32	A	B	C	D	E	57	A	B	C	D	E	82	A	B	C	D	E
8	A	B	C	D	E	33	A	B	C	D	E	58	A	B	C	D	E	83	A	B	C	D	E
9	A	B	C	D	E	34	A	B	C	D	E	59	A	B	C	D	E	84	A	B	C	D	E
10	A	B	C	D	E	35	A	B	C	D	E	60	A	B	C	D	E	85	A	B	C	D	E
11	A	B	C	D	E	36	A	B	C	D	E	61	A	B	C	D	E	86	A	B	C	D	E
12	A	B	C	D	E	37	A	B	C	D	E	62	A	B	C	D	E	87	A	B	C	D	E
13	A	B	C	D	E	38	A	B	C	D	E	63	A	B	C	D	E	88	A	B	C	D	E
14	A	B	C	D	E	39	A	B	C	D	E	64	A	B	C	D	E	89	A	B	C	D	E
15	A	B	C	D	E	40	A	B	C	D	E	65	A	B	C	D	E	90	A	B	C	D	E
16	A	B	C	D	E	41	A	B	C	D	E	66	A	B	C	D	E	91	A	B	C	D	E
17	A	B	C	D	E	42	A	B	C	D	E	67	A	B	C	D	E	92	A	B	C	D	E
18	A	B	C	D	E	43	A	B	C	D	E	68	A	B	C	D	E	93	A	B	C	D	E
19	A	B	C	D	E	44	A	B	C	D	E	69	A	B	C	D	E	94	A	B	C	D	E
20	A	B	C	D	E	45	A	B	C	D	E	70	A	B	C	D	E	95	A	B	C	D	E
21	A	B	C	D	E	46	A	B	C	D	E	71	A	B	C	D	E	96	A	B	C	D	E
22	A	B	C	D	E	47	A	B	C	D	E	72	A	B	C	D	E	97	A	B	C	D	E
23	A	B	C	D	E	48	A	B	C	D	E	73	A	B	C	D	E	98	A	B	C	D	E
24	A	B	C	D	E	49	A	B	C	D	E	74	A	B	C	D	E	99	A	B	C	D	E
25	A	B	C	D	E	50	A	B	C	D	E	75	A	B	C	D	E	100	A	B	C	D	E

SECTION 3

1	A	B	C	D	E	26	A	B	C	D	E	51	A	B	C	D	E	76	A	B	C	D	E
2	A	B	C	D	E	27	A	B	C	D	E	52	A	B	C	D	E	77	A	B	C	D	E
3	A	B	C	D	E	28	A	B	C	D	E	53	A	B	C	D	E	78	A	B	C	D	E
4	A	B	C	D	E	29	A	B	C	D	E	54	A	B	C	D	E	79	A	B	C	D	E
5	A	B	C	D	E	30	A	B	C	D	E	55	A	B	C	D	E	80	A	B	C	D	E
6	A	B	C	D	E	31	A	B	C	D	E	56	A	B	C	D	E	81	A	B	C	D	E
7	A	B	C	D	E	32	A	B	C	D	E	57	A	B	C	D	E	82	A	B	C	D	E
8	A	B	C	D	E	33	A	B	C	D	E	58	A	B	C	D	E	83	A	B	C	D	E
9	A	B	C	D	E	34	A	B	C	D	E	59	A	B	C	D	E	84	A	B	C	D	E
10	A	B	C	D	E	35	A	B	C	D	E	60	A	B	C	D	E	85	A	B	C	D	E
11	A	B	C	D	E	36	A	B	C	D	E	61	A	B	C	D	E	86	A	B	C	D	E
12	A	B	C	D	E	37	A	B	C	D	E	62	A	B	C	D	E	87	A	B	C	D	E
13	A	B	C	D	E	38	A	B	C	D	E	63	A	B	C	D	E	88	A	B	C	D	E
14	A	B	C	D	E	39	A	B	C	D	E	64	A	B	C	D	E	89	A	B	C	D	E
15	A	B	C	D	E	40	A	B	C	D	E	65	A	B	C	D	E	90	A	B	C	D	E
16	A	B	C	D	E	41	A	B	C	D	E	66	A	B	C	D	E	91	A	B	C	D	E
17	A	B	C	D	E	42	A	B	C	D	E	67	A	B	C	D	E	92	A	B	C	D	E
18	A	B	C	D	E	43	A	B	C	D	E	68	A	B	C	D	E	93	A	B	C	D	E
19	A	B	C	D	E	44	A	B	C	D	E	69	A	B	C	D	E	94	A	B	C	D	E
20	A	B	C	D	E	45	A	B	C	D	E	70	A	B	C	D	E	95	A	B	C	D	E
21	A	B	C	D	E	46	A	B	C	D	E	71	A	B	C	D	E	96	A	B	C	D	E
22	A	B	C	D	E	47	A	B	C	D	E	72	A	B	C	D	E	97	A	B	C	D	E
23	A	B	C	D	E	48	A	B	C	D	E	73	A	B	C	D	E	98	A	B	C	D	E
24	A	B	C	D	E	49	A	B	C	D	E	74	A	B	C	D	E	99	A	B	C	D	E
25	A	B	C	D	E	50	A	B	C	D	E	75	A	B	C	D	E	100	A	B	C	D	E

FOR TPR USE ONLY	V1	V2	V3	V4	M1	M2	M3	M4	M5	M6	M7	M8

MORE EXPERT ADVICE

from

THE PRINCETON REVIEW

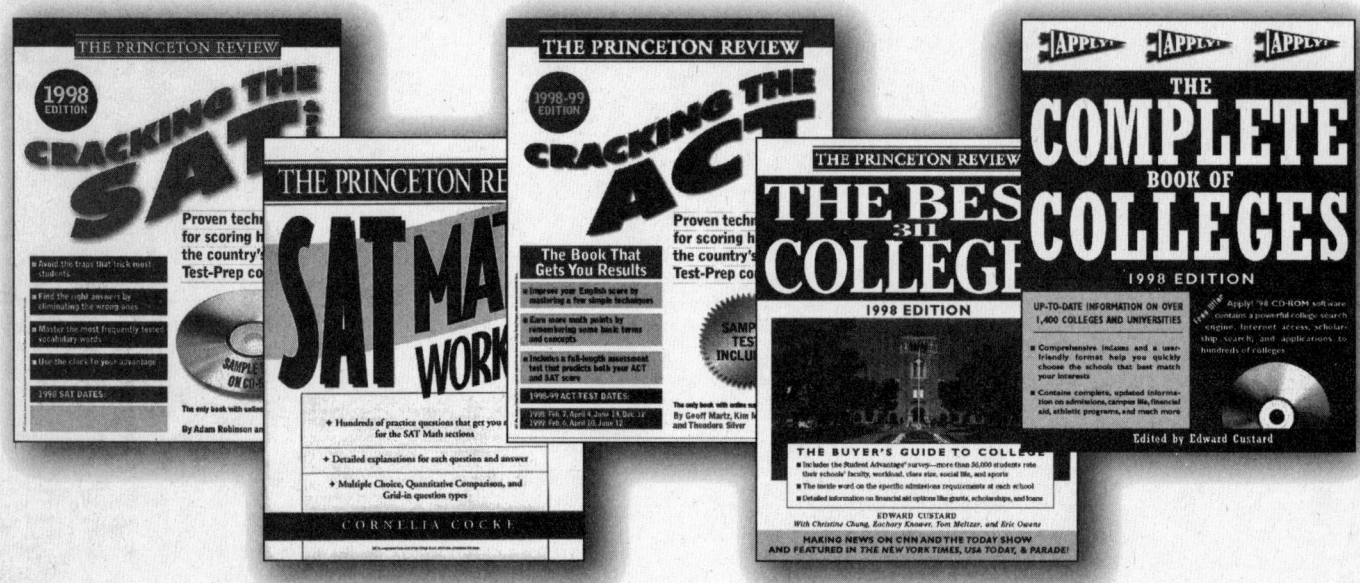

We help hundreds of thousands of students improve their test scores and get into college each year. If you want to give yourself the best chances for getting into the college of your choice, we can help you get the highest test scores, the best financial aid package, and make the most informed choices with our comprehensive line of books for the college-bound student. Here's to your success and good luck!

CRACKING THE SAT & PSAT 1998 EDITION
0-679-78405 $18.00

CRACKING THE SAT & PSAT WITH SAMPLE TESTS ON DISK 1998 EDITION
0-679-78404-7 • $29.95 with Mac and Windows compatible disks

CRACKING THE SAT & PSAT WITH SAMPLE TESTS ON CD-ROM 1998 EDITION
0-679-78403-9 • $29.95 Mac and Windows compatible

SAT MATH WORKOUT
0-679-75363-X • $15.00

SAT VERBAL WORKOUT
0-679-75362-1 • $16.00

INSIDE THE SAT BOOK/CD-ROM
1-884536-56-5 • $34.95
(Windows/Macintosh compatible interactive CD-ROM)

CRACKING THE ACT 1998-99 EDITION
0-375-75084-3 • $18.00

CRACKING THE ACT WITH SAMPLE TESTS ON CD-ROM 1998-99 EDITION
0-375-75085-1 • $29.95 Mac and Windows compatible

COLLEGE COMPANION
Real Students, True Stories, Good Advice
0-679-76905-6 • $15.00

STUDENT ADVANTAGE GUIDE TO COLLEGE ADMISSIONS
Unique Strategies for Getting into the College of Your Choice
0-679-74590-4 • $12.00

PAYING FOR COLLEGE WITHOUT GOING BROKE 1998 EDITION
Insider Strategies to Maximize Financial Aid and Minimize College Costs
0-375-75008-8 • $18.00

BEST 311 COLLEGES 1998 EDITION
The Buyer's Guide to College
0-679-78397-0 • $20.00

THE COMPLETE BOOK OF COLLEGES 1998 EDITION
0-679-78398-9 • $26.95

WE ALSO HAVE BOOKS TO HELP YOU SCORE HIGH ON
THE SAT II, AP, AND CLEP EXAMS:

CRACKING THE AP BIOLOGY EXAM 1998-99 EDITION
0-375-75108-4 $16.00

CRACKING THE AP CALCULUS EXAM AB & BC 1998-99 EDITION
0-375-75107-6 $16.00

CRACKING THE AP CHEMISTRY EXAM 1998-99 EDITION
0-375-75109-2 $16.00

CRACKING THE AP ENGLISH LITERATURE EXAM 1998-99 EDITION
0-375-75096-7 $16.00

CRACKING THE AP U.S. HISTORY EXAM 1998-99 EDITION
0-375-75097-5 $16.00

CRACKING THE CLEP 1998 EDITION
0-679-77867-5 $20.00

CRACKING THE SAT II: BIOLOGY SUBJECT TEST 1998-99 EDITION
0-375-75101-7 $17.00

CRACKING THE SAT II: CHEMISTRY SUBJECT TEST 1998-99 EDITION
0-375-75102-5 $17.00

CRACKING THE SAT II: ENGLISH SUBJECT TEST 1998-99 EDITION
0-375-75099-1 $17.00

CRACKING THE SAT II: FRENCH SUBJECT TEST 1998-99 EDITION
0-375-75103-3 $17.00

CRACKING THE SAT II: HISTORY SUBJECT TEST 1998-99 EDITION
0-375-75104-1 $17.00

CRACKING THE SAT II: MATH SUBJECT TEST 1998-99 EDITION
0-375-75100-9 $17.00

CRACKING THE SAT II: PHYSICS SUBJECT TEST 1998-99 EDITION
0-375-75106-8 $17.00

CRACKING THE SAT II: SPANISH SUBJECT TEST 1998-99 EDITION
0-375-75105-X $17.00

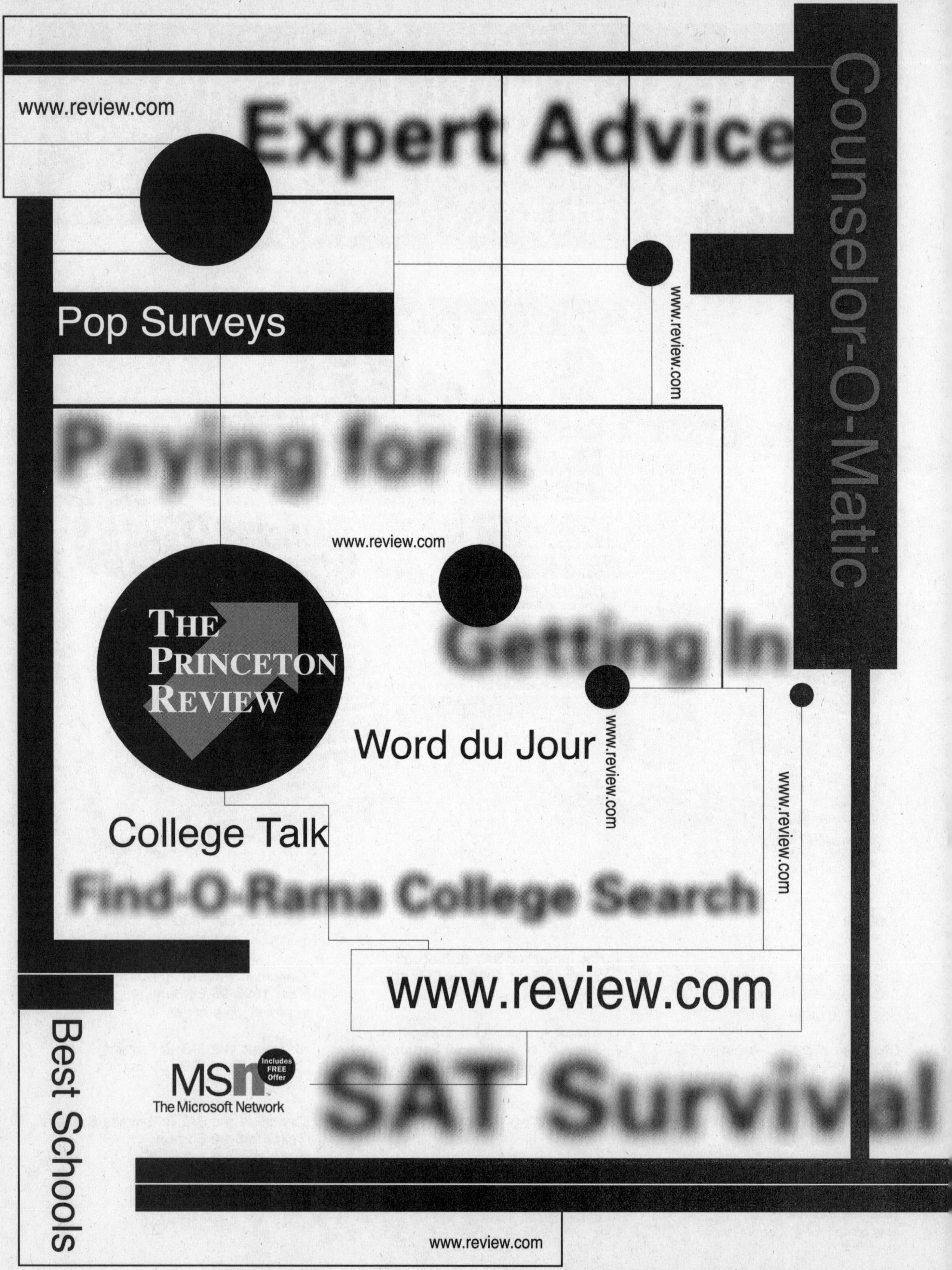

www.review.com

Expert Advice

Counselor-O-Matic

Pop Surveys

www.review.com

Paying for It

www.review.com

Getting In

THE PRINCETON REVIEW

Word du Jour

www.review.com

www.review.com

College Talk

Find-O-Rama College Search

www.review.com

Best Schools

MSn The Microsoft Network
Includes FREE Offer

SAT Survival

www.review.com

Free!

Did you know that The Microsoft Network gives you one free month?

Call us at 1-800-FREE MSN. We'll send you a free CD to get you going.

Then, you can explore the World Wide Web for one month, free. Exchange e-mail with your family and friends. Play games, book airline tickets, handle finances, go car shopping, explore old hobbies and discover new ones. There's one big, useful online world out there. And for one month, it's a free world.

Call **1-800-FREE MSN**, Dept. 3197, for offer details or visit us at **www.msn.com**. Some restrictions apply.

Microsoft® Where do you want to go today?®

©1997 Microsoft Corporation. All rights reserved. Microsoft, MSN, and Where do you want to go today? are either registered trademarks or trademarks of Microsoft Corporation in the United States and/or other countries.

FIND US...

International

Hong Kong
4/F Sun Hung Kai Centre
30 Harbour Road, Wan Chai,
Hong Kong
Tel: (011)85-2-517-3016

Japan
Fuji Building 40, 15-14
Sakuragaokacho, Shibuya Ku,
Tokyo 150, Japan
Tel: (011)81-3-3463-1343

Korea
Tae Young Bldg, 944-24,
Daechi- Dong, Kangnam-Ku
The Princeton Review- ANC
Seoul, Korea 135-280,
South Korea
Tel: (011)82-2-554-7763

Mexico City
PR Mex S De RL De Cv
Guanajuato 228 Col. Roma
06700 Mexico D.F., Mexico
Tel: 525-564-9468

Montreal
666 Sherbrooke St.
West, Suite 202
Montreal, QC H3A 1E7 Canada
Tel: (514) 499-0870

Pakistan
1 Bawa Park - 90 Upper Mall
Lahore, Pakistan
Tel: (011)92-42-571-2315

Spain
Pza. Castilla, 3 - 5° A, 28046
Madrid, Spain
Tel: (011)341-323-4212

Taiwan
155 Chung Hsiao East Road
Section 4 - 4th Floor,
Taipei R.O.C., Taiwan
Tel: (011)886-2-751-1243

Thailand
Building One, 99 Wireless Road
Bangkok, Thailand 10330
Tel: (662) 256-7080

Toronto
1240 Bay Street, Suite 300
Toronto M5R 2A7 Canada
Tel: (800) 495-7737
Tel: (716) 839-4391

Vancouver
4212 University Way NE,
Suite 204
Seattle, WA 98105
Tel: (206) 548-1100

National (U.S.)

We have over 60 offices around the U.S. and run courses in over 400 sites. For courses and locations within the U.S. call 1 (800) 2/Review and you will be routed to the nearest office.